Upper Palaeolithic Faunas from South-West France

A Zoogeographic Perspective

Katherine V. Boyle

BAR International Series 557
1990

B.A.R.

5, Centremead, Osney Mead, Oxford OX2 0DQ, England.

BAR -S557, 1990: 'Upper Palaeolithic Faunas from South-West France'

Price £27.00 post free throughout the world. Payments made in dollars must be calculated at the current rate of exchange and $8.00 added to cover exchange charges. Cheques should be made payable to B.A.R. and sent to the above address.

ISBN 0 86054 704 3

For details of all new B.A.R. publications in print please write to the above address. Information on new titles is sent regularly on request, with no obligation to purchase.

Volumes are distributed from the publisher. All B.A.R. prices are inclusive of postage by surface mail anywhere in the world.

Printed in Great Britain

"The raw material of biogeography is the
distribution of species in time and space."
(Myers & Giller 1988:15)

"The distribution map lies behind some of
the most central themes in archaeology".
(Hodder & Orton 1976:1)

ACKNOWLEDGMENTS

This work is essentially my Ph.D. thesis, awarded in 1988 by the University of Cambridge. The reason for publication is to make available to others a larger body of French Upper Palaeolithic faunal data in one volume than has previously appeared.

I should like to thank my supervisor Dr. P.A. Mellars, who, after encouraging my interest in the European Palaeolithic whilst still in Sheffield, kept me on the relatively straight and narrow during three years of data collection and examination. I would also like to thank Dr. C.S. Gamble (Southampton University) who, during the academic year of 1982/1983 showed me that there is more to Palaeolithic archaeology than a mere 'laundry list' of species. I need also to thank Dr. P. Callow (Cambridge University) for his help with the multivariate analysis of data.

Most of the published data used here was collected between January 1984 and December 1985 with the assistance of library and museum staff in Cambridge (Haddon, Sedgwick, Botany and Zoology), London (University College, Museum of Man, British Museum Natural History), Paris (Musee de l'Homme, La Sorbonne) and Bordeaux (Institut du Quaternaire, Bibliotheque Universitaire, Musee d'Aquitaine). I am grateful to the above for their assistance, especially in France where my reasons for collecting anatomical element counts were initially met with some degree of scepticism.

Two extended visits to Bordeaux, when I was welcomed to the Institut du Quaternaire by Francoise Delpech, Francois Prat and Andre Gilbert, yielded data unavailable elsewhere. During my first visit (1985/86) further literature was examined, especially a series of Doctorat d'Etat theses. In the summer of 1987 Natalie Memoire (Museum d'Histoire Naturelle) allowed me to examine material from the site of Pair-non-Pair, while Alain Roussot (Musee d'Aquitaine) let me spend June examining material from a number of sites. I should like to thank them and all the staff at both museums for their welcome. The first visit was funded by Corpus Christ College, the second by a C.N.R.S. Visiting Research Fellowship.

Additional data were made available in London by Andrew Currant of the Department of Palaeontology (B.N.N.H.), where I examined material from a number of Perigord sites. Several people assisted in my search for relevant background information. I should like to thank Graeme Barker (Leicester University), Robin Kay (Rowett Research Institute), Robin Putman (Southampton University), and H.A. Tyler (Ministry of Agriculture, Fisheries and Food); also Geoff Bailey (Cambridge) for advice, not always heeded! Meanwhile Dr. Bryan Gordon (Ottawa) showed a keen interest in the research, for which I am grateful.

Closer to home, I owe a debt of gratitude to members of College who helped maintain my sanity whilst battling with the University Computer Service - in particular Mark West, an 'informed layman' who asked questions - not always helpful. Finally, to Father J.C. Berry I dedicate this volume in thanks for the support given when we were both completing doctoral research.

TABLE OF CONTENTS

LIST OF TABLES

LIST OF FIGURES

Chapter One

INTRODUCTION.

When, in October 1983, the research reported here commenced, the general purpose of the study was a relatively straightforward one. It was felt that the time had come to make an attempt to rework the Upper Palaeolithic faunal data available from sites in South West France, especially that part which has yielded abundant lithic and artistic material in addition to faunal data. A useful starting point was provided by the work of Delpech (1975; 1983), although the number of sites which she considered was relatively small - given the overall size of the available database, although some "fouilles anciennes" were included. She does not, however, consider any sites which yield only qualitative data. As a review of the literature showed, the region has attracted much attention during the last century. It is in the Perigord that we find one of the densest European Palaeolithic records; sites in the area are rich in art and have yielded material upon which the definition of several cultural periods has been based, eg: La Gravette (Gravettian), La Madeleine (Magdalenian). There are 88 Magdalenian sites alone listed by White (1985). Add to these the earlier deposits and the total reaches 269. Furthermore, if we add those sites to the north, west and immediate south of the region covered by White, a total of 597 sites is available for study, of which more than 300 yield Magdalenian deposits (Banque Nationale de Paris 1979; Champagne & Espitalie 1981; Delpech 1975; Lenoir 1983; Leveque & Miskovsky 1983; Rigaud 1982; White 1985). It should be noted that these take the form of caves, rockshelters (abris sous roche), open air sites and a handful of 'surface scatters'.

From the time Lartet and Christy (1877) and de Mortillet and de Mortillet (1900) discussed the Palaeolithic archaeology of the Perigord, the period has been characterised as one which is associated with a hunter-gatherer way of life. The Upper Palaeolithic has long been termed the 'Age du Renne', (de Mortillet & de Mortillet 1881) during which specialised reindeer-hunting has been inferred based upon the abundance of the species in assemblages resulting from human activity at sites in the area.

A great deal of work has been undertaken concerning South West French Palaeolithic faunas over the years. However, most data have been presented in the form of faunal appendices to site reports. Only in the last twenty years have we seen a determined effort by some specialists to consider faunal material as a worthwhile database in itself. Thus, French archaeozoological research has culminated in significant studies by Delpech (1983, 1984), in which the meaning of faunal associations over somewhat wider geographical areas is often considered. However, while methodological and theoretical developments abound in the Anglo-American world, approaches adopted by French specialists have in general changed little, remaining essentially palaeontological in nature.

Even in the recent publications of Delpech (1983, 1984) faunal material is still generally viewed as a means through which to establish a relative chronology and the nature of the surrounding environment and associated climate. With a few tentative, descriptive exceptions, the role of human behavioural factors has received scant attention, the unwritten assumption seeming to be that the proportion in which each species occurs in an assemblage represents the proportion of that species in the natural environment, although the mere presence or absence of taxa is often the base upon which discussions are based.

In the light of current approaches to the study of French palaeolithic faunas the decision was made to examine available data in order to determine whether any patterning could be recognised and explained. The aim of the research was thus twofold: firstly, the identification of recurrent geographic and chronological patterning, and secondly, the search for possible causes of such patterns. It became necessary to ask whether it were possible that factor A, B or C ... explained the regularities and variations observed, and almost inevitably, two principal types of causal factor emerged: environmental (non-human) determination and behavioural (human) causation.

In order to decide which factors were operating, the potential role of ecological variables (climate, snow depth, vegetation etc.) was considered, in the belief that it is first necessary to establish the nature and role of the environment or to eliminate this variable in the consideration. The environment surrounding any locality provides the bulk of the natural resource base which man, by means of an adopted foraging strategy, may exploit. Even in a situation in which a non-environmental interpretational framework is adopted (eg: in the case of possible nutritional preferences or social/ideological taboos), it is impossible for a group of hunter-gatherers to regularly consume a prey species which is not present on a regular basis. It may therefore be assumed that any species regularly found at sites within an area will have occurred naturally in the vicinity of those sites. The occurrence of a very small number of anatomical elements at a single site may, on the other hand, be explicable in terms of transportation from elsewhere. If a large quantity of this additional prey is required hunters may move closer to their quarry or transport quantities of the material, probably already processed and thus avoiding the hauling of bone and unusable material. If we cannot explain recurrent patterning in terms of such factors as natural resource availability then attention must be turned to other (perhaps human) causal factors. In cases in which reindeer comprises over 90% of the herbivore material yielded by a site, should we not perhaps infer some form of specialised subsistence strategy rather than simply assume a severely restricted (narrow) environment in which few resources are available?

Several assumptions were made while undertaking the research; they are stated here in conjunction with associated objectives of the research:

1. We are looking for recurrent geographic and sequential patterning in species representation, based on the fact that biogeographical studies have shown that we can expect to observe regularities in community structure over time and through space. It is assumed here that some structure may be seen in faunal assemblage composition due either to natural or human causes.

2. In an attempt to explain the co-occurrence of various taxa in differing proportions in terms of ecology and exploitation strategies employed, it is assumed that species occur in regions which offer certain preferred conditions. The persistent co-occurrence of horse and antelope indicates, for example, the existence of cold, dry steppe conditions. It is assumed that, in order to exploit species, certain hunting techniques are more appropriate than are others eg: stalking v. driving in forests and open plains. Furthermore, it is assumed that such was the case during the Upper Palaeolithic and that some techniques were appropriate for more than one species.

3. When attempting to examine and account for element representation at those sites which have been found to yield suitable data it is necessary to have a standard point of reference. It is therefore assumed that behavioural patterning observed among the contemporary Nunamiut by Binford (1978) is of relevance to our data, given that reindeer is the major prey species in each case. There is reason to believe that this assumption is justified, for studies by Thomas (1982), Legge and Rowley-Conwy (1988) and Speth (1983) have applied Binford's (1978) methodology in other contexts either directly or in modified form.

When determining the region from which our data were to be taken, an area larger than that of the Perigord proper was chosen, both in order to increase our database and to consider a region of such size as recent research has shown hunter-gatherers to occupy (Binford 1983). Binford (1983:110) defines "residential core areas" as one in which base camps are located during the course of one year, albeit with seasonal adjustments. This area may exceed 5,400 square kilometres. The area covered by foraging trips is closer to 25,000 square kilometres. For this reason the southern parts of the Charente and Vienne, the western portion of the Massif Central, the Lot and the Lot et Garonne were considered. In addition, to the West, that part of the Gironde through which the Dordogne flows was included. The data available from the latter region are severely limited chronologically but supplement the Perigord material geographically, providing a contrasting resource base and record of exploitation. Thus, the larger region selected for study is one which is based upon the broader Dordogne river drainage basin, allowing us to adopt a regional perspective in addition to the more common site orientated report (see fig. 1.1)

1.1 The database

Any attempt to identify patterning in faunal assemblages rests on the basic data available - in the present study these are the temporal and spatial distribution of faunal resources, the condition of the remains and their association with alternative data types. The

S O U T H   W E S T   F R A N C E

Fig. 1.1

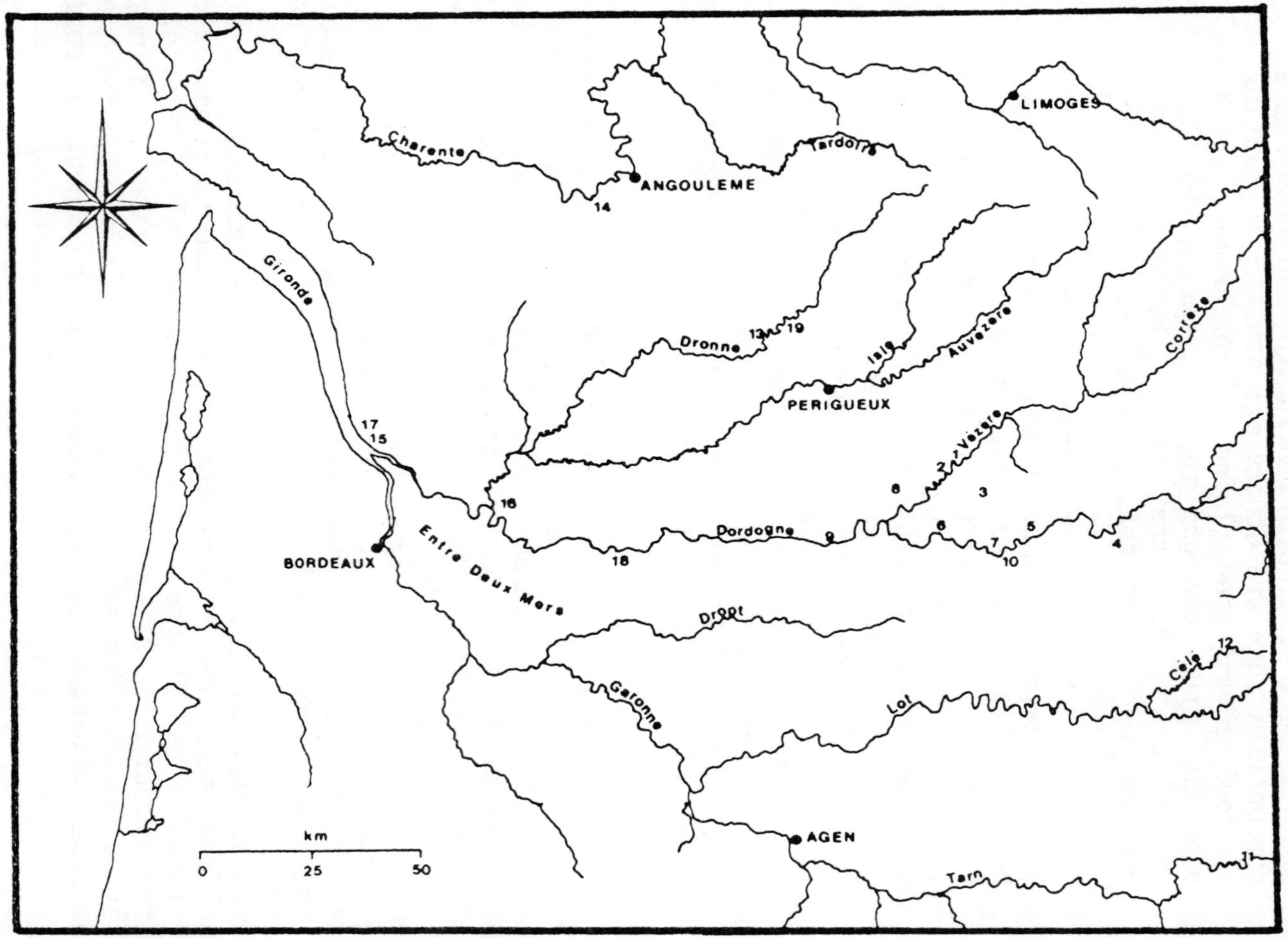

Map showing the Area of Study and major Upper Palaeolithic Sites which yield quantitative data

data, faunal and other, employed in the present study derive primarily from archival sources (site reports etc) and thus we have substantially less say regarding the form these data take than would be the case if we were working with primary sources (the faunal material itself).

The initial intention in the present, archaeozoological study was to consider only the quantitative faunal data available in literature sources. However, as data collection progressed it became clear that by completely omitting non-numerical material, a large amount of potentially useful information was being overlooked. We are not yet in a position to judge the degree to which the available quantitative samples are representative of the faunal picture as a whole, and thus the decision was made to collect both qualitative and quantitative data. In this way we are not omitting an important element of the potential record. For the present study, these data were available mainly for the Chatelperronian, Early Aurignacian (Basal & I), Later Aurignacian (II-IV), Upper Perigordian / Protomagdalenian, Solutrean (Upper), Early Magdalenian (I & II), Middle Magdalenian (III), Upper Magdalenian (IV,V & VI) and Azilian contexts.

In the present study, the majority of the data with which we are concerned comprises relative indications of the frequency of animal species, expressed either in qualitative or quantitative terms. In addition to this there is a more limited body of data concerning the frequency of anatomical elements, usually quantitative in nature which have been collected and summarised in tabular form where available (see appendix).

1.1.1 Qualitative Data.

Traditionally, the aim of faunal analysis has simply been to provide a list of those species which are present in an assemblage. Such presence/absence data are of limited value but do provide an indication as to the range of species available at the site, and the associated "faunal environment". Although the presence of a species is commonly taken to imply that the species existed in the vicinity of the site (despite the fact that either complete animal carcases or parts thereof may have been transported some distance from the kill), the absence of the species from an assemblage need not imply the complete absence of that species from the environment: it merely means that remains of that animal did not accumulate at the location in question. However, it is possible to employ this non-numerical material in inter-site comparisons, involving the total number of species (N) encountered, or the regularity with which certain species occur throughout a range of sites. We are also able to employ such nominal-scale (presence/absence) data in some multivariate statistical classification techniques, eg: using Jaccard's Coefficient of Similarity in cluster analysis (see chapter 7 below).

Key to fig. 1.1: 1. La Madeleine 2. Laugerie Haute 3. Cap Blanc 4. Combe Cullier 5. Caminade 6. Le Flageolet 7. Maldidier 8. La Ferrassie 9. Gare de Couze 10. Le Piage 11. Battuts 12. Sainte Eulalie 13. La Chevre 14. Bourgeois-Delauny 15. Pair-non-Pair 16. Fongaban 17. Roc de Marcamps/Grotte des Fees 18. Le Morin 19. Pont d'Ambon

Ranked data, in which species are listed in ascending or descending order of abundance, provide a marked improvement on simple nominal scale data. With such information we are able to employ further statistical techniques in pattern recognition studies, such as those described by Meddis (1984). Descriptive qualitative data are perhaps the most difficult with which to deal; there are a great many terms applied to species frequency. Having to distinguish between such terms as 'rare', 'a few' and 'not common' is a frequent problem. The terms encountered most frequently in the literature have been standardised in the present study as shown below:

Absent	0	
1 case	1	
rare	2	(french = un peu, rare)
common	3	(french = commum, plusieurs)
abundant	4	(french = abondant)
very abundant	5	(french = tres abondant)
dominant	6	(french = la plupart)

Clearly the recording of data in this form can be highly subjective and, for the majority of the present discussion the classification has been reduced to more basic categories: present, absent, dominant.

All the qualitative data collected prior to 1 January 1987 have been recorded spatially in the form of presence/absence maps, and chronologically in terms of the present / absent / dominant categories as discussed in chapters 6 and 8. In order to ensure that the data were as comparable as possible, this simpler level of recording non-numerical material was selected.

1.1.2 Quantitative Data.

Only rarely can interesting and reliable behavioural or environmental conclusions be drawn from the comparison of the simple presence or absence of taxa making up archaeozoological assemblages. For most forms of interpretation - both palaeoenvironmental and subsistence-related (Grayson 1984:16) - we need some indication of the relative abundance of the species observed (Klein & Cruz-Uribe 1984:24).

The consideration of numerical data allows us to make more detailed statements regarding taxonomic abundance. As pointed out earlier we should not, of course, assume that the set of bones recovered at the site mirrors the originally deposited material let alone reflects the faunal resource base which we are trying to reconstruct. Clearly we can not assume that fluctuations in taxonomic abundance provide a direct reflection of changes in the environment. However, a direct correlation between the relative abundance of characteristic species and the palaeoclimate has often been assumed, and forms a common feature of many palaeolithic faunal reports.

For the majority of the analysis reported in this study fragment counts (N.I.S.P. - Number of Identified Specimens per Taxon) were selected in preference to M.N.I. (Minimum Number of Individuals) for several reasons. N.I.S.P. is the most readily obtained index of species abundance, comprising the total number of bones and fragments which can be assigned to the species in question, and as such is the more commonly available form of data. There is no standard way in which to calculate M.N.I. counts and it is only rarely that we find details of the method adopted in each case. The various formulae which may be used yield different results even when considering the same assemblage. Problems of comparability are much greater for M.N.I. counts than for N.I.S.P. totals due to the variety of formulae and the way in which fragmentary bones are treated. Furthermore, the use of M.N.I. counts in considerations of subsistence patterns implies the assumption that animals were consumed in 'living-animal units' (Binford 1982). N.I.S.P. counts can be calculated as the basic identifications are done, and, as such, are additive in that figures can be easily updated.

However, problems arise when using N.I.S.P. totals:

1. This procedure fails to take account of the fact that the skeletons of some taxa have a larger number of elements than do those of others. Ideally, such species should not be compared.

2. N.I.S.P. overemphasises those species whose skeletons reach the site intact. Smaller animals are more likely to be brought back from the kill in one piece than are larger animals. The latter may be brought in the form of meat parcels (Binford 1978).

3. N.I.S.P. is sensitive to bone fragmentation. A heavily butchered carcase will be represented by a larger number of bones and fragments than will one which is relatively unprocessed. A carcase subjected to heavy post-use taphonomic processes will also be over-represented.

4. Differential preservation means that the number of bones yielded by a deposit may bear little relationship to those initially deposited. Such is also true of M.N.I., but the problem here is less critical.

5. Chaplin (1971) maintains that N.I.S.P. counts from different sites are not comparable and do not lend themselves as readily to subsequent analysis as does M.N.I.. The point remains however, that there should be little, if any, discrepancy between totals obtained by different analysts - unlike the M.N.I.

Given the comparability and relative abundance of N.I.S.P. totals over M.N.I. counts, the decision was made to consider the former, increasing our overall data-base and, relatively speaking, the size of each assemblage, for by definition, N.I.S.P. exceeds M.N.I., usually by a substantial margin.

Other means of quantification have recently been proposed in an attempt to improve our ability to gauge the relative abundance of faunal species. These have, as yet, had little impact upon the analysis of French Palaeolithic faunas. That which will concern us here is the concept of 'Fractional Minimum Numbers of Individuals' (henceforth fM.N.I.) (Binford 1978:71), the calculation of which requires quantitative anatomical element frequencies. Such data were sought whilst collecting species frequency material. The concept and use of fM.N.I. will be considered in detail in Chapter 9, given that the technique applies solely, in the present study, to the behavioural analysis of element representation reported in that chapter.

In the present study, detailed quantitative data have been considered primarily for samples totalling at least 100 identifiable fragments, although, due to the relative lack of large assemblages we have also been forced to consider smaller samples. Provided that the reader is aware of the fact that, when sample size only marginally exceeds 50 bones an increase of approximately 2% may represent only a single element, these smaller assemblages may be included (as they have been, for example, by Delpech 1983). In a regional survey of material it is important that a 'representative' sample of quantitative data be employed, and thus it was decided that a total of at least 10 sites was necessary in the present study, the major exception being the Later Aurignacian. In this case, however, the non-numerical data form a useful supplement. Where quantitative material totalling fewer than 50 bones occurs the site/level was added to the qualitative data base, so as to avoid complete omission of the occurrence. The data are presented in Appendix I, which contains those employed in the quantitative analysis, Appendix IV comprising the material which it was felt better not to include therein, but which has here been noted for the sake of completeness.

Having selected our means of quantification and minimum sample size, figures were converted into percentages in order to aid comparisons and were recorded on data sheets in the form shown in Appendix I. In addition to the relative frequencies of the major herbivore species note was made of the total number of herbivore taxa (N). Included in this figure were the mammoth, woolly rhinoceros, ass, hare and rabbit in addition to the major species upon which attention is concentrated. The major species with which we are concerned are reindeer, red deer, bovids, horse, ibex, roe deer, chamois, boar and antelope and they were recorded in this order. Only the major species were recorded in terms of percentage frequencies: the others are scarce, rarely totalling more than 1%.

1.1.3 'Supplementary' Faunal Data.

Taphonomy has taught us that apparent changes in faunal assemblage composition reflect a combination of various factors such as the genuine disappearance or appearance of a taxon in the environment, a shift in hunting/scavenging strategies or changes in preservation conditions. Therefore, in addition to the qualitative and quantitative indications of species abundance outlined above, various forms of 'taphonomic' data were collected where possible. Shipman (1981:6) emphasises that the study of taphonomy concentrates upon events occurring between death and fossilization and the role of these in data retrieval. We therefore find ourselves collecting data concerning the frequency of bone surface cut-marks, fracture types etc. These data were collected, both in London and Bordeaux and will form the basis of work presented elsewhere. In the present study, our consideration of butchery and carcase management strategies depends largely upon variations in the relative frequency of different anatomical elements.

1.2 Summary

Chapter 2 presents a review of literature reporting research previously undertaken into the Upper Palaeolithic of South West France. In particular we see a marked lack of regional synthesis, for French Palaeolithic research has, until quite recently, been orientated, almost exclusively, towards the site-by-site record of artefactual and environmental data.

Chapter 3 comprises a summary of the physical geology of South West France, environmental changes characterising the course of the Upper Palaeolithic and a brief survey of the cultural chronology of the area. No attempt is made to include a consideration of faunal data in the environmental reconstruction (as is often done) since these data are our major concern later on. By employing large herbivore remains in climatic reconstruction we should become engaged in circular arguments - especially if faunal material is then examined in terms of these reconstructed prevailing environmental conditions.

Chapter 4 contains a summary of the basic ecology and ethology of the major herbivore species encountered. This is based primarily upon modern ecological data, but does include an attempt to furnish information concerning two of the extinct species encountered which appear to have been of some importance in the Upper Palaeolithic economy (the mammoth and woolly rhinoceros). The chapter concludes with a brief discussion of modern and historic hunting techniques employed in the exploitation of the herbivores in question in the northern hemisphere. Meanwhile, Chapter 5 summarizes relevant seasonality indications; it is historical in nature, showing the developments and change in our understanding of material in question.

Four chapters describing the research results follow: Chapter 6 discusses the nature and 'causes' of chronological variation in assemblage composition and structure; in Chapter 7 the results of bivariate correlation and multivariate statistical analyses are described; Chapter 8 considers spatial (zoogeographic) patterning, while in Chapter 9 a 'behavioural' consideration of element representation is presented. It is this last chapter, in which an attempt is made to identify activities taking place at and the function of various sites, which, I believe, should provide much information, but which is severely limited by both the quality and quantity of available data. A plea is made for detailed faunal reports in which the frequencies of all element types should be recorded in addition to other basic taphonomic information. The behavioural and taphonomic studies which have had great impact upon Anglo-American zooarchaeological research are impinging, only very slowly, on studies in Western Europe. In the final chapter summary and conclusions are provided and, based upon these conclusions etc., some suggestions are made regarding future research.

In view of the range and abundance of faunal material derived from sites in S. W. France it is clearly impossible to attempt to re-analyse more than a minute part of the material and it is therefore almost inevitable that our data are primarily derived from published sources. Such was the case in the present study. Those faunal samples examined at first hand were those which are available for study at the British Museum (Natural History) (Laugerie-Haute, La Madeleine, Bruniquel and Chatelperron). Also studied was the material from Grand Moulin, Grotte des Fees, Jolias, Roc de Marcamps and Pair-non-Pair in the Gironde, and Reignac (Dordogne), courtesy of the Museum d'Histoire Naturelle and Musee d'Aquitaine although not all this material is included in analysis here - due to sample size problems.

Problems necessarily arise when the major data comprise material derived from published site reports. It is not possible to specify the nature of the data reported; of particular note is the lack of anatomical element counts, although such lists do occasionally appear (Capitan et. al. [1906]; Delpech [1972]). In addition the means of quantification is often not stated. Therefore, given a percentage value for a selected species, one is not informed as to whether figures refer to Minimum Number of Individual (M.N.I.) or Number of Identifiable Specimens per Taxon (N.I.S.P.) frequencies. The way in which M.N.I. counts are calculated is rarely explained and thus care must be taken when comparing these figures for contrasting totals are obtained from the alternative formulae available. For this reason N.I.S.P. counts are here employed when available, allowing the calculation of proportional (%fM.N.I.) figures in the way advocated by Binford (1978:69) for consideration of behavioural patterns.

When the analysis of assemblage data commenced there appeared to be no standard way in which archaeozoologists were prepared to gauge 'intra-assemblage diversity'. A search was made of the available, extensive ecological literature, yielding a collection of pertinent indices and coefficients, some of which could be calculated given our data. Approximately nine months after I first calculated these figures Grayson's (1984) 'Quantitative Zooarchaeology' appeared in which taxonomic diversity was discussed. The appearance of this and subsequent 'manuals' (eg: Klein & Cruz-Uribe 1984) concerning relevant methods and techniques of quantification and analysis indicates that we are now in a position to begin to rework our existing data, collecting and presenting new material in new ways.

Chapter Two

HISTORY OF UPPER PALAEOLITHIC ARCHAEOZOOLOGICAL RESEARCH IN SOUTH WEST FRANCE.

This chapter reviews the history of research into the Upper Palaeolithic faunas of South West France, aiming primarily to illustrate the ways in which data have been described and analysed over the last one hundred and fifty years but also to show the ways in which various, recurrent, questions have been, and continue to be, raised. The discussion will follow a historical rather than thematic pattern although problems of seasonality, climatic and behavioural (ie: human) interpretation are considered. They remain the major issues with which we are still concerned, having arisen initially in the mid-nineteenth century. In France research has largely centred on climatic interpretation, Anglo-American research betraying anthropological origins, focusing on behavioural aspects of the data.

2.1. Early Research: pre-1914.

Some of the earliest research undertaken in South West France was conducted by MM. Chaubard and de Reignac in 1834 at the Grotte de Ratus, Commune de Gavaudun (Morala 1985) in the Agenais. J.-.L. Corubes continued their work in the area between 1850 and 1865. It is during the 1870's however that major research is reported by Chauvet (1872) and Lartet and Christy (1877), by which time the Upper Palaeolithic was commonly called the 'Age du Renne'. Attention will be focused on the second of these.

2.1.1. Lartet and Christy (1877): Reliquiae Aquitanicae.

The account presented by Lartet and Christy provides a clear indication of the state of research in the 1870's. Both environmental and behavioural (human) interpretation of observed phenomena are provided and, although the palaeontological record is poorly represented in the discussion its importance is emphasised. Lartet and Christy attempt to impress upon the reader the importance of faunal material as part of the record - both archaeological and ethnographic:

> "It is not so much the existence of the multitudinous implements in stone and bone, with the evidence of their manufacture on the spot, which invests these deposits with their chief interest, but the even more multitudinous examples of bones, broken up by man, of animals extinct in that part of Europe, out of all record of history or tradition, and the failure as yet to detect amongst them any undoubted indication of the early domesticated animals" (Lartet and Christy 1877:21).

The homogeneity of the fauna is noted throughout the district. Reindeer is almost everywhere predominant; in some areas horse follows, in others bovids. The presence of the ibex and chamois, now found only in the Alps and Pyrenees is considered, as is that of the wild boar. In addition peculiarities are identified, including, for example, large felid metacarpals bearing scraping-marks at Les Eyzies, two molars of the Great Irish Deer at Laugerie Haute and great bear phalanges with notches produced by cutting tools at Laugerie Basse.

As regards large herbivore populations, two conclusions are reached. Firstly, the faunal assemblages yielded by Dordogne sites are seen to be composed of two groups of species. The first of these comprises mammoth, woolly rhinoceros, horse and bovid, the second - reindeer, horse and aurochs, however, the mammoth and hyaena are often associated with the latter group. Secondly, when considering the fauna collected from Cro-Magnon, the distribution of saiga is discussed. Lartet notes that it is only in stations with barbed arrow-heads and with reindeer predominating that saiga occurs. The discussion implies that this species may have been used in chronological considerations in the Dordogne. Its expanding range is associated with certain artefact categories. This work has continued until the present day and the discussion of saiga populations remains one of the main issues in Upper Palaeolithic faunal studies in South West France.

As already indicated, many of the questions still asked today were posed in the 1870's by Lartet and Christy. Issues which have only recently begun to be considered in a systematic manner were touched upon by the authors.

There seems to have been a general taphonomic awareness through which natural bone-cave deposits were distinguished from archaeological deposits. This distinction has recently been emphasised by Binford (1981:18) in his distinction between 'geological' and 'archaeological' deposits. The influence of the likes of Buckland (1823) is apparent in the discussion of hyaenas and other carnivores as agents of accumulation. An alternative mode of formation is also considered; C. Prevost (in Lartet & Christy 1877) suggests that running and torrential waters were responsible for many of the cave bone deposits, especially in the case of underground caverns. It is interesting to note that in recent years taphonomic studies have gained ascendancy once again.

Similarly, human behaviour was often inferred from the state of recovered bones: at Les Eyzies reindeer vertebrae found in articulation are considered in terms of kill and butchery strategies. Based upon both faunal material and ethnographic data, conclusions are reached regarding the consumption of the brain as an immediately-consumed delicacy.

Seasonality issues are also raised, based, to a large extent, upon 'modern' reindeer behaviour and ecology (Anderson 1877; Austen 1877). Lartet and Christy remain unconvinced regarding year-round occupation of the Perigord and infer reindeer migrations, prompted by insect harassment. In order to explain the presence of antlers indicating otherwise, they suggest that reindeer were killed elsewhere throughout the year by humans, their antlers being brought to the Perigord by migrating human populations.

Climatic inferences are drawn, based upon the presence and association of certain species, eg: the co-existence of the reindeer and hippopotamus. It is this aspect of faunal assemblage analysis which has continued to be of greatest importance in French archaeology. Perusal of the literature appearing over the following century shows that the climatic significance of faunal species takes priority in analysis and interpretation, although some vague indication of human-causation occasionally appears.

2.1.2. Developments.

The general picture of 'Old Stone Age' life in France was envisaged as one of harsh conditions and an intense struggle to survive. Faunal species were commonly considered in terms of whether or not they are now extinct in France. For example, Chauvet (1897) points out that at the Abri en Face de Fieux (Charente) there are no species which have subsequently left the region. The assemblage comprises boar, rabbit and roe deer, indicating that conditions, at the time of formation, were mild. He tells us that in Upper Palaeolithic levels at La Quina (Gardes, Charente) reindeer predominated, followed by horse and bovids. Attention is also drawn to the ways in which teeth and antler have been modified to form pendants. Such objects are rare and hence of particular interest. The way in which bones are broken is also noted, as is the association of bone material with hearths (eg: Vire (1905) at Lacave, (Lot). In such cases faunal data are considered in terms of butchery and culinary practices, the discussion showing an awareness of the contrasting roles of factors now considered in 'taphonomic' studies.

By the early twentieth century we are beginning to find details of both numbers of individuals and element frequencies, as the need to quantify assemblages is recognised. Element frequencies are often provided for reindeer, it being the commonest species. Some indication of the age of the animals at death is also given (Capitan et. al. 1906) and in association with anatomical details, the form in which major species were conveyed to sites is postulated. At the Abri Mege (Teyjat, Dordogne) for example, Capitan et. al. (1906) suggest that individuals were transported as complete carcases, for all elements of the body are deemed to be represented. However, a climatic interpretation is also sought. The rarity of horse and bovids, and the absence of red deer are seen to indicate cold conditions rather than a failure to exploit these other available resources. The concept of an essentially specialist reindeer kill-site does not appear. Similar analysis and results are reported by Breuil (1906) at Les Cottes (Haute-Vienne).

1907 saw the publication of Hue's "Osteometrie des Mammiferes", a two-volume study of quaternary fauna presented from a palaeobiological viewpoint. The structure of skeletons is discussed, including variations in the frequency with which certain elements occur while the volumes provide a guide to identification of skeletal parts. Meanwhile, when considering the bone material recovered at the Grotte de la Mairie (Teyjat, Dordogne) the significance of microfaunal remains is considered by Capitan et. al. (1908).

During the subsequent decade site reports follow a relatively standard pattern, providing details of site location and an account of the recovered lithic material. In a level-by-level (chronological) discussion, faunal material is then outlined. The degree of precision in quantification varies from report to report, from vague indications such as 'rare', 'abundant' or 'common', to precise individual or fragment counts. These are often mixed within one faunal list, and several means of quantifying remains are regularly found in one site report. Such is the case in Capitan and Peyrony's (1912) discussion of La Ferrassie (Savignac-de-Miremont, Dordogne). Once again however, attention is concentrated upon climatic conditions reconstructed from the material available.

In summary, by 1914 the most commonly found interpretational framework is that relating to prevalent climatic conditions. This framework has remained dominant until the present day, and has only recently felt the impact of a behavioural approach.

2.2. <u>1918 - 1939.</u>

The period between 1918 and 1939 is one during which numerous site reports appeared. There is however a noticeable lack of attempts to discuss general patterns. Data collection rather than synthesis seems to have been the rule.

The Limeuil (Dordogne) report by Capitan and Bouyssonie (1924) provides an account typical of its time. Site details are followed by a discussion of lithic and faunal material before proceeding to a description of the engraved stone for which the site is renowned. Art-work takes priority over other material; such site reports are commonly accompanied by detailed descriptions and sketches of the **gravures.** It must, nevertheless, be said that the discussion of animal bones recovered is comparatively detailed. We are provided with mammalian bone counts and an indication of which bird species were present is added. When describing the state of preservation of the reindeer present we are told that all parts of the skeleton are represented but that the long bones predominate, usually broken in such a way as to indicate that the marrow was extracted. Those parts of the carcase yielding most meat are the most commonly found.

In 1930 Bouyssonie and Delsol discussed the predominance of reindeer among 'edible' species at Jolivet (Terrasson, Dordogne). It is the first time one encounters a division of animal types in terms of criteria which may be of value to the archaeologist when considering the palaeoeconomic significance of assemblages. Species are commonly considered as herbivores, carnivores, rodents, birds etc. with a further division according to size. The authors comment upon the representation of element-types and suggest that the site was not a true habitation site, based upon the lack of what are considered to be base-camp-associated stone tools. No attempt is made to judge site function from the faunal material available. Finally they suggest that the location of the site was chosen due to the nature of the physical environment at that point. The area is likely to have been one to which game herds came in order to graze and drink.

During the 1930's several sites were described and their implications considered. Peyrony's 1932 publication covers the Bourdeilles prehistoric sites in the Dordogne (Abri des Francillous, Abri des Bernous, Grotte des Bernous, Trou de la Chevre, Fourneau du Diable and Grotte de Pey de l'Aze). In the course of this report the reindeer remains from Fourneau de Diable are discussed: articulations, phalanges, vertebrae and teeth are well represented, belonging, it is noted, to individuals of all ages. In 1934 Peyrony presents material recovered from Magdalenian and Azilian levels at Longueroche (Dordogne) while 1935 sees his publication of researches concerning the Abri Castantet in the Sergeac valley (Dordogne).

By 1936 the persistent climatic framework of interpretation, so characteristic of French faunal research, is well established (Peyrony 1936), the presence of reindeer being equated with cold, tundra conditions, the saiga with cold, dry steppe, red deer with mild, temperate conditions and the boar with woodland or temperate forest.

An account of research undertaken at Grotte des Fees a Marcamps (Gironde) was published in 1939 (Lacorre 1939). It remains today one of the better site reports and represents a landmark in faunal studies in South West France. An awareness of the potential of fishing, hunting and gathering is apparent and the positioning of the site is considered in terms of the advantages it offers for hunting game herds. Stalking, driving and trapping are seen as possible hunting strategies. Descriptive details of element representation of the major species are provided, with comments concerning what Lacorre (1939) considers (rightly so) to be two interesting observations - namely 1) the importance of saiga and 2) the co-existence of saiga and reindeer.

The importance of saiga is considered by Lacorre to represent a phase of the late Wurm during which the reindeer was eclipsed in the west of our region. During such periods he suggests that we commonly see an increase in horse frequencies and, although more rarely, bovids. Unfortunately, such fluctuations are considered solely in terms of climatic factors, omitting reference to possible shifts in hunting strategies. Early Magdalenian deposits at the Grotte des Fees (Marcamps, Gironde) are thus interpreted climatically; the predominance and consistent presence of spermophiles are considered to imply interstadial conditions. The discussion is based upon modern ecological conditions in the grass-steppe of Russia where saiga survive in January temperatures of -13° to -37°C and July ranges of +21° to +39°C. An average range of -20° to +24°C is inferred. The result of the discussion is the use of the presence of saiga to invoke the existence of Russian steppe vegetation in the Gironde region. It is now known that such conditions are drier and a little colder than most of the course of the Wurm. The discussion concludes with a list of 26 sites at which both saiga and spermophiles are encountered, implying the prevalence of cold steppe.

The co-existence of Saiga and reindeer at the Grotte des Fees is considered incongruous (Lacorre 1939) and a mixture of habitats is therefore inferred. Based upon the facts that in Russia reindeer and saiga migrate seasonally in such a way that they never co-inhabit an area, that vegetation required for forage is different as is dental morphology. Lacorre invokes the existence of varying microclimates in the surrounding area. Presumably reindeer occupy patches of tundra, the saiga inhabiting dry, cold, grassy steppe in the valleys and on isolated slopes (Lacorre 1939).

However, despite the fact that the site is important for its saiga and reindeer deposits, the horse is listed as the major prey species, represented in particular by long bones and crania. It is the former which make up the greater part of the bone count, having been uniformly broken for marrow extraction. Even phalanx I of the bovids has been broken in this way, suggesting that maximum yield was sought from the kill.

2.3 Early Post-War Developments (1945 - 1965).

The immediate post-war period was largely dominated by Bouchud's discussion of fauna at various South West French sites (1952a; 1953; 1959a; 1959b). However, one major volume appeared soon after the war: Malvesin-Fabre's (1948) "Essai sur la faune pleistocene de la Gironde: paleobiologie et paleoclimatologie". The report serves as a descriptive summary of the then current knowledge regarding palaeolithic faunas. Emphasis is placed upon broadly climatic interpretations and their implications. Malvesin-Fabre concludes, for example, that reindeer populations arrived in the area during the autumn in order to over-winter in the region. This is a theory which was shortly to come under attack by Bouchud. The major feature of the palaeoclimatic interpretations prevalent at that time was the apparent consideration of the whole system. Malvesin-Fabre, for example, discusses Aurignacian assemblages in the Gironde as a reflection of vegetation (forage) conditions which are considered to be caused largely by changes in sea-levels and atmospheric humidity/aridity.

Site reports of the period continue to follow the standard pattern in which fauna appears to take second place to stone tool assemblage analysis. In many ways it is surprising that archaeozoological research did not decline in a more significant way, due to the increased interest in stone-tool assemblage variation led by Bordes through the 1950's into the 1960's and 1970's. It is to the likes of Bouchud that we owe the survival and development of our subject through that period.

In 1953 two papers concerned with 'methodological' issues appeared (Leroi-Gourhan 1953; Bouchud 1953). The first of these concerns the interpretation of bone remains. In it, Leroi-Gourhan points out four areas in which our understanding of palaeolithic archaeology can benefit from zoological study. Our knowledge of both human and animal environments is improved, accompanied by an increase in our appreciation of the general climatic conditions of the time. We can begin to gain a better understanding of dietary, social and religious practices and shed light upon aspects of bone technology. A major criticism of much faunal analysis, both in the past and today, is the apparent failure to include bone art works in faunal considerations. Such material is usually separated from the rest of the faunal evidence, being placed in a mobiliary art category and as such may never reach the faunal specialist. This is not a problem unique to Palaeolithic studies.

Leroi-Gourhan (1953) stresses that, in palaeoclimatic considerations, assemblages with large numbers of bones are desirable; he talks of thousands rather than hundreds.

In addition to the above, the author suggests that attempts should be made to determine whether assemblages reflect predation by hyaenas or other carnivores, thus advocating a taphonomic approach similar to that proposed by Binford (1981, 1983). Butchery data should be detailed as well as indications of the 'state of preservation'.

At the same time Bouchud (1953) considers the climatic significance of Palaeolithic faunas. After more than fifty years of prehistoric research in the area by various analysts, Bouchud presents a justification of the use of fauna in palaeoclimatic reconstruction. He argues that the Palaeolithic distribution of fauna is of climatic significance due to its mixed nature; we encounter species occurring together which one would expect under contrasting regimes. The presence of local-scale micro-climates is inferred as is the role of seasonal migrations, for, although species may be able to live in various climates, migration may be necessary in order for reproduction to be possible (Bouchud 1953:432). Bouchud acknowledges that the problems involved in climatic reconstruction are substantial, but maintains that attempts at such reconstruction are of value. Basing his argument upon modern ecological data it is essential that the discussion be of detailed nature. Given the heterogeneous nature of the assemblages, it is to the details of each species' range-ecology that we must turn.

In 1954 Bouchud raises the problem of reindeer migrations and their implications for Palaeolithic archaeology. His review is based upon research undertaken by him which was later to have a considerable impact on our knowledge of the Palaeolithic in S.W.France.

The 1950's see the publication of many papers recording details of specific assemblages (Bouchud 1952a, 1952b, 1956, 1957; Delporte 1956). In addition 1956 sees Lacorre's account of reindeer migrations in the Les Eyzies area, maintaining, contrary to Bouchud's findings, that the reindeer herds migrated through the area twice a year (Lacorre 1956:302).

2.4. 1966: Essai sur le Renne

In 1966 the published version of Bouchud's doctoral thesis concerning the palaeobiology and climatic implications of reindeer, rodents and birds appeared, in which he emphasised the existence of two varieties of reindeer in South West France (forest and tundra). In addition, based upon biometrical analysis of reindeer astragali from 32 levels at 19 sites, a third type is identified. Thus, during the Upper Pleistocene he identifies three subspecies co-inhabiting our region (Delpech 1983:146):

Tundra:
- small   *Rangifer tarandus tarandus* var. minor
- larger   *R. t. arcticus* (= *R. t. groenlandicus*)

Forest:

R. t. caribou (= R. t. terranovae)

The division is substantiated by variations in antler structure. Furthermore, climatic considerations based upon biometric analysis are discussed, especially chronological trends traced through animal size variations.

Perhaps Bouchud's (1966) "Essai sur le Renne et la Climatologie du Paleolithique moyen et superieur" is best known as a source of early seasonality-of-occupation indications. He considers the evolution and attrition of teeth during the life of the reindeer, which allows him to date both mandibles and isolated teeth in terms of months of the year. He concludes that the study of teeth and antlers from various archaeological levels demonstrates year-round continuity of human occupation and the permanent presence of reindeer in South West France. However, the use of antlers is minimal and barely signifies in his discussion.

By discussing the year-round occupation of Perigord by reindeer taking part in small-scale limited migrations, Bouchud rejects the idea that long distance movements occurred and states that some herds may even have stayed in one small area. He thus disagrees with suggestions made by Obermaier (1939) that herds wintering in the Dordogne spent the summer in the Alps (Bouchud 1966:243). In order to move from one range to another, it is necessary only to go to the Massif Central. Suitable forage etc. conditions could be found within 80 Km of Les Eyzies. In Canada larger-scale migrations are necessary in order to locate desired conditions.

Both reindeer frequencies and microfaunal analysis were used by Bouchud to trace the climatic development of the Middle and Upper Palaeolithic. He (Bouchud 1966:243) employs the standard Wurm I, II, III and IV divisions, the last two of which equate with the Upper Palaeolithic in his discussion. He points to the fact that climate varies spatially as well as chronologically (Bouchud 1966:245) according to several factors: latitude, altitude, distance from the sea and orientation. Such local variation allows us to explain seemingly incongruous faunal associations, for species known to occupy different ecological niches (both today and in the past) occur in a single assemblage.

In 1973 Binford strongly criticised the methods employed by Bouchud (1966) and the assumptions under which he worked; Bouchud is forced to assume that births occurred simultaneously at a regular time each year and that attrition occurs at a constant rate. He also assumes that complete eruption of the premolars takes 27 months, whereas, as Binford points out (1973:239) eruption may take between 22 to 26 months from birth. Thus, unknowingly, Bouchud reduces four months to one month. Simultaneity of births, constant attrition and a standard duration of maturation should not be assumed. The highlighting of such weaknesses inevitable leads us to question all Bouchud's conclusions regarding Perigord Palaeolithic seasonality. What is more, recent work has shown that winter occupation is more commonly observed in the area (Gordon pers. comm.).

Despite the claim that man was permanently in the Dordogne, Bouchud (1966:240) emphasises the non-sedentary nature of human groups. However, he maintains that shelters and caves did not remain unoccupied for long, while efforts were concentrated on killing sufficient game.

In addition to discussing the implications of his research for palaeoclimatic reconstruction and migrations, Bouchud briefly considers the possibility of reindeer domestication. He lists four indications of domestication (Bouchud 1966:240):

1. The abundance of juveniles,
2. The presence of all parts of the skeleton,
3. Evidence of castration as seen in antlers, and
4. Relevant art-work.

Comparisons are made between the percentage of young in modern wild reindeer herds and faunal assemblages and the conclusion reached is that hunting rather than domestication was the norm.

Despite the fact that each part of the carcase is commonly represented in most deposits, the frequency (relative and absolute) thereof varies considerably. Accompanied by the dubious evidence available, Bouchud (1966:241) discusses possible castration of reindeer, the conclusions reached concerning element representation being interpreted as an indication of some degree of domestication. However, to assume that the presence of all types of element at a site indicates domestication is a dangerous thing. No account is taken of the extent to which large carnivores may have been responsible for part of the assemblage - in terms of accumulation or subsequent destruction.

Finally mobiliary, and other, art is examined to find signs of control of animals by man, eg: harnesses. As Bouchud himself says (1966:242) such work must be carried out with extreme caution and he concludes by stressing the role of hunting as opposed to herding during "even the Upper and Final Magdalenian"(1966:241).

We have almost ten years to wait until a volume of comparable importance is produced, namely Delpech's (1975) doctoral thesis, subsequently published in modified form (1983). However, several noteworthy publications did appear during the late 1960's and 1970's, each of which made its own contribution to our understanding of Palaeolithic faunal assemblage variation.

2.5. Preliminary to Change: 1967 - 1973

Following the publication of the 'Essai sur la Renne' several papers appear either supporting or refuting Bouchud's 1966 conclusions (eg: Guillien and Henri-Martin 1968). Some of these are discussed below in chapter V. Bouchud attributes the reindeer to a different species, *Rangifer tarandus : Rangifer guettardi* (1967, 1968). In 1972 he re-iterates this belief, claiming that all Wurm and post-Wurm reindeer fall into the *Rangifer tarandus* group; post Riss-Wurm forms are called *R. t. guettardi*. However, Bouchud's work has recently been called into question (Delpech 1983:147).

In 1968 Prat completed a thesis on the remains of Megaceros sp. in Palaeolithic deposits in South West France. We are presented with a site-by-site consideration of the archaeozoological material, including occasional element counts. Much of the discussion is centred upon the ecological significance of occurrences. Climatic and topographic factors are seen to cause the distribution of the species. Its presence is considered to indicate cold, humid conditions (Bouchud 1965) leading to the development of more temperate open pastures, very open forests and marshy depressions. Megaceros sp. is essentially an open plains inhabitant and the rugged relief of the Dordogne is invoked to explain the scarcity of the species in that region, in contrast to the Charente and Gironde.

In 1968, in 'Les Apports de la Paleontologie du Quaternaire a la Connaissance de l'Homme Fossile' Bouchud (1968b) isolates three problems inherent in much archaeological research. These include the facts that much of the data is to be found in widespread, scattered sources, that there are many sites and that complicated statistics are often involved.

Bouchud goes on to summarise findings and hypotheses originating in the previous decade. Man was not permanently in the Perigord. He invokes evidence relating to the discovery of both Atlantic and Mediterranean shells, indicating relatively long-distance movements. Humans did not hunt selectively and did not domesticate the reindeer. In many cases, carcases were transported to the site complete; all parts of the skeleton are expected to be recovered. It implies that no part of the carcase was rejected, and, as we examine the ways in which bones were butchered and broken, we may begin to identify the ways in which non-meat yields, eg: sinew, marrow and hides were used. In the Aurignacian levels of the Abri Facteur (Dordogne) further non-food yield is, Bouchud believes, indicated by the large quantity of passeriforms, probably, he suggests, killed for their feathers.

Various accounts of sites and assemblages continued to appear throughout the 1960's and into the early 1970's (Delpech 1967, 1968, 1970; Prat & de Sonneville Bordes 1969; Daniel 1969; Bertouille & Bertouille 1969; Vezian & Vezian 1970). There seems, by 1970, to be no detectable influence of the so-called 'New Archaeology' in evidence in the U.S.A. and U.K.. The standard approach to faunal analysis lay in the measurement of selected elements in order to determine size, sex and age of the animals and the reconstruction of palaeoclimatic conditions from faunal species associations. This palaeontological as to opposed archaeological framework had developed over 50 - 60 years and has shown considerable resistance to change.

By 1970 we begin to detect slight changes in approach. Delpech provides an account of faunal material from Flageolet II (Bezenac, Dordogne), couche IX. A minimum number of individuals (M.N.I.) count of 12 is given for reindeer; 5 of these are under 3 years of age. She employs the techniques developed by Bouchud (1966) to determine the season of death, and thus her results (March, April, June, September and December) may be questioned. The report is however, of value in providing a discussion of butchery patterns and carcase usage. The almost complete absence of vertebrae and the small number of ribs, scapulae and coccyges indicate that man was not bringing the complete reindeer back to the site. The nature of the topography in the area further supports this view. 13 proximal metapodia (5 metacarpals, 8 metatarsals) show traces of butchery at the articulation. These traces run perpendicular to the main axis. Delpech (1970) concludes from this that the hunters occupying Le Flageolet II severed the metapodia from the carpals/tarsals and discarded that part of the leg which bears least meat, extracting the tendons from the articulation.

Delpech (1970) then goes on to list four criteria which she considers to be of importance when selecting prey species: availability, technology and ease of hunt, nutritional considerations and culinary preferences. It is the latter to which she attributes unusual variations. However, despite these signs of the adoption of a 'behavioural' approach, the author concludes with a discussion of climatic conditions as revealed largely by the large herbivore remains.

The early 1970's see a large number of publications, many of which are either dedicated to microfauna (Chaline 1972; Seronie-Vivien 1971) or contain a full discussion of the large herbivore material (Delpech 1972), while Delpech sets out to consider the palaeoclimatic significance of the macrofauna. The use of microfauna is more likely to produce reliable results than will the consideration of large herbivores and carnivores, for the reason that human foraging decisions are much less likely to have influenced microfaunal assemblages. The examination of microfauna gains in importance, much of which may be attributed to Mourer-Chauvire (Mourer-Chauvire 1975; Desbrosse & Mourer-Chauvire 1973).

In 1973 Delpech questions the use of fauna in climatic interpretation, based upon modern animal ecology and ethology, given that no parallels with the Pleistocene environment exist today. In the course of her discussion of Hokr (1951) and Fabre (1964) on which much palaeoclimatic faunal analysis has been based, Delpech makes the assumption that herbivore frequencies and associations depend more upon climate than carnivore predation. She goes on to reiterate and develop much of what was said in 1970.

2.6. New Approaches: 1974 - 1982.

The late 1970's are perhaps the most significant years for faunal analysis in South West France. During this period we see the appearance of several studies, each making its own significant contribution to our understanding of the data.

In 1974 Delpech and Rigaud reported on their investigation of fragmentation patterns at Le Flageolet I, couche VII. Much of this work is the result of Rigaud's period of study in the U.S.A., working with Binford in Alaska among the Nunamiut.

Seven types of bone fragment are identified and the proportion of each in 8 zones is calculated. The analysis results in suggestions as to the use of bones in parts of the site, eg: grease extraction, butchery, marrow extraction and processing. The similarity between butchery techniques employed by Palaeolithic occupants of Le Flageolet I and the Alaskan Inuit is noted. Their preferences are identical as regards their choice of marrow-yielding bones and the procedures employed.

The publication in 1975 of the Abri Pataud Excavation report includes a detailed account of the fauna by Bouchud (1975). Delpech's Doctorat d'Etat thesis (1975) provides a data-base collated from various sources, reflecting both traditional and newer approaches to faunal analysis, although the former predominate. Bay-Peterson's (1975) thesis concerning Pre-Neolithic Faunal Exploitation in Southern France and Denmark is an attempt to progress beyond climatic interpretation and, incorporating auxiliary data, aims to model the Palaeolithic / Mesolithic economic system.

The aim of the research undertaken by Delpech (1975) was to interpret the variations in faunal associations in terms of the changes in climatic conditions which allow the establishment of a chronological sequence. The volumes reflect the general research interests of the Bordeaux 'Institut de Quaternaire': stratigraphy, chronology and palaeoclimatology. The palaeontological discussion concerns the microevolution of species and other causes of variation among the animal species concerned.

Delpech (1975) presents a discussion of the methods employed in her investigation of faunal patterns. The importance of biometric analysis is evident, as are the problems of interpretation. Only the reindeer receives detailed attention. She considers Bouchud's (1966) claim for the co-existence of sub-species and proceeds (after carefully examining cranial remains, antlers, carpals, tarsals and phalanges) to dispute his hypotheses (Delpech 1975:184). Differences in size of individuals are attributed to climate and forage rather than species-type or micro-evolution. We are thus presented with a case in which we have only two types of reindeer occupying South West France during the Upper Palaeolithic (Gordon 1980) - the forest and tundra varieties.

We are provided with a useful account of the ecology of each species upon which much of the discussion of palaeoclimatic reconstruction is based. Climatic provinces are identified, based upon the calculation of three indices:

1. Total percentage of species attributed to cold conditions (reindeer, mammoth, woolly rhinoceros, saiga, horse, ibex and chamois).
2. Percentage of bovids.
3. Percentage of more temperate and humid species (elk, *Megaceros* sp. red deer, ass, roe deer, boar).

Four environmental groups are identified: cold and dry, cold and humid, temperate and humid, and temperate and dry. However, there are considerable problems in attributing species to ecological groups for 'mixed' assemblages are usually found.

Finally, Delpech (1975) considers palaeoethnographic aspects of selected faunal assemblages. Bone material, in this way, is employed as evidence of human activity. For example, at Fongaban

(Saint-Emilion, Gironde), where bovids represent 98% of the fauna, hunters preferentially selected juveniles. Lower limb bones predominate and are usually complete, whereas the bones yielding more meat are rarer and commonly broken.

Seasonality indications are also considered. Following Bouchud's procedures, conclusions of year-round occupation are often reached and once again can be called into question.

Bay-Peterson sets out to model faunal exploitation systems in Southern France (1975). It is a thesis in which attempts are made to extend the discussion traditional French approaches, and as such represents a landmark in French Palaeolithic studies. Various principles of animal ecology and ethology are applied and their implications studied. The concept of territoriality is considered as is the 'generalised' v. 'specialised' nature of assemblages, while the overall approach strongly reflects the 'Palaeo-economic' school of thought developed in Cambridge during the late 1960's and early to mid 1970's.

Bay-Peterson stresses the role of faunal species as resources. They are not considered as climatic indicators. By exploiting major herd species (eg: reindeer) only seasonally and switching to alternative resources at other times, the author maintains that hunters can survive in markedly more compact animal territories. Thus, the importance of alternative resources is stressed, encouraging us to consider the whole resource base rather than only the main species.

The model presented is one of a broad-based economy, in which foragers exploited several complementary resources. Four exceptions to the broad-base, which, we are told, is more apparent in South West France than in the South East, are identified:

1. Intensive reindeer exploitation in the Dordogne/Vezere region.

2. Specialised ibex exploitation in the Ariege (Grotte de la Vache).

3. Extensive saiga exploitation during the Magdalenian in the Gironde and Charente.

4. Specialised culling of horse at Solutre.

Throughout the discussion one is aware of the marked difference in approach in comparison with that of Delpech (1975), there being a general awareness of relevant ecological and anthropological literature.

The other major 1975 publication is the report of investigations carried out at the Abri Pataud (Movius 1975). Attention will here focus on Bouchud's analysis of the faunal material. He provides a summary of each level of the stratigraphy, each one representing, perhaps, several hundred years of deposition. He presents 'evidence' of the year-round occupation of the rock shelter, based upon the antlers recovered from the site.

Concentrating attention upon palaeobiology and palaeoclimatic conditions, Bouchud (1975) discusses the probability of finding different parts of the carcase. He selected reindeer for consideration due to the abundance of the species. He identifies this species as Rangifer tarandus L. and later on, in his discussion of climatic conditions and herbivore associations, refers to forest and tundra types.

Presenting a species-by-species discussion of ecological implications, Bouchud (1975) considers three major groups of animal:

1. Tundra, cold steppe and cold, temperate steppe species (reindeer [tundra], mammoth, woolly rhinoceros, muskox)

2. Taiga, forest steppe (and mountain) species (reindeer [forest], elk, red deer, roe deer, boar, bison, Bos sp., horse, ass).

3. Mountain species (ibex, chamois).

The percentages of groups of species (excluding reindeer) are then calculated and their values plotted chronologically: horse, bovids, cervids, ibex, chamois and extinct species. The horse and bovid sequences are very similar, while the cervid curve is the reverse of these.

Bouchud's (1975) discussion sets out to consider the palaeontology and implied climates, and indeed succeeds in doing this. In so doing it represents a traditional approach which in the latter half of the 1970's has begun to be supplemented by alternatives, much of which has been influenced by recent archaeozoological research in the U.S.A..

Several major publications of the late 1970's represent French approaches to archaeological and zoological material. In both 'La Prehistoire Francaise' (de Lumley 1976) and 'La Fin des Temps Glaciaires' (de Sonneville-Bordes 1979) we are presented with a regional and chronological account of Upper Palaeolithic cultures and fauna, in addition to descriptions of climate and the environment, including vegetation.

In 1978 we see the publication of 'Les Derniers Chasseurs' (Rozoy 1978) once again taking the form of a synthesis of recent work. Rozoy does, however, make clear the fact that he can find no trace of change between the Upper Palaeolithic and Mesolithic. In order to avoid the implication of the lack of such change he employs the term 'Epipaleolithique'. In the discussion of faunal material we encounter, for the first time, the conversion of bone material to meat weights. A relatively detailed account of meat yields at Rouffignac (Delpech & Suire 1976) is thus provided. We are told of the frequent use of young animals, implying that there was little or no concern for the continuing availability of game; the degree of resource management was small. Possible hunting techniques are assessed, including the likelihood of trapping and snaring.

In 1979 Spiess presents a study of the Abri Pataud faunal assemblages in a way which is substantially different from that in which Bouchud (1975) considered the material. He studies the material from the viewpoint of an environmentalist, concentrating upon the interaction of culture and environment. Modern and recent caribou hunters from North America are considered in the context of caribou ethology. In so doing he discusses Burch's (1972) conclusions (concerning the reindeer as an available resource) and various assumptions commonly made concerning the interaction between animal herds and human hunters. He is aware of the importance of prey behaviour to predation success, but assumes that reindeer are easy to kill (1979:138). Both Burch (1972) and Spiess (1979) make this assumption, the latter claiming that "Caribou do seem to be 'stupid' when compared with other deer species". This is unproven.

As for his discussion of Aurignacian levels at the Abri Pataud, the account is, in general, informative, being a marked contrast to previous accounts. However a few weaknesses are apparent.

Firstly, despite the presence of Delpech's (1975) thesis in his bibliography, Spiess discusses Burch's acceptance of two of Bouchud's multiple species of reindeer and fails to make any comment upon her suggestion that a large part of the variation in species is due to forage and climatic condition.

Several of his conclusions are based upon somewhat scanty data. For example, in mentioning the lack of evidence for large scale drives, Spiess (1979:185) fails to acknowledge that such activities would have taken place at some considerable distance from the site. The faunal assemblage from the abri itself will only represent (to some extent) the on-site activities and therefore the material brought back to camp. Studies need to be undertaken of the material, as yet to be recovered, from the area between the site itself and the Vezere. It is worth noting that the Abri Pataud is not the only site at which this is the case.

In order to determine the season of occupation, Spiess uses tooth eruption and annuli, plus foetal bones. Bone material is employed in M.N.I. calculations from which he progresses to estimates of length of occupation in terms of man-days, a concept fraught with problems! The use of annuli is, as Gordon (1980:372) points out, remarkably unsuccessful, using only 11 of 171 teeth. However, as in the case of foetal bones, antler shedding and tooth eruption, his conclusion is one of winter occupation.

When discussing his conclusions regarding seasonality issues, Spiess (1979:192-97) appears to be unaware of Bouchud's (1977) claim for summer occupation of the rockshelter and winter occupation of the area outside the site. In stressing winter occupation attention is not given to autumn indications; the presence of salmon bones alone would indicate late summer use of the site.

Finally, in discussing the diversity of fauna at the site Spiess (1979) attempts to rid us of the opinion commonly held that the environment was cold, harsh and sparse. The boggy tundra or mossy forest-tundra of the present-day Arctic was not encountered in the region of Les Eyzies (1979:254).

1979 also sees the appearance of two contrasting doctoral these - one from the French (Bordeaux) school, the other from the British (Cambridge).

Le Tensorer's thesis concerning the Agenais area studies stratigraphy, palaeoclimates and the Palaeolithic material from the Lot-et-Garonne. The thesis is notable for one thing (in zoological terms) - the identification of purely palaeontological deposits in which there is no evidence of human occupation. As in the case of most French theses, the volume concentrates, however, upon stratigraphic, sedimentological and lithic material. Faunal considerations are sparse.

Meanwhile, in setting out to examine Upper Pleistocene horse assemblages Levine (1979) considers the ethnography of hunters and herders and the ethology of equids to "establish behavioural parameters prerequisite to the construction of interpretational models" (1979). She discusses the state of preservation of the material and proceeds to describe the four major hunting techniques which she considers feasible. In adopting a 'behavioural' approach, the author then considers butchery techniques. These she views as a function of the distance of kill-site from base-camp and the type of animal killed. Each site should be interpreted in the light of the environmental and behavioural agents acting upon it at the time. No standard formula can be adopted in the consideration of every site. Thus Levine (1979) reminds us of the complexity of the situation which we are attempting to model. For our own purpose her study of the Jaurens natural deposits is of primary importance, applying, as she does, Cluster Analytical techniques in a study of population structure.

Jaurens lies in the Correze (Commune of Nespouls), the fossil bed varying in thickness from 10cm to 1.20m. The bed is rich, particularly so in elliptical lenses. Many of the bones which were discovered in articulation, were accompanied by intact skulls. The deposit, formed by flood water sweeping carcases into the cave, has been dated to about 29 300 B.P. (Ly - 359). The assemblage is examined in the belief that it is of value to have examples of age distributions which are unaffected by man. These may be compared to similar distributions from archaeological sites.

Levine (1983) presents several population structure models which are applied to various archaeological sites, and in so doing provides us with a standard method which may be applied elsewhere. The value of such theses as these lies in the degree to which reported results may be applied to others.

Cluster analytical techniques are also applied by Hemingway (1980) in a consideration of French Initial Magdalenian faunas. In this case the results are discussed in terms of associated ecological conditions. Four groups are identified:

1. Reindeer

2. (A) Woodland temperate group, in which temperature requirements are constant but with a wide tolerance range in precipitation. Only tundra zones appear to be avoided (Hemingway 1980:69).

3. (B) Mountain and cold species, including the more tolerant carnivores.

4. (C) Large herbivores of open steppe and grassland.

Applying these results and a knowledge of the then available associated seasonality data, Hemingway suggests a model in which winter occupation sites (1 and 4) are to be found in Western France and summer occupation sites farther to the East.

The major publication of 1980 is that of Laville et. al. 'Rockshelters of the Perigord'. In this, faunal material is considered, as is pollen analysis, as 'auxiliary evidence'. It is considered of value in the establishment of chronologies due to the fact that extinctions and replacement of one species by another occurred and can be identified from the faunal assemblages. The claim is that the real value of palaeontological data lies in the reconstruction of climate and complements the sedimentological results on which climatic frameworks are based (Laville et. al. 1980:96). However, despite the predominance of climatic considerations the authors are well aware of the role of cultural variables in assemblage formation, and state that these assemblages do not necessarily represent a 'representative cross-section of the potentially exploitable animal populations that occupied the Perigord at the time' (Laville et. al. 1980:351).

In 1982 Le Gall presented the results of his analysis of fish assemblages from South West France. His consideration of seasonal occupation of sites presents a picture somewhat different from recent reindeer based discussions (Gordon 1982). Supporting Bouchud's (1966) conclusions, evidence of year-round occupation is provided.

2.7 The Current Situation.

The publication of Delpech's doctoral thesis, in modified format (1983) has provided archaeologists with a synthesis of some of the available faunal data of Upper Palaeolithic age in South West France. It is a useful summary of the chronological and climatic significance of various species. In addition the author considers the spatial distribution of reindeer frequencies at selected sites, during Dryas I (Delpech 1983:173) but fails to discuss the human behavioural implications of assemblages. She follows this with a relatively detailed geographic discussion of the saiga, building on Lacorre's (1939) report.

The importance of three ungulate groups is considered by Delpech et. al. (1983) in an account of Quaternary palaeoclimates.

These groups consist of species associations commonly encountered in deposits in the region:

1. Open range, Arctic group (reindeer, ibex, chamois)
2. Open non-Arctic group (saiga, horse, bovids)
3. Woodland group (red deer, doe deer, boar)

The frequency of each group is considered at various sites and observed fluctuations therein are noted.

1983 also saw the publication of a volume of papers produced in homage to Jean Bouchud. These papers represent many of the current research interests of French specialists, ranging over the total length of the Palaeolithic and covering the (usually climatic and chronological) significance of various species (eg: de Sonneville-Bordes & Laurent 1983; Poplin 1983; Guerin & Faure 1983).

In 1984 we have a volume dedicated to the important site of La Ferrassie (Delporte 1984). In it we are presented with an account of the faunal material recovered during excavations. Size variations in the individual animals and chronoclimatic implications thereof are discussed (Delpech 1984). The volume, as a whole, represents a schematic account in which data are considered under standard headings for example, rodents, birds and lagomorphs.

If we turn now to analysis undertaken by Anglo-American specialists in the last few years, three general trends are noticeable. Firstly, the significance of carnivore populations is now being considered (Straus 1983). Carnivores do not represent a large part of recovered faunal assemblages in South West France, unlike their counterparts in Northern Europe and Northern Spain. However, the existence of such species should not be forgotten as they represent man's competitors for available food.

Secondly, element representation and butchery data are now being considered (Olsen 1987). In a discussion of the use of reindeer carcases by the occupants of the Grotte des Eyzies, she concludes that entire carcases were brought to the site for dismemberment. In addition to treating the animal as a food source Olsen discusses the potential non-food yield. She concludes that, with a short hunting season, meat should have been stored and alternative resources used. We can see here the substantial influence of Binford's work among the Nunamiut (1978) enabling us at last to progress beyond palaeoclimatic reconstructions. Similar work has also been undertaken by Chase (1986) although it is Middle Palaeolithic material from Combe Grenal which forms the database under consideration.

The third and final development has been the marked increase in attempts to describe and explain subsistence and social systems in South West France. Models are being put forward in which various species such as salmon (Jochim 1983) or reindeer and other large herbivores (Mellars 1985) are looked upon as important resources. The extent of dependence upon such species as salmon has caused some dispute in considerations of the 'social complexity' of South West France as revealed, for example, by the quantity of and variety in art forms in the region.

When employing an ecological framework to explain complexity in the Perigord, Mellars (1985:273) remarks upon the existence of a relatively stable climate for 20 000 years and the simultaneity of changes in climate and cultural material at the end of the last glacial. He also claims that the region in question is a "well-defined, ecologically uniform geographic area" (1985:273). However, he goes on to emphasise the high diversity of resources, the broad diversity representing a patchy environment in which varying habitats abound. For example, the possible closeness of winter and summer ranges may preclude the necessity for long distance migrations.

The existence of a mixture of biotopes and the accompanying diversity of resources provide various alternatives in the event of a shortfall in the preferred items. The implications of this and the associated high overall carrying capacity are considerable for local population density. It is suggested that much of the archaeological record is the result of population concentrations rather than an evenly spread, relatively high population density (Mellars 1985). Studies of this nature are increasing in frequency, in which the discussion of implications of faunal analysis forms a major part.

2.8. General Conclusions

At the point at which the present research was undertaken there seemed to be two generally accepted approaches to the study of French Upper Palaeolithic fauna. The first of these is represented by the work of Delpech (1983, 1984) and Laville et. al. (1980, 1983). In these, faunal assemblages are largely seen to represent the climatic conditions prevalent at the time of formation, in conjunction with local topographic factors. Fauna are considered in terms of various combinations of conditions. Hence we find reference to 'cold, dry', 'cold, humid', 'temperate, dry' and 'temperate, humid' regimes. Variations on these climatic themes allow the consideration of various forms of vegetational environment:

1. Open tundra
2. Forest tundra
3. Steppe
4. Temperate woodland
5. Open woodland
6. Others

Fluctuations in environmental conditions can be identified and the recognition of such changes from site to site and level to level may lead to the establishment of a relative chronology. Faunal extinctions and replacements, reflecting the effects of palaeoclimatic changes, allow us to construct such chronologies.

The second approach stems from archaeozoological research recently undertaken in the U.S.A. and the implications thereof. Attempts are being made to use faunal data as a means of determining the nature of the social and economic system. The data take the form of element counts and butchery data (Chase 1986) and frequency of species types (Straus 1983, Mellars 1985). In addition attention is being placed on the type of information which faunal material can yield. Seasonality issues are now being considered in detail the results of Gordon's research being discussed in chapter V. Researchers are beginning to focus, once again, upon the possible links between the environment and art works which have been recovered both in S.W. France and elsewhere (Jochim 1983).

We are no longer concerned with only one animal species - the reindeer. The existence of a broad economy is often postulated in which several alternative, albeit secondary, resources were available. The impact of ecological and socioecological (Winterhalder and Smith 1982) studies by American and British specialists is beginning to impinge. It could be said that current research is somewhat reminiscent of the Palaeoeconomic Cambridge School, with the addition of some mathematical input and a characteristically American modelling approach.

In summary, the whole system is now being considered in both cultural and environmental terms, and each approach should complement the other. It is only by looking at each aspect of the organisation of assemblages (environmental, chronological, spatial, cultural etc.) that we can hope to eventually determine the nature of Upper Palaeolithic adaptation.

Chapter Three

CHRONOLOGY AND ENVIRONMENT.

The aim of this chapter is to present an outline of the environmental and chronological background of the Upper Palaeolithic of South West France, against which is set the subsequent discussion of our faunal data.

The chapter begins with a summary of the physical geology of the Aquitaine and surrounding regions, continuing with a discussion of the Perigord chronoclimatic framework established primarily by Laville (1973) on the basis of geological (sedimentological) (Laville 1975, Laville et. al. 1980, 1983), palynological (Leroi-Gourhan 1984, Turner & Hannon 1988) and isotopic evidence (Laville et. al. 1983). The changing climatic conditions which led to the patterning in the environmental data are then considered. There then follows an outline of the cultural chronology of South West France which incorporates recently obtained radiocarbon dates. The chapter finishes with a discussion of environmental conditions which might be considered to have been "typical" of cold (stadial/glacial) and warm (inter-stadial/inter-glacial) episodes of climatic history.

3.1 Physical Geology.

Aquitaine is a tertiary basin of Molasse (Oligocene and Miocene sandstones and sandy marls) of fluvio-lacustrine origin (Durand-Delga 1980), triangular in shape with the Bay of Biscay forming the base, the Col de Naurouze the apex and the Pyrenees and Central Highlands the arms (Laborde 1961). The region was formed by the progressive drying of marine sediments during the Oligocene. Subterranean fluvial erosion led to the formation of caves and underground galleries, while surface water carved out the early stages in the development of the valley-interfluve landscape typical of parts of the region today.

The basement rocks of the Aquitaine Basin are of Precambrian to Palaeozoic age, the floor of the region dipping southwards to the Pyrenees and westwards to the coast. This is due to the Hercynian Orogeny which produced a marked NE/SW dip in the originally horizontal deposits (White 1985). The maximum depth exceeds -7000m on the Atlantic coast, south of Bordeaux.

A major south-south-west trending fault from the northern edge of the Montagne Noire provides the main complication, forming the southern boundary of the Mesozoic plateau which dominates the northern half of the region.

It is the Quaternary period which provides the majority of the deposits in the region, largely as a result of the erosion of older land surfaces and the redeposition of derived material. Deposits occur in an alluvial fan (as gravels) at the foot of the Central

Pyrenees. They form river silt along the flood plain of the Garonne and the course of the Dordogne and Vezere (Vita-Finzi 1978) while, although the sand dunes along the present Atlantic coastline are thought to be strictly Holocene in age (Vigneaux 1975), much of the surface deposits inland are considered to be of substantially earlier date (post-dating 20 000 B.P.), for example, the sediments comprising the Montignac deposit (the major Quaternary sedimentary unit in the Dordogne region) date to between 22 000 and 9 000 B.P. (Vita-Finzi 1978:44). In the Charente abundant evidence of quaternary deposits is found in the form of river terraces, slope deposits, cryogenic features and cave/rockshelter deposits (Vigneaux 1975, see p.62, below).

The region with which we are concerned provides a wide variety of landscape types, varying from the granite plateaux of the Limousin Montagne, the limestone plateaux and river valleys of the Perigord, the deep cut Jurassic limestone valleys of Quercy and alluvial plains of the Garonne in the Agenais area, and Gironde.

Eastern Limousin forms the western bastion of the Massif Central (see fig. 3.1) a survival from the Hercynian mountain system. The Massif Central comprises granite and crystalline schists which were uplifted by the tectonic movements which caused the Alpine folds. Faulting occurred along the eastern and southern left escarpments facing the Rhone and Mediterranean, while the plateau itself was subjected to minor faulting resulting in rifts (now the courses of the Allier and Upper Loire) and the rejuvenation of the drainage system. At the same time rivers cut deeply into the rock, especially in the Jurassic limestone of the south west, eg: Dordogne, Lot and Tarn.

The Limousin itself is an area of plateau which cuts across the crystalline rocks and Jurassic sediments (Sparks 1972:471). The plateau surface tilts down in a north westerly direction and is drained by the Charente and feeders of the Dordogne and Loire (Laborde 1961).

Throughout the Aquitaine, from north to south and east to west, as one meets more recent deposits and landscapes, the relief becomes more gentle. The plateaux west of the Limousin Montagne are low-lying, reaching altitudes of 400 - 500 m: the Ambazac Hills are a part of the crystalline formation of the Massif Central, and like the Bas-Limousin plateau between the Vezere and Dordogne rivers have a relatively mild climate.

The Brive basin, situated in the east of our region, is a warm, sheltered sandstone depression in which topographic features reinforce the mild aspect of the climate and are likely to have had a similar effect 18 000 years ago (Demars 1982). Meanwhile the Perigord and Quercy plateaux are quite different. The former comprises Cretaceous limestone whereas the latter is of Jurassic limestone bedrock and yields classic karstic features in the Causses area eg: underground rivers of the Martel, Gramat and Limagne Causses.

In both areas the relief owes something to changing river erosion. Downcutting into deposits, which vary horizontally as regards their degree of resistance, has led to changes in river courses, although in general, it is true to say that river courses remained much the same, during the Wurm, as they are today in the confined gorges of the Vezere and Dordogne. Upper Cretaceous

Fig. 3.1

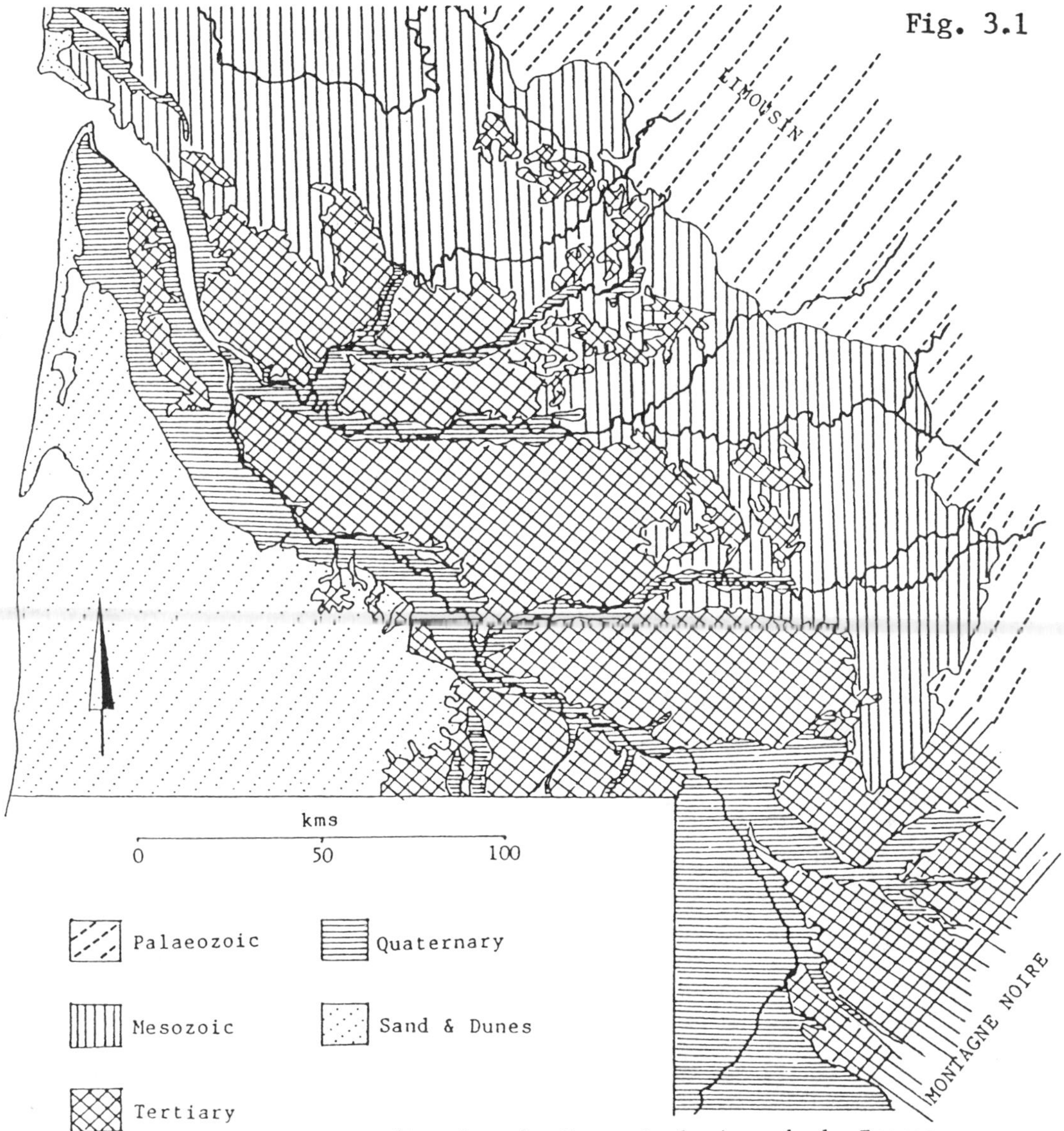

(Based on La Carte Geologique de la France,
Bureau de Recherches Geologiques et Minieres.)

Geological Map of South West France

limestone ranges from the dolomitic to marlaceous varieties (White 1985) and have considerable influence on the width of the valleys (Fenelon 1951). While marls and less resistant limestones may see changes in river course direction (and are often witness to wide, sloping river valley profiles), Perigord rivers change direction and cut deep into the rock at points where water courses cross synclines and anticlines bringing harder material to the surface. The valleys of the Dordogne, Lot and Vezere cut deeply into the Limestone plateaux, resulting in a landscape which is associated specifically with the Perigord region.

The area can be divided into two parts: the Perigord Blanc lies in the northern half of the region, associated with the Dronne, Isle and Auvezere rivers as they run through crystalline, Jurassic and Cretaceous bedrock (Cauvin 1971). The area is one of frequent outcrops of limestone, characterised by coppices rather than forests. Perigord Noir is a more forested area especially along the Dordogne itself and south into the alluvial plain. The area is centred upon the Dordogne-Vezere courses.

In the Quercy region, karstic features are observed. The dry limestone country between the Dordogne and Aveyron yields bare plateau surfaces known as the Causses de Quercy (Laborde 1961:91). Distinctive relief is a result, fluvial erosion and weathering varying with the permeability, jointing and solubility of the bedrock. Two types of solution hole are observed - (a) Creux: a funnel-shaped depression with a hole at the centre, (b) Gouffre or Aven: shaft-like holes (often natural animal traps, such as La Grotte Bernard (Bertouille et. al. 1969)). River level springs rather than surface tributaries are a common feature of karstic landscapes. They are more commonly observed in the Quercy region, surface tributaries such as those in the Roque St. Christophe area being the Perigord norm. However, in pockets of karstic relief, springs are occasionally observed, eg: the Sergeac valley, although these are in addition to surface water.

The geological feature most commonly associated with the Perigord area of France is the cave / rockshelter. These are not variations on one theme defined purely by size. Caves are formed as water permeates through joints and along bedding planes until it reaches impervious rock within or beneath the limestone formation. The water follows natural drainage directions (Holmes 1969) until reaching an exit. This may be many miles from the input point. When, and only when, the drainage course is established, material (dissolved in CO_2-charged water) is transported away, and fresh water, continuing the dissolution process, enters the system. In this way a series of channels and caves is produced. As the surface openings are enlarged cave mouths develop although the size thereof is often a poor guide to the extent of the formation inside.

Rockshelters are not necessarily the direct result of dissolution. Several modes of development have been postulated including the following which Laville et. al. (1980:46-52) consider to be of importance in S.W. France:

(a) Frost action cuts away at a cliff face

previously eroded by running water, eg: Abris at river level along the Vezere.

(b) Found above river level, most begin as niches resulting from chemical and mechanical weathering.

(c) Also above river level, caves may be exposed at the surface by cliff face retreat or tunnelling from the cave to the cliff face.

Thus, an 'abri sous roche' is a rockshelter upon the floor of which rock debris (eboulis), produced by frost shattering, is deposited. Such deposits eventually form much of the matrix comprising the stratigraphic levels from which artefactual and zoological material is recovered.

Quaternary deposits are to be found in the form of river sediments and cave fills inland, while the Landes is a region of widespread Quaternary alluvia and aeolian sand deposits laid down during the course of the last two stadials of the Wurm Glaciation (Vigneaux 1975:25).

Between 18 000 B.P. and 10 000 B.P., sea-level rose from -120 metres to -50 metres. The Garonne-Gironde eroded and incised the bedrock encountered, resulting in large quantities of fluvial sand and gravel. The river took the form of a typical periglacial river network with an average gradient of 5.10^{-4}m (Vigneaux 1975) carrying a large volume of deposits, subsequently deposited as the Wurm terrace. At the time of the last glacial maximum (ca. 18 000 B.P.) sea-levels were substantially lower than they are today (Laville et. al. 1983) due primarily to the increased volume of land-locked water (ice) and as a result, the area of additional exposed land was considerable.

Between 10 000 B.P. and 6 000 B.P. sea-levels rose from -50m to -10m while the riverbed slope declined, transporting substantially less material - most of which was granulometrically fine. At this stage the river developed a series of meanders (Vigneaux 1975). The following table (table 3.1) shows the gradual rise in sea-levels since 18 000 B.P.:

Table 3.1 Sea-level Changes since 18 000 B.P.

B.P.	Sea level (O.D.)
18 000	- 120 m
13 000	- 90 to - 80 m
11 000	- 60 m
10 000	- 50 m
9 000	- 40 m
7 000	- 12 m
6 000	- 10 m

(Source: After Thibault 1979;
Castaing et. al. 1971)

Examination of aerial photographs, topographic and geological maps of Perigord shows that the area has not, and never could have had, a uniform or homogeneous environment. Topographic variations force us to consider the region as a 'complex network of microenvironments' (Laville et. al. 1980:101). The rich mosaic of vegetation and landscape types throughout the course of the last 35 000 years provided a wide range of potential resources in an environment which will be considered later.

3.2 Chronology of the Upper Palaeolithic in South West France.

The establishment of both chronology and climate of the Upper Palaeolithic of South West France has been based largely on the sequence of stadials (cold) and interstadials (temperate) and their associated fluctuating environmental conditions. Traditionally, two main data-types are available for study, each of them having made substantial contributions to our knowledge of the Palaeolithic of the Perigord. In the discussion that follows we shall firstly consider the results of analysis of these databases, namely the sedimentological and palynological records. After this, an attempt will be made to integrate the available 'industries' and radiocarbon dates, especially those recently produced by the Oxford Accelerator Unit.

3.2.1 Sedimentology

Much of the sedimentological research of the 1960s and 1970s emerged from Laville's monumental study of the rockshelter and cave deposits of the Perigord (1975). In it he produced a scheme which has been widely applied in studies of the Palaeolithic in S.W.France and elsewhere. It is important to remember, however, that Laville's chronoclimatic scheme was devised solely for the Perigord, so that differences may be observed in other regions. (For example, see fig. 3.2 in which differences are observed between the Perigord (Dordogne - Vezere) and the Agenais (Lot-et-Garonne) at a distance of 90 - 100 km).

After systematic fieldwork, careful recording (including photographic) of the type section to be considered, detailed description of sediments as they occur in the section and a tentative description of the depositional history of the site (Laville et. al. 1980), summaries of archaeological occupation levels and associated cultural debris and pedological horizons are made. The latter, as is pointed out, are not true stratigraphic units in the strict sense, but are included for the sake of completeness. They result from post-depositional biological and chemical weathering of already existing sediments.

At least one (but preferably more) characteristic sampling column of sediments is taken from each sedimentary unit. Blocks of diameter over 100mm are removed prior to transfer to the laboratory, morphological examination of these larger elements being completed on site so as to establish whether the slabs result from roof collapse or strictly climate-related processes.

In the laboratory the sediment from each separate sampling column is dried. The artefacts, bones etc are removed. The sediment

The Upper Palaeolithic Climatic Stratigraphy of the Lot-et-Garonne and Perigord.

Fig. 3.2

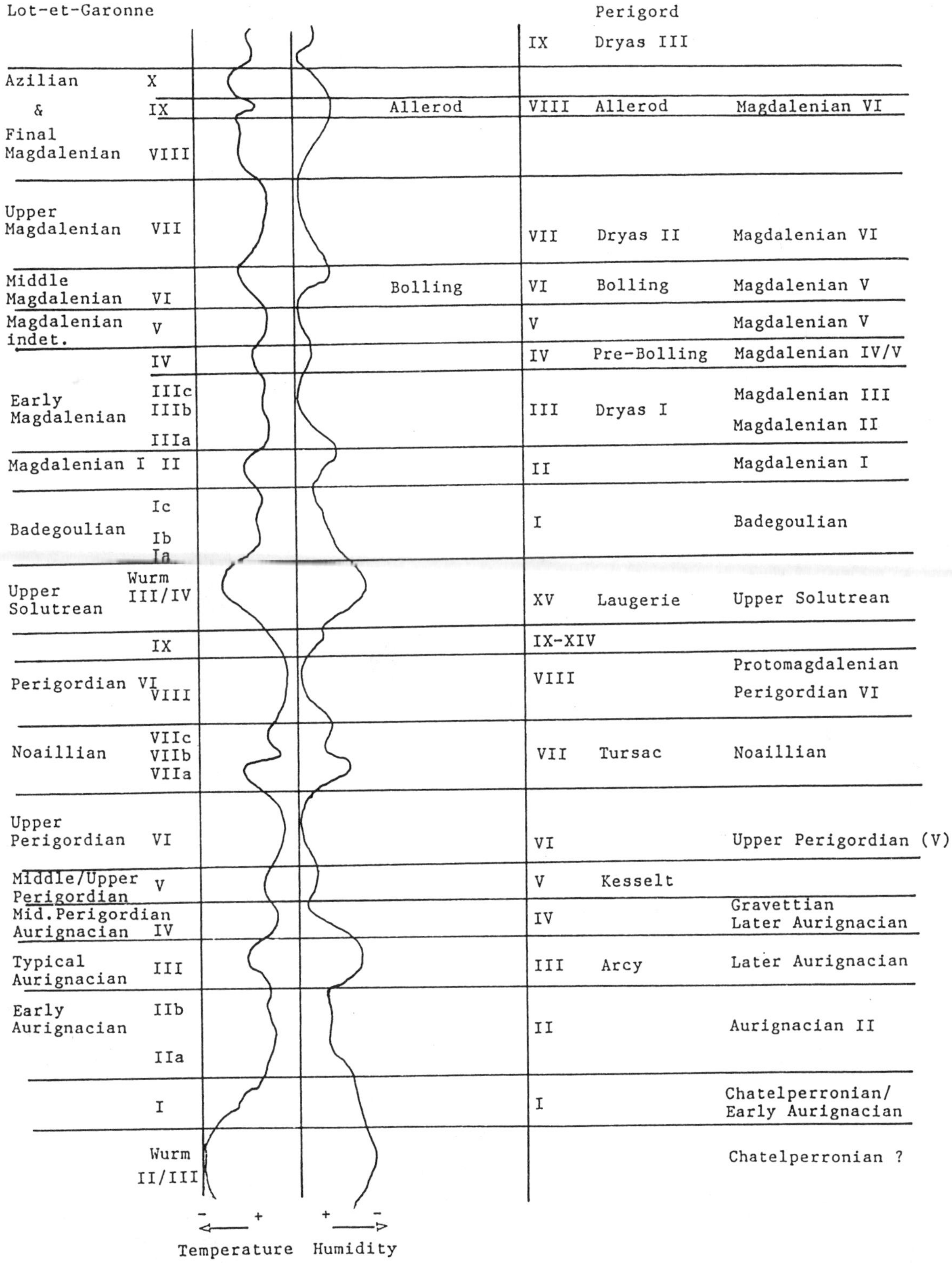

(Source: after Laville et. al. 1980, Le Tensorer 1981)

is then divided into a spectrum of size classes by granulometry, the categories being listed below. The percentage value of each is recorded and presented graphically in the form of a frequency polygon (Laville et. al. 1980:85).

Table 3.2. Major Granulometric Classes.

Size (mm)			
100^{+}	blocs	COARSE	
100-10	cailloux/ eboulis		rock fragments
10-5	granules	MEDIUM	
5-2	graines		
2-0.05	sables		
0.05-0.002	limons	FINE	matrix
0.002-	argiles/ colloides		

(Source: after Laville et. al. 1980:83)

The term (eboulis) is preferred in the description of rock fragments in general. No attempt is made here to translate the terms employed in the belief that much of their precise meaning is lost in so doing.

At this stage the coarse, medium and fine deposits are subjected to global granulometric, calcimetric and micromorphological analysis, once again being recorded and presented graphically. The choice of technique for each size category is shown in table 3.3. As a result of such analysis, based upon the composition of the sediments of each stratigraphic unit, cold (rigorous) and warm (mild) phases are identified (see table 3.4). The former are broadly associated with coarser sediments, cryoclastic debris resulting from ice- and frost-shattering, and decalcification. Milder episodes are characterised by fine deposits, sols d'alteration stalagmitic formations and calcareous 'crusts' on sediments, resulting from redeposition of carbonates.

It is not the intention to discuss in detail the results presented by Laville et. al. (1980). Suffice it to say that Wurm III is subdivided into fourteen phases, Perigord I - XIV, a scheme we see repeated by Le Tensorer (1981) in his discussion of the Agenais region where Wurm III, Lot-et-Garonne I - IX are identified.

Laville et. al. (1980) divide the Upper Palaeolithic into two groups of sites, groups II and III, group I representing the earlier, Lower and Middle Palaeolithic deposits. Group II is divided into nine subphases, some of which are further subdivided. The broad scheme is outlined in figure 3.2, showing its structure in both the Perigord (after Laville et. al. 1980) and Lot-et-Garonne (after Le Tensorer 1981). Group III comprises the later Wurm III deposits (X - XIV) and all of Wurm IV, making a total of sixteen phases, including the Wurm III - IV interstadial. The distinction between these phases rests largely upon the alternation of cold and dry (rigorous), and mild and

Table 3.3 Laboratory Analysis of Rockshelter Deposits.

COARSE FRACTION	MEDIUM FRACTION	FINE FRACTION
Sieved into 9 size categories from 10-20mm to 90-100mm. Weighed, 'percentaged', graphed. Larger categories more important in cases of frost shattering in cold, rigorous conditions.		Screening through 12 sieves of progressively smaller mesh size. Silts and clay divided by densimetry. Repeated after acid treatment. Difference reflects chemical weathering. Proportions of sands, silts and clays shown as %s. Sands/silt may be divided into 23 groups.
Variations in form exhibited by stones, reflecting their origin & type of secondary post-deposition modification. Frost weathering: frost slabs, frost-cracked stones and fissured stones. Identified & counted in main size group. Primary cryoclastic weathering of rockfaces produces frost slabs. Frost cracked & fissured stones result from secondary weathering of éboulis after deposition. Fissures reflect milder conditions. Blunting & porosity: degree of secondary alteration. 4 categories of bluntness, summed into a Bluntness Index. Bluntness caused by mechanical & chemical agents. Carbonate concretions: illuvial concretions due to leaching.	Secondary frost weathering on 'granules' & leaching may be observed. Allocthonous material may be identified. Petrographic examination of mineralogical component recommended.	Calcimetry indicates loss of calcium carbonate to chemical leaching. Degree of leaching determined by remaining quantity of carbonate. Other techniques : X-ray diffraction, pH (acidity) comparison of samples. Morphoscopy: examination of grains under magnification.

(Source: after Laville et. al. 1980:83-92).

Table 3.4 Chronostratigraphic Scheme for Würm III & IV

B.C.	Phase	Climate	
9000	X	Temperate	Azilian
	IX	Rigorous	Magdalenian VI
10000	VIII	Mild & humid	Magdalenian VI
	VII	Rigorous	Final Magdalenian (Magdalenian VI)
	VI	Mild & humid	Magdalenian V
	V	Rigorous	Magdalenian V
11000 12000	IV	Mild or temperate & humid	Magdalenian IV
14500	III	Rigorous	Magdalenian III Magdalenian II
16000	II	Mild & humid	Magdalenian I
	I	Rigorous	Badegoulian
18000	Würm III/IV	Temperate Very humid	Final Solutrean Upper - Final Solutrean
	XIV	Colder and relatively dry	Upper Solutrean Middle Solutrean
	XIII	Mild & humid	Middle Solutrean
	XII	Colder and relatively dry	Lower Solutrean
	XI	Mild & humid	Lower Solutrean
	X	Very cold & dry	Protosolutrean
20000	IX	Mild & humid	Aurignacian V
	VIII	Rigorous	Protomagdalenian Perigordian VI
	VII	Moderate & humid	Noaillian
	VI	Rigorous	Perigordian V (Font Robert & Noailles)
26000 27000	V	Mild & humid	Perigordian IV Evolved Aurignacian
	IV	Rigorous	Aurignacian III (Evolved Aurignacian)
	III	Mild & humid	Aurignacian II (Later Aurignacian)
31000	II	Very rigorous	Aurignacian II & I (Early Aurignacian)
	I	Moderately cold & very humid	Aurignacian I Chatelperronian

(Source: after Laville et.al. 1980)

humid (temperate) conditions, the characteristics of which were outlined very briefly above.

Since the publication in 1980 of Laville, Rigaud and Sackett's "Rockshelters of the Perigord", Laville has proposed a slightly modified framework for the chronostratigraphy of the Palaeolithic of Perigord (1988). Only the Upper Palaeolithic will concern us here. In the new scheme only one interstadial is recognised during the course of the Wurm, instead of three. Traditionally interstadials have been placed between Wurms I and II, II and III, and III and IV. Only that between Wurm II and III is retained in the new scheme with the name "Wurmian Interstadial". It is seen to be composed of three parts:

> 1. Alteration of sediments and formation of soil horizons during the climatic optimum.
>
> 2. A phase of erosion.
>
> 3. Slope wash and colluvial deposit accumulation in the final stages of the interstadial (Laville 1988:156).

By removing the Wurm III - IV interstadial we are left with the "Late Wurm", a period comprising twenty four climatic phases, of which phase XV corresponds to the Wurm III - IV interstadial (see table 3.5, in which the main features of the scheme are summarised). The fourteen climatic phases of Wurm III which saw the development of the Aurignacian, the Perigordian sequence and the early Solutrean remain unaltered. The later period's (Wurm IV) nine phases associated with the Magdalenian and Azilian become phases XVI to XXIV.

3.2.2 Palynology

The reconstruction of vegetational and associated climatic history is among the applications of pollen analysis listed by Moore & Webb (1978:2). The study of climatic change based upon palynological evidence depends upon the assumption that "variations in climate influence the constitution of vegetation " (Woodward 1987:1) and thus the assumption that a rise in tree pollen reflects a warming of climate and an associated increase in humidity (Leroi-Gourhan 1984) is one commonly made. Climate does play an important role in the determination of vegetation patterns, but is not alone in controlling the plant communities existing in an area. Other factors, such as the nature of the bedrock and thus the soil have determining roles. In order to employ vegetational changes as a means of dating deposits, a practice which is often adopted, it is important that the main changes in relative pollen frequency employed when determining pollen zones be contemporaneous and identifiable over large geographic areas (Moore & Webb 1978:5).

No two pollen samples derived from deposits representing the same time period furnish the same results. Laville et. al. (1980:97) use this as a reason to advocate the adoption of sedimentology as a means of reconstructing climatic history. They point out, for example,

Laville's Traditional and Revised Chronological Schemes

Table 3.5

SCHEME						
Old	B.P.	New				
IX		XXIV		rigorous		
VIII		XXIII				
VII		XXII				
VI	14000	XXI			thermophiles reappear	Absence of Thermophilous Trees
V		XX	Pleniglacial	rigorous		
IV		XIX		mild & humid		
III		XVIII		rigorous		
II	17000	XVII		mild & humid		
I		XVI		rigorous		
Würm III/IV	20000	XV		mild & humid		
XIV		XIV				
XIII		XIII		mild & humid	(not seen in pollen)	
XII		XII				
XI		XI		mild & humid	(not seen in pollen)	
X		X				
IX		IX		mild & humid	(not seen in pollen)	
VIII		VIII	Climatic Instability	rigorous	thermophiles disappear	
VII	23000	VII		mild & humid	TURSAC	
VI		VI		rigorous		
V		V		mild & humid	KESSELT	
IV		IV		rigorous		
III	30000	III		mild & humid	ARCY	
II		II				
I		I				
	35000	Würmian Interstadial				

(Source: after Laville 1988:149).

that the composition of a pollen sample is determined, or at least influenced, by the topographic position of the site, especially in relation to such factors as the prevailing winds. It is important to note that site position will also influence the sedimentological record at each location. Thus geological and botanical evidence have to be treated with equal respect.

The preparation of pollen samples for analysis and subsequent interpretation entails the division of samples taken from each level of a site into a range of plant forms (Laville et. al. 1980). On microscopic identification arboreal and non-arboreal pollen are distinguished, each group then being further divided according to species and finally their associated environmental requirements (for example, the absence of frosts, humidity). Each pollen type is counted and converted to relative frequency, the sample from each level being placed in relative chronological position so as to produce a composite profile.

The techniques, both physical and chemical, which are commonly employed in the laboratory preparation of pollen samples will not concern us here. It is the results obtained which will form the focus of our attention, for the relevance of techniques depends largely upon the mineralogical composition of the associated sediment (Renault-Miskovsky 1986:60) and the potential survival rating of pollen grains in the sediment. As a result of preservation problems, pollen transport within the soil/sediment profile, and other taphonomic factors, the size of pollen samples from cave and rockshelter deposits is often small - a problem which has recently been emphasised (Turner & Hannon 1988) and to the consequences of which we shall return later.

The reader concerned with the methodologies commonly adopted is referred to Birks & Birks (1980), Bradley (1985) and Leroi-Gourhan & Renault-Miskovsky (1977).

Much of the palynological research undertaken relating to later Wurm (Upper Palaeolithic) deposits in South West France has been undertaken by M.-M. Paquereau (1970a). In a detailed synthesis of the climatic history of the Aquitaine and Bay of Biscay, she and several others presented a summary of the state of research in 1983 (Laville et. al. 1983). This and the scheme presented by Paquereau (1980) will form the basis of our discussion.

Table 3.6 shows, in detail, the chronological changes identified in the vegetational cover of South West France between approximately 35 000 and 10 000 B.P. The predominant ground flora and tree cover are listed, as is the associated (landscape) type. The terms have been translated and, where possible, equated with those employed by Paquereau in 1970 and summarised by Laville et. al. (1980).

Three phases of the Wurm II-III Interstadial have been identified from the sedimentological data available: a climatic optimum, a period of erosion and a phase of colluviation and run-off. At La Ferrassie the last stage of the Interstadial is identified and dated to 36 000 - 34 000 B.P., the vegetation being characterised by a strong woodland component (A.P. = 48 %) and an established temperate undergrowth with grasses, sedges and ferns (implying the existence of humid conditions).

Table 3.6

Chronological Changes in Upper Palaeolithic Vegetation Cover in South West France

Stage	Ground flora	Tree cover	Landscape	Climate	Site / C14 date
IX	Grass, sedge hygrophiles	Pine, birch	Cold meadowland	rigorous	
VIII	Hygrophiles	Oak, hazel, elm, lime, alder, willow, pine, birch	Mixed oak forest	warm and humid	
VII	Grass, Compositae	Pine, hazel, birch, alder, willow	Open parkland	rigorous	Pont d'Ambon c.3 10 350 B.P.
VI	Grasses, sedges, hygrophiles	Hazel, alder, oak pine	Open woodland*	mild & humid	
V	Grass, Compositae	Pine, birch, willow hazel	Cold meadowland	rigorous	Le Flageolet II c.11 12 870 B.P.
IV	Grasses, sedges	Hazel, alder, willow pine, birch	Cool parkland	mild & humid	
III	Compositae, grass	Pine	Cold steppe	rigorous	Laugerie Haute Est c.2 13 990 B.P.
II	Grass, sedges hygrophiles	Alder, willow, hazel pine	Cool parkland	mild & humid	Laugerie Haute Est c.19 17 040 B.P.
I	Grass, Compositae	Pine	Cold steppe	rigorous	Laugerie Haute Est c.12-18 18 260 B.P.
Würm III/IV = XV	Grass, sedges hygrophiles	Alder, willow, hazel pine, oak	Closed parkland	mild & humid	Laugerie Haute Ouest c.5 19 600 B.P.
XIV	Grass, Artemisia heliophiles	Pine	Cold meadowland	rigorous	
XIII	Grass, sedges	Hazel, pine, willow birch, alder	Cool parkland	mild & humid	
XII	Grass, <u>Artemisia</u> sp. heliophiles	Pine	Cold meadowland	rigorous	Laugerie haute Ouest c.12a 20 180 B.P.
XI	Grass, sedge Compositae	Pine, hazel, alder birch	Cool parkland	mild & humid	Laugerie Haute Ouest c.12d 20 750 B.P.
X	Grass, Cichoriae	Pine	Cold steppe	very rigorous	Abri Pataud c.2 22 000 B.P. Laugerie Haute Ouest c.15 21 980 B.P.
IX	Grass, sedge	Pine, hazel, alder birch, willow	Cool parkland	mild & humid	
VIII	Cichoriae, grass	Pine	Cold steppe	very rigorous	
VII	Grass, sedge hygrophiles	Hazel, alder, willow oak	Closed parkland	mild & humid	Abri Pataud c.3 23 000 B.P.
VI	Cichoriae, grass	Pine	Cold steppe	very rigorous	
V	Grass, sedge hygrophiles	Hazel, pine, alder oak	Closed parkland	mild & humid	
IV	Cichoriae, grass heliophiles	Pine	Cold steppe	very rigorous	Abri Pataud c.7 32 800 B.P.
III	Grass, sedge heliophiles	Hazel, pine, alder willow, oak	Cool parkland	mild & humid	
II	Grass, Compositae	Pine, birch	Cold meadowland	rigorous	Abri Pataud c.12 33 000 B.P.
I	Grass, Compositae sedges	Pine, willow, birch hazel, alder	Cool parkland	unstable cold, mild humid	

* Temperate woodland (Paquereau 1970)

(Source: after Paquereau 1980)

Wurm III, phase I sees a fall in arboreal pollen (henceforth A.P.) totals, although pine (dominant), willow and birch remain. Hygrophiles are locally abundant, steppic elements being absent. In phase II A.P. totals are low, the landscape being open and dry with an important Cichoriae component.

Phases III, V and VII are all characterised by improved conditions. A.P. rises to 25 - 40 % with hazel as the predominant deciduous element. Lime, alder, willow and hornbeam are identified although the existence of hornbeam has recently been questioned (Turner & Hannon 1988). There is a large open woodland or closed parkland cover, although the more thermophilous species are lacking. These phases (III, V and VII) have been termed the Arcy, Kesselt and Tursac Interstadials respectively (Laville et. al. 1980).

Phases IV, VI and VIII meanwhile see a return to cold and dry conditions, a phenomenon reflected by both pollen and sedimentological data. Arboreal pollen reaches only 5 - 10 %, accompanied by important frequencies of xerophilous species such as <u>Ephedra</u> and <u>Galium</u>. We see a steppic environment (Laville et. al 1983).

The Tursac Interstadial (phase VII) is identified from both the pollen and sediments available. The period is traditionally dated to approximately 24 000 to 23 000 B.P., three subdivisions being recognised:

- VIIa, mild and humid. A.P. - ca. 30 %
- VIIb, colder. A.P. falls, pine predominates
- VIIc, warmer and humid. A.P. reaches 40 %, with numerous thermophilous species.

Phases IX to XIV, at the end of Wurm III, are known only from Laugerie Haute and Le Malpas (Laville et. al. 1983). Phases IX, XI and XIII represent the milder, more temperate periods according to the sedimentological data. The pollen however is less clear. A.P. never exceeds 22% of the total, comprising primarily pine with some birch and hazel. Thermophilous species, such as oak, are often completely absent (Laville et. al. 1983:228) while the occurrence of other deciduous trees can be described only as sporadic.

Of phases X, XII and XIV, the first is the most rigorous, being a true steppe. XII and XIV are characterised by a limited quantity of birch, hazel and willow, while steppic elements are rare.

The Wurm III - IV Interstadial is less marked than is its predecessor (Wurm II - III). A.P. reaches only 30%, being composed primarily of pine. Hazel, alder and willow do however exist, although in low frequencies. The rarity of the thermophilous species which might be expected during the Interstadial implies that temperatures remained low, although the abundance of hygrophiles highlights the degree of humidity.

Wurm IV, phase I is characterised by low A.P. frequencies, most of which are composed of pine with some birch, while heliophiles and Compositae are of importance. Phase II is more humid, seeing the return of oak, although in low frequencies. During phase III humidity and temperature again decline, a cold steppic environment predominating. Phase IV (the Pre-Bolling) sees a rise in A.P. (pine, birch, willow and alder) but no broad-leaf thermophiles are present. The Dryas I of phase V which follows is cold, but concludes with a rise in temperature and humidity. In the course of phase VI A.P. reaches 30%, comprising - in addition to the usual species - oak, lime, buckthorn, box and spindle (Laville et. al. 1983:231). During phase VII (Dryas II) we see once again the return of cold, steppic conditions.

The Allerod (phase VIII) is milder and more humid than other Wurm IV episodes, despite the interruption of a brief cold snap. Towards the end of phase VIII we see typical 'mixed oak forest', with hazel, elm, lime, alder and oak. Numerous hygrophiles comprise the ground flora.

Finally, in phase IX there is a fall in A.P. to ca.25% prior to the onset of Holocene conditions. Oak and other thermophiles disappear, grasses dominating the ground flora which still maintains a strong sedge and hygrophile component.

In 1988 a report concerning the vegetational history of S.W. Europe was published (Turner & Hannon 1988), based upon recent analysis of cave/rock shelter pollen data from S.W.Europe (Hannon et. al. 1987). Consideration of one type of palaeoenvironmental data can of course be hazardous. Ideally, several data types should be studies together. However, several important points are made, of which those of relevance to our area will be considered here.

The identification of the interstadials commonly depends upon fluctuations in the relative frequencies of pine and other more temperate arboreal species. Turner & Hannon begin their discussion of the cave and rockshelter deposits of South West France by contesting the existence of various interstadials mainly on the grounds of downwash through deposits. Thus the Arcy, Kesselt and Tursac Interstadials are rejected, despite the fact that both the pollen and sedimentological records provide firm evidence for, not only the earlier Cottes but also the Arcy Interstadial.

Turner & Hannon (1988:471) also point out that the Laugerie and Lascaux Interstadials, traditionally dated to between approximately 19 500 - 18 500 and 18 000 - 16 500 B.P. respectively, occur at a time at which oxygen-isotope analysis in the N. E. Atlantic (see below) suggests that conditions off the west coast were exceptionally cold. Rejecting the existence of the Laugerie and Lascaux Interstadials on the grounds of small sample size (in some cases fewer than 50 pollen grains), potential mixing by downwash from higher levels and the failure of pollen to agree with other forms of data, the interpretation of pollen data associated with these interstadials make, we are told, "no ecological sense" (Turner & Hannon 1988:475).

The existence of _Carpinus_ (hornbeam) and _Fagus_ (beech) in late Wurm assemblages, like that of _Pinus pinaster_ (maritime pine), is generally held to indicate warm, potentially interstadial conditions. Turner & Hannon (1988) point out, however, that according to Huntley & Birks (1983) _Carpinus_ and _Fagus_ fail to reach South West France until approximately 2000 years ago. Downwash from Holocene levels or laboratory contamination are proposed as possible sources of error.

The hornbeam associated with the Lascaux Interstadial should, Turner & Hannon claim (1988), be dismissed as irrelevant.

Melting of mountain glaciers was well underway by 16 000 B.P. following the glacial maximum at 20 000 to 18 000 B.P. Restrictions are placed on vegetation growth as much by the general aridity of the the climate as by fluctuating, low temperatures. We see a resulting arid steppe between 16 000 B.P. and 13 000 B.P.

No evidence is found among continental pollen profiles for the "Younger Dryas" (Dryas III), traditionally associated with harsh, rigorous conditions, although some evidence of Younger Dryas cooling is seen in marine core CH6719. Basing their argument on data from N.Spain Turner & Hannon find no sign of climatic deterioration between 11 000 and 10 000 B.P. However, they do acknowledge that such cooling may be possible at Le Moura (Pays Basque) close to the Atlantic (1988:480). Climatic improvement at about 10 000 B.P. is acknow-ledged throughout S. W. Europe, marked by an increase in birch, pine, hazel and oak pollen. However, the claim for Dordogne refugia and survival of thermophiles is refuted.

3.2.3 The Marine Record

The marine climatic record has also been considered recently both in terms of pollen content and oxygen-isotope concentration. The marine pollen diagrams available from the Bay of Biscay are problematic (Turner & Hannon 1988:469). Exceedingly high pine frequencies (80-95% total pollen) and very low Graminae bias the samples, implying a large degree of sorting and degradation. The profiles are not directly comparable with those from terrestrial sources due to the over-representation of pine. Furthermore, the cores available have not been radiometrically dated.

Turner & Hannon (1988) conclude their discussion of results reported by Duplessy et. al. (1981) and Pujol & Turon (1974) by warning the reader that only broad correlations between marine and terrestrial sequences may be expected.

The oxygen-isotope record as summarised by Laville et. al. (1983) reveals a more concise picture of climatic events during Wurm III and IV.

The end of isotope stage 3 coincides with the earliest part of Wurm III, assuming that the Cottes and final stage of the Wurm II/III interstadial are equated with one another. All of stage 2 and part of stage 1 then fall within the Upper Palaeolithic time span.

Between 29 000 and 18 000 B.P. the isotope record reflects the rigorous nature of much of the climate. Ice volume reaches a maximum at ca. 18 000 B.P., when summer surface-water temperatures fall below 6°C, arctic water entering the Gulf of Gascony during early parts of isotope stage 2. At 18 000 B.P. we see the first sign of ice volume melt. This process has been divided into four parts which are summarised in conjunction with the major pollen changes in table 3.7.

Termination IA, beginning soon after 18 000 B.P., marks the major phase of deglaciation between 15 000 and 13 500 B.P. The isotope record does not allow us to identify the Laugerie and Lascaux Interstadials. During Event I summer surface-water temperatures rise to ca. 20°C, almost equalling those of today. Event 2 temperatures fall to 11 - 12°C.

Laville et. al. (1983) also consider the pollen associated with the isotope record, and, in contrast to Turner & Hannon, conclude that the results compare well with those of the terrestrial pollen record.

Table 3.7 Deep Sea Record and Vegetation Change.

Isotope stage	B.P.	Marine Record		Terrestrial Record	
1	10 K	Termination IB	↑ Oak; Artemisia ↓	Mixed oak forest	Dryas III
	11 K	Event 2		steppe	
		Event 1	↓	Open Woodland	Allerod
					Dryas II
	13 K				Bolling
2		Termination JA	↑ Oak		Glacial
	15 K	. . . . . . glacial	↑ Birch	steppe	

(Source: after Laville et. al 1983:234)

3.3 The Climate in Western Europe.

We have established, from sedimentological, palynological and isotope analyses, that the climate of the Upper Palaeolithic was one of fluctuating temperatures (cold - temperate) and humidity (dry - humid). The discussion thus far has briefly considered the nature of the climate and vegetation, but not the causes. This remains to be done.

Between 30 000 and 16 000 B.P., the climate of Europe was essentially cold and dry, but often interspersed with shorter, milder episodes while the period between 20 000 and 16 000 B.P. is commonly regarded as the glacial extreme in Western Europe, European and Laurentide ice sheets being at their greatest extent.

Much of the understanding of the climatic conditions prevalent at 18 000 B.P. in Europe is the result of the deep sea core analysis and computer simulation undertaken (Ruddiman and McIntyre

1981; McIntyre et. al. 1976; Bradley 1985; Lowe and Walker 1984). Reconstructions are based upon the premise that the continental ice-sheets and sea-ice cover modified surface temperature patterns, and hence the location and intensity of surface winds and jet streams (Kutzbach and Wright (1985). Employing CLIMAP (1976, 1981) findings, temperature, winds and precipitation patterns are discerned. The series of results is of particular value because simulation studies, oxygen-isotope analysis and palynological conclusions are often seen to agree, at least to some extent, in terms of climatic history (Eicher et.al. 1981).

The Atlantic yields cores particularly conducive to the study of palaeoclimates at the time of the last glacial maximum (18 000 B.P.). The shape of the ocean is such that currents are forced into meridianal alignments; here watermass interchange occurs freely - between the inner Arctic and other latitudes. This is significant in considering atmospheric circulation, for polar water can flow southwards through the Norwegian Sea and lead to ice sheet formation over Northern Europe.

Lamb (1977:289) attributes much of the climate at 18 000 B.P. to easterly surface winds (which he suggests were light and variable in summer) and northerly winds over the sea between Greenland and North and West Europe.

At 18 000 B.P. the Arctic Polar Front (centred on lat. 42°N), provided the dividing line between northern 'dynamic' regimes and relatively stable, milder southern conditions (McIntyre et.al. 1976). The greatest temperature difference between the present and 18 000 B.P. is to be found between 42°N and 60°N latitude. North and south of this region differences are smaller. Average surface air temperatures in unglaciated Britain, far from the Atlantic, were ca. 10 - 12°C lower than at present, 8 - 10°C in summer, 12 - 14°C in winter. In continental Europe temperatures were 16°C lower. However, warm days of up to 30-35°C may well have occurred. Warm month temperatures may have been as high as, if not higher than today, for lower pressure centres deviated only slightly from their modern positions.

Seasonal temperature variations are significant. Kutzbach and Wright (1985) show that, south of the North European ice-sheet, January temperatures were 20 - 30°C lower than today, while summer temperatures (July/August) were lower by only 5°C. Based upon the fact that surface temperatures in August were substantially warmer than those in February, McIntyre et. al. (1976) suggest that along the western coast of Europe there was some northward flow of warmer water. November temperatures south of the Polar Front indicate that this was the season which was most similar to present conditions. Precipitation was lower in both seasons, air streams being cold and dry.

By 13 000 B.P. half of the northern hemisphere ice volume of the last glacial maximum had disappeared, despite the fact that the ice-sheets were still extensive in area (75-80 %). Ice disintegration by marine calving and downdraw was rapid, caused by a rise in summer insolation, rising sea-levels, a chilling of the sea surface (by meltwater, thus lessening atmospheric moisture extraction and transport to the ice-sheets) and winter-freezing of low salinity meltwater (Ruddiman and McIntyre 1981).

At approximately 13 000 B.P. the Polar Front moved northwards to the northwest section of the subpolar gyre. At 11 000 B.P. it moved 800km south, almost to its glacial maximum. Temperatures in summer fell from 14° to 7.5°C and in winter from 8° to 1.5°C during this, the 'Younger Dryas'. This was followed by retreat to the mouth of the Labrador Sea and close to the Greenland coast at 10 000 B.P.. Warming continued until ca. 6 000 B.P. It can be seen therefore, that deglaciation lasted 10 000 years.

Environmental changes after 10 000 B.P. owe much to the increased warmth of the Late Glacial (Dennell 1983:104) although rarely did variations in the various components thereof coincide. At this point, Europe was dominated by anticyclones and warm air from the South, with easterly winds, (normally resulting in cold winters). Prevailing circulation patterns continued to change (Lamb 1977). As American ice disintegrated, an east-west zonal course was adopted by winter circulation. Between 8 000 and 6 000 B.P. summers in Western Europe may have been 2°C warmer than today, associated with mild winters and increased precipitation totals.

3.4. Upper Palaeolithic Cultural Chronology in the Perigord.

In S.W.France the establishment of a cultural chronology has depended almost entirely upon the recognition of defined fossiles directeurs, namely stipulated stone and bone tools, considered characteristic of stages in the cultural development of the region.

During the 1930's Peyrony systematically defined the Early Upper Palaeolithic of S.W.France in terms of five Perigordian industries (Delporte 1976). His conclusions were based on work at such sites as Laugerie Haute, La Gravette and La Ferrassie. He described the Perigordian as one characterised by prismatic nuclei, abundant blades and bladelets (often modified and used as endscrapers), and a poor but developing bone industry. In 1936 he identified five types of Perigordian industry, terms which, to some extent, are still employed: Perigordian I (type de Chatelperron) or Chatelperronian; Perigordian II (type de Bos-del-Ser); Perigordian III (type de Laugerie Haute) now Perigordian VI; Perigordian IV (type de Gravette) or Gravettian); Perigordian V (type de la Font Robert). This scheme has subsequently been modified to take account of results of more recent excavations and associated analysis and forms the basis for the current discussion.

The Chatelperronian (Perigordian I or Lower Perigordian) is traditionally considered to be the earliest of the Upper Palaeolithic cultures. It is broadly characterised by Chatelperron points, endscrapers, burins, truncated and retouched elements, and a variety of Mousterian-type tools such as sidescrapers, notches and denticulates (Harrold 1988:157). A high proportion of these types are produced from blades derived from prismatic cores. The extent to which Chatelperronian industries are associated with Wurm II/III Interstadial deposits in S.W.France is debatable, for only at La Ferrassie do we find such a level and here its attribution is questioned (op.cit).

Intense erosion is often blamed for the apparent lack of Hengelo/Les Cottes Interstadial deposits (Leroyer & Leroi-Gourhan 1983:41), which, based upon two occurrences known to exist in France and a third in Northern Spain, one might expect to be associated with Chatelperronian industries. At Arcy-sur-Cure, St.Cesaire and Cueva Morin Lower Perigordian levels are known to date from this interstadial (see fig. 3.3). Chatelperronian industries from La Quina and Chatelperron itself (Allier) date to the earliest Wurm III, Perigord I deposits as described by Laville et. al. (1980).

The Aurignacian complex is traditionally seen to follow the Chatelperronian. It is characterised by more tool types than the former and a developing split-base bone point industry. For example, in the earliest Aurignacian ("0") keeled and nosed endscrapers develop, as do Dufour bladelets, while throughout the Aurignacian we find retouched blades and abundant bone tools (Straus & Heller 1988). The belief that the Lower Perigordian precedes the Aurignacian is based upon the fact that Aurignacian levels usually directly overlie Chatelperronian deposits. As Harrold points out, of the 22 Chatelperronian occurrences in France, 17 are known to have Aurignacian levels overlying them, either immediately or separated by sterile sediments. At 2 of the remaining sites however, Chatelperronian and Aurignacian industries interstratify, although in some cases the lithic assemblages are too small to allow detailed analysis.

At Roc de Combe and Le Piage we find the sequence outlined below (see table 3.8). Based on these examples and the degree of inter-site overlap demonstrated by Laville et. al. (1980) and Leroyer & Leroi-Gourhan (1983), the Chatelperronian and Aurignacian industries can not be viewed simply as successive stages in the Upper Palaeolithic chronological sequence. They are instead generally considered to represent contrasting cultural traditions, resulting from the activity of two distinct ethnic groups. However, despite the common assumption that the Aurignacian and Perigordian are "parallel phyla" of independent origin (Laville et. al. 1980:283), one is forced to conclude that the Chatelperronian does, in general, precede the Aurignacian, although with a period of overlap lasting between 1000 and 1500 years.

Table 3.8 <u>The Sequence of Chatelperronian and Aurignacian at Roc de Combe and Le Piage.</u>

Roc de Combe.		Le Piage.	
Aurignacian I	(level 7)	Aurignacian	(level F)
Chatelperronian	(level 8)	Chatelperronian	(level F1)
Aurignacian	(level 9)	Aurignacian	(level G-I)
Chatelperronian	(level 10)	Aurignacian	(level J)
		Aurignacian	(level K)

Based on these examples and the degree of inter-site overlap demonstrated by Laville et. al. (1980) and Leroyer & Leroi-Gourhan (1983), the Chatelperronian and Aurignacian industries can not be viewed simply as successive stages in the Upper Palaeolithic chronological sequence. They are instead generally considered to represent contrasting cultural traditions, resulting from the activity

Fig. 3.3
Figure showing the Relationship between Chatelperronian and Aurignacian Industries based upon Results of Sediment and Pollen Analyses.
(After Leroyer & Leroi-Gourhan 1983:42)

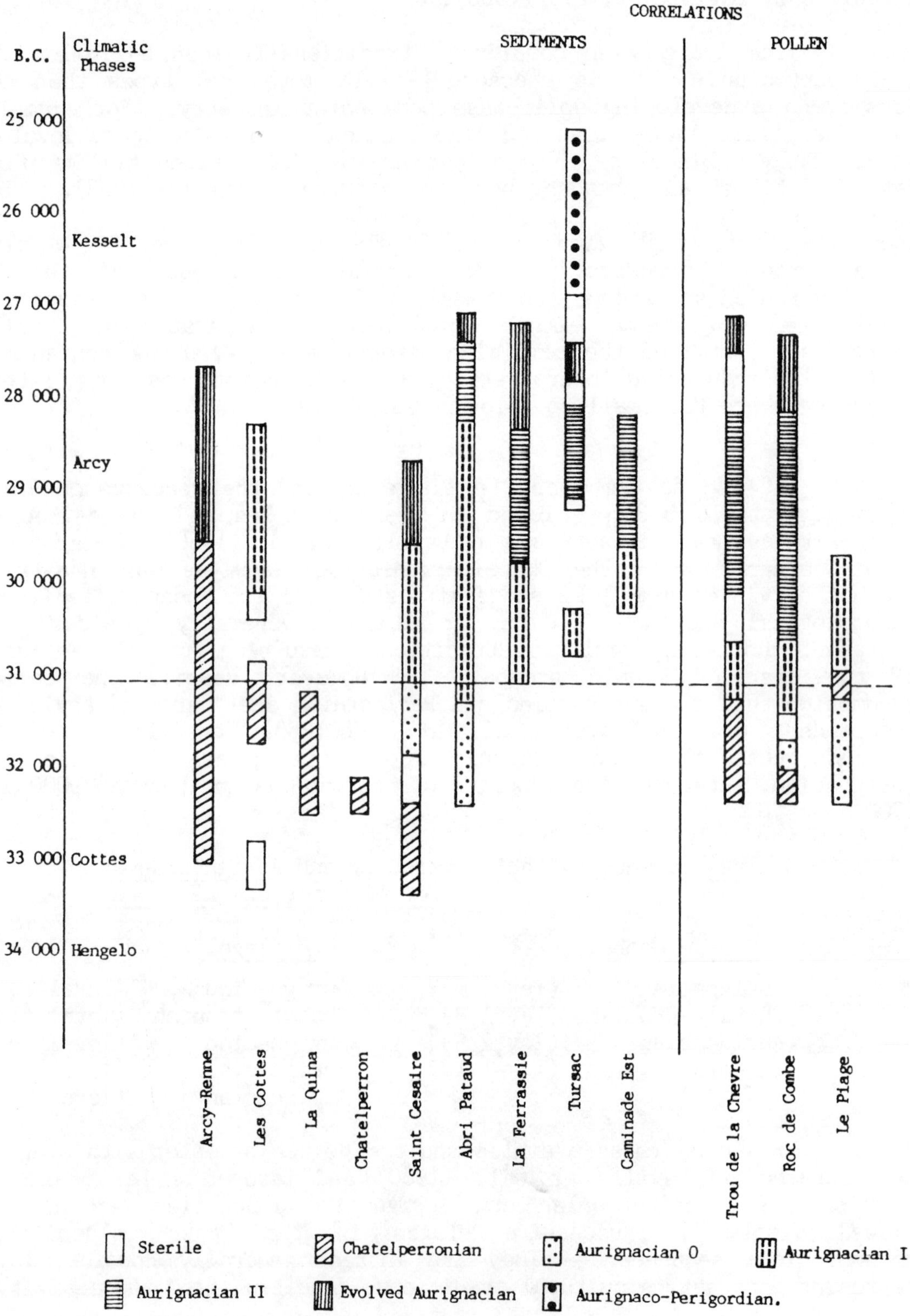

of two distinct ethnic groups. However, despite the common assumption that the Aurignacian and Perigordian are "parallel phyla" of independent origin (Laville et. al. 1980:283), one is forced to conclude that the Chatelperronian does, in general, precede the Aurignacian, although with a period of overlap lasting between 1000 and 1500 years.

The Chatelperronian of the Perigord is, in general, associated with deposits of Wurm III, Perigord I age (33 300+/-500 B.P. (GrN-4333) at Les Cottes, level G) with associated Early or Archaic Aurignacian deposits developing simultaneously. At Le Flageolet I, for example, level XI yields a date of 33 800+/-1800 B.P. (OxA-598). By Wurm III, Perigord II we are left with only Aurignacian deposits, with the exception of Chatelperron itself to the immediate east of our region (30 800+/- 500 B.P. (GrN-4258) at Les Cottes level E, Aurignacian I).

With the disappearance of the Chatelperronian the Aurignacian flourishes. Shortly after 30 000 B.P. it is well established (La Ferrassie K4, 28 600+/- 1050 B.P. (OxA-409), Aurignacian II [Mellars et. al. 1987:129]) to be replaced at ca. 28 000 B.P. by the Middle Perigordian (IV).

Between 28 000 and 27 000 B.P. the Aurignacian disappears to be replaced by the Gravettian or Perigordian IV - a period which is relatively poorly represented in comparison with others. It is characterised by varying abundance of truncation and dihedral burins (Rigaud 1988). Simple endscrapers on blades and flakes, and numerous backed pieces - including Gravette and microgravette points (which vary substantially in number) also characterise the period. Radiocarbon dates provided by the Groningen laboratory for Perigordian IV levels at the Abri Pataud date the industry to between 28 150+/-225 B.P. (GrN-4634) and 26 600+/-200 B.P. (GrN-4477), both from level 5, lens K-1 (Bricker & Mellars 1987:229). Two more recently obtained dates from Perigordian IV levels at Pataud agree in essence:

Level 5 (lens K-1)	28 400+/-1100 B.P.	OxA-169
Level 5 (lens R-3)	26 000+/-1000 B.P.	OxA-581

Perigordian V is more complex. Based upon a larger database the Upper Perigordian (V) has been divided into three successive stages, although the extent to which unilinear evolution is demonstrated is highly questionable (Laville & Rigaud 1973). The division was the result of work conducted at La Ferrassie by D. Peyrony who believed that a clear temporal sequence could be demonstrated. He found three levels containing Perigordian V material, each significantly different from the others.

Level L: directly superimposed on K. Contains Perigordian V^3 with Noailles burins.

Level K: directly above J. Contains Perigordian V^2 with truncated elements.

Level J: directly overlying Aurignacian IV in H'' (or separated by a layer or eboulis). Contains Perigordian V^1 with Font Robert points.

Subsequent analysis of the La Ferrassie deposits has allowed a reconsideration of the Upper Perigordian (Delporte & Tuffreau 1973; Delporte 1984) as revealed by a more detailed stratigraphy than that recorded by Peyrony (1934).

Table 3.9 La Ferrassie Stratigraphy, after Delporte (1976) & Mellars et. al. (1987)

	B3-B1	sterile	moderate & humid
	B4	Perigordian V^3	
	B5	eboulis	cold
OxA-401: 23 800+/-530 B.P.	C4-B6	sterile	moderate & humid
	D1	Perigordian V^2	climate improving
OxA-403: 27 530+/-720 B.P	D2	Perigordian V^1	cold
Gif-2698: 22 520+/-550 B.P.			
OxA-402: 27 900+/-770 B.P.			
Gif-2696: 23 900+/-550 B.P.			
	D3	Perigordian V^1	cold
Gif-2700: 22 520+/-550 B.P.	E1	Perigordian V^1	cold
Gif-2701: 23 580+/-550 B.P.			

Delporte (1976:23) identifies two facies of the Upper Perigordian with Font-Robert points (Perigordian V^1) and distinguishes between three subphases of the Perigordian with Noailles burins (or Noaillian). Perigordian V^{1a} (type Ferrassie) comprises numerous scrapers and Font-Robert points, gravette and microgravette points. Burins, including those on truncations, are rare. In Perigordian V^{1b} (type Roc de Combe, Le Flageolet) assemblages burins outnumber scrapers, dihedral burins exceeding those on truncation. The division of Perigordian V^1 is not supported by Rigaud (1988) on the grounds of both typology and practicality. As Rigaud says, there seems little point in identifying a group based solely on one occurrence, as in the case of Perigordian V^{1a}.

It is the Noaillian, or Perigordian V^3, which has received most attention (David 1973; David & Bricker 1987), a period which David interpreted as a distinct cultural entity - quite separate from contemporary Upper Perigordian industries in the Perigord. It is worth noting however, as Rigaud (1988) has subsequently pointed out, that the "gravettian character" is consistently present throughout the Upper Perigordian, as are Noailles burins, although Delporte (1976:27) does isolate three groups within the Noaillian complex which he sees as a result of functional or activity-related variability:

Perigordian V^{3a} as seen in the upper level at the Abri Labattut, comprises numerous Noailles burins, backed blades and points, and a regular gravette component.

Perigordian V^{3b} as seen at Le Facteur (level 10-11), has numerous Noailles burins but lacks backed blades and points.

Perigordian V^{3c} at La Rochette and Bassaler-Nord has fewer burins, backed blades and points.

Comparison of the chronological spread of Perigordian V^3 industries at various sites during Laville's (1975) Wurm III, Perigord VI-VIII (see table 3.10) leads Delporte to discard a chronological explanation of the variability which he identifies (1976:29) in favour of a functional alternative, the contemporaneity of the Noaillian and Perigordian 'cultures' thereby being 'confirmed'.

Table 3.10 <u>The Chronology of Upper Perigordian Industries.</u>

Wurm III Perigord VIII	Laugerie Haute, Abri Pataud	Protomagdalenian
	Laugerie Haute	Perigordian VI
	Roc de Combe	Final Perigordian
	Laugerie Haute	Perigordian VI
	Le Flageolet I, Trou de la Chevre	Perigordian V^{3c}
	La Ferrassie	Perigordian V^{3a}
Wurm III Perigord VII (a - c)	Roc de Combe	Final Perigordian
	Abri Pataud	Perigordian VI
	Le Flageolet I	Perigordian V^{3c}
	Les Jambes	Perigordian V^{3c}
	Le Flageolet I, Trou de la Chevre	Perigordian V^{3b}
	Roc de Combe, Le Facteur	Perigordian V^{3a}
	La Ferrassie	Perigordian V^2
Wurm III Perigord VI	Les Jambes	Perigordian V^{3c}
	Le Flageolet I	Perigordian V^{3b}
	Roc de Combe	Perigordian V^{3a}
	La Ferrassie	Perigordian V^1
Wurm III Perigord V		Perigordian IV
		Aurignacian IV

(Source: after Delporte 1976)

David's belief that the Noaillian and other Upper Perigordian industries are in fact separate cultural traditions, each with its own history, is based upon detailed attribute analysis of selected assemblages. The former is seen to originate outside S.W.France, arriving from Provence, eg: La Bouverie, Var (David & Bricker 1987). The Noaillian is divided into two major parts, Early and Late, characterised as follows:

Early Noaillian	Late Noaillian
Higher %s of Noailles burins	Higher %s of regular blades
Burins made by dihedral technique	Truncation burins exceed dihedral burins
	Raysse burins of importance
	Marginal retouch important
Abri Pataud, 4 lower	Abri Pataud, 4 middle
Abri Facteur, c.10-11	La Raysse, c.4
Roc de Gavaudun, C.2	Bassaler-Nord, c.4

A poorly defined Final Noaillian is also found, according to David & Bricker (1987) at the Abri Pataud and Les Jambes (Dordogne).

The parallel development of the Upper Perigordian and Noaillian traditions is described by David & Bricker (1987) in terms of Laville's (1975) Wurm III climatic phases. The Early Middle Perigordian (IV or Gravettian) began towards the end of Laville's cold, humid phase ending Wurm III, Perigord IV. The transition to the Late Middle Perigordian occurred during Perigord V (mild and humid).

Early Upper Perigordian industries equate with Wurm III, Perigord VI and continue into the tripartite climatic phase VII. Meanwhile the Noaillian appears in the Perigord during Perigord VI, the change from David & Bricker's Early to Late Noaillian co-inciding with the Tursac Interstadial of Perigord VII, traditionally dated to 24 000 to 23 000 B.P., but recently redated to approximately 27 000 to 24 500 B.P. (Bricker & Mellars 1987). The Noaillian (Perigordian V^3) disappeared by the end of Wurm III, Perigord VII, the colder conditions of phase VIII being associated with Final Upper Perigordian levels (Perigordian VI).

The Final Perigordian (VI) follows directly after the Upper Perigordian (V) industries (Rigaud 1976:59), a sequence clearly demonstrated at the Abri Pataud and Roc de Combe. The industry was first defined by Peyrony as Perigordian III but later reassigned by Sonneville-Bordes (Delporte 1976). It is characterised by the presence of many burins on truncations, numerous gravette and microgravette points and truncated and backed blades (Straus & Heller 1988). Many types of scraper are absent as, in general, are those elements typical of the Upper Perigordian - with the exception of some Noailles burins. There is, however, a relatively rich bone industry.

Continuity from Perigordian IV to VI is seen by Rigaud (1988:394), although David & Bricker (1987) do not agree. In the latter case, a break is seen between Perigordian V and VI.

Dates recently obtained from the Oxford University Radiocarbon Accelerator Unit place Perigordian VI between 24 600 and 21 740 B.P.. At the Abri Pataud dates listed by Mellars et. al. (1987) and Bricker & Mellars (1987) include both 24 500+/-600 B.P. (OxA-686) and 21 740+/-450 B.P. (OxA-599) for level 3 (lens 3), 24 440+/-740 B.P. (OxA-165), 24 250+/-750 B.P. (OxA-164) and 23 180+/-670 B.P. (OxA-163) for level 3 (lens 2A).

Perigordian VII (or Protomagdalenian) is similar in nature to Perigordian VI, with dihedral burins outnumbering truncation burins. The is an important backed bladelet component. In the Perigord stricto sensu the Protomagdalenian is known only from Laugerie Haute and the Abri Pataud. In our wider study-area we may also add the site of Blot (Haute-Loire) where the industry is dated to 21 700+/-1200 B.P. (Gif-564) and 21 500+/-700 B.P. (Gif-565) (Delporte 1976:34). Based on further dates from the Abri Pataud and Laugerie Haute the Protomagdalenian appears to range from ca. 23 000 to 21 500 B.P.

Aurignacian V assemblages were first discovered and defined by Peyrony. The assemblage was uncovered above Perigordian III (now VI) at Laugerie Haute where no Perigordian IV and V were found. Aurignacian V is therefore deemed to be separated from its earlier phases, by the course of the Middle and Upper Perigordian (Demars 1985:328).

At Laugerie Haute, the only Perigord site at which Aurignacian V is found, the industry is characterised by thick endscrapers, burins and simple bevel-base bone points (Straus & Heller 1988). Apart from the large number of crude carinate and nosed scrapers, a feature which it shares with the earlier 'Typical Aurignacian', Aurignacian V is largely defined on the basis of its lack of Perigordian tool types (Laville et. al. 1980).

In recent years however, Demars (1985) has suggested the possible existence of a small number of open-air sites in S.W.France which might be attributed to Aurignacian V. Two of these fall within our area of study - Cublac and Bombetterie, both in the Correze (Demars 1985:330). He points out that it is possible that in Spain Aurignacian V has been found at El Pendo and in S.E.France, at La Salpetriere (Gard), although he admits that this is questionable.

Demars rejects the possible link between Aurignacian V and Typical Aurignacian industries. Likewise, he does not accept the idea of a new human population responsible for a new lithic culture, for a change in lithic technology need not imply a new population - merely a shift in manufacturing technique (1985:332). He instead suggests that the Aurignacian V of Laugerie Haute, and possibly elsewhere, represents just one stage in a manufacturing cycle in which an idea or technique develops, is perfected and eventually disappears (often rapidly), to be replaced by another which develops, is perfected and in turn disappears.

The Solutrean represents perhaps the most distinctive Upper Palaeolithic cultural tradition in S.W.France. Most of the research into the Solutrean was undertaken by Smith (1966, 1969) and the reader is referred to his account thereof, Le Solutreen en France (1966).

Named after the site of Solutre (Saone et Loire) where it was first recognised, the Solutrean is best represented at the site of Laugerie Haute where Peyrony's pioneering research was undertaken. The period begins toward the end of Wurm III and finishes during the early phases of Wurm IV (Laville et. al. 1980). The stone tool inventory is typically Upper Palaeolithic, with abundant endscrapers, numerous perforators but few burins. There appears to be no distinctive microlithic element in assemblages and bone work is poor in terms of both quality and quantity. The major feature of Solutrean assemblages is the foliate or leaf-shaped point produced by unifacial and bifacial retouch. Such artefacts make their appearance during the third of five Solutrean phases (Laville et. al. 1980):-

> Protosolutrean- directly overlying Aurignacian V with unifacial and plane-faced points.
>
> Lower Solutrean - unifacial points increase in frequency. Solutrean retouch becomes more 'elegant' and characteristic. At the Abri Pataud (level 1) this is dated to 20 400+/-450 B.P. (OxA-373)(Mellars et. al. 1987).
>
> Middle Solutrean - bifacial pressure flaking results in the production of elegant laurel-leaf points, some of which, it is believed, served ceremonial and/or symbolic rather than functional purposes (due to their thin and delicate form).
>
> Upper Solutrean - laurel-leaf points are present. Unifacial points continue. The diagnostic foliate points begin to resemble willow leaves rather than laurel leaves. This stage is dated to 19 230+/-300 B.P. (Gif-3609) at Le Roc de Sers.
>
> Final Solutrean - characterised by shouldered foliate points and the disappearance of unifacial retouch. This phase is dated to ca. 19 700 B.P. at Laugerie Haute, traditionally coinciding with the Laugerie Interstadial.

This classic Solutrean development applies primarily to the Perigord where the industry is well represented. Elsewhere distinct regional variants occur (Smith 1966).

Some degree of overlap between the Upper/Final Solutrean and the Early Magdalenian is seen in much of France on an inter-site basis. For example, while the Final Solutrean persists at Le Malpas, the Lower Magdalenian develops at Laugerie Haute. Laville et. al (1980) suggest that the frequent claim for overlap be due to the occurrence of mixed deposits or the collection of artefacts made by earlier human groups by later populations.

The Magdalenian, named after the site of La Madeleine (Tursac), at which only the Later Magdalenian is found, was initially divided into six successive stages by Henri Breuil, phases I-III being the Lower Magdalenian, IV-VI the Upper Magdalenian (Breuil 1912). This scheme, in which definitions were based upon changes observed in the bone/antler/ivory industry, has subsequently been modified although much of it remains unaltered.

Laville et. al. (1980) divide the Magdalenian into two groups of stages. Magdalenian "0" was found underlying Magdalenian I (which developed during the Lascaux Interstadial) at Laugerie Haute and named appropriately, dated to 18 260+/-360 B.P. (Ly-972). These two phases are grouped together, characterised by abundant raclettes, scaled marginal retouch, carinate scrapers, small flakes and blade segments.

Magdalenian II-VI is relatively homogeneous, rich in burins and characterised by backed bladelets. Geometric microliths develop, taking the form of triangles, rectangles and semi-lunates. The scalene triangle, retouched on one or two sides (Joffroy & Theriot 1984) develops throughout the colder Magdalenian II and increases in frequency during Magdalenian III. The bone industry associated with these, and the Early Magdalenian, comprises primarily the sagaie, as opposed to the harpoon, characteristic of the later stages. During Magdalenian IV various types of burin are known from stone tool assemblages, while in Magdalenian V contexts burins and double-sided scrapers are important.

Laville et. al. (1980) isolate five tool types which are considered characteristic of the Late Magdalenian. All are to be found in Final Magdalenian contexts (dated to 12 370+/-220 B.P. (Ly-2700) at the Abri Vidon (Cessac), 12 130+/-160 B.P. (Gif-3739) at Pont d'Ambon, level 3b (Bourdeilles) and 11 750+/-310 B.P. (Ly-976) at Gare de Couze, level C) together with an increase in microliths, but some are known to occur much earlier. The five categories are:

1. Parrot-beak truncation burins,
2. Teyjat points (a tanged point),
3. Laugerie-Basse points (a foliate piece),
4. Azilian points and
5. Thumbnail scrapers.

It is the true harpoon which is the determining criterion in distinguishing between Upper Magdalenian traditions.

During Magdalenian IV (13 440+/-330 B.P. (Ly-922) at La Madeleine, couche J, niveau 14) reindeer antler harpoons first appear in prototypical form (Laville et. al. 1980) with only tentative barbs. A single row of barbs characterises the harpoon during Magdalenian V (13 070 - 12 750 B.P. at La Madeleine, Delibrias & Evin 1980) while during Magdalenian VI those with two rows of barbs appear.

Finally, the Azilian emerges from the Terminal Magdalenian between 11 000 and 10 000 B.P. (see table 3.11). Reindeer antler tools are replaced by those formed from red deer bone/antler. Only the Azilian point and thumbnail scrapers increase in frequency, the Magdalenian inventory being replaced by one markedly smaller and less diverse.

Table 3.11 <u>Azilian Radiocarbon dates from Pont d'Ambon and Penne.</u>

Pont d'Ambon	c.2	Gif-3561	9 990+/-250 B.P.
	c.2	Gif-3740	9 640+/-120 B.P.
	c.3a	Gif-3368	10 350+/-190 B.P.
	c.3a	Gif-2570	9 830+/-180 B.P.
Penne		Ly-1175	10 110+/-440 B.P.

(Source: after Delibrias & Evin 1980)

3.5 Environmental Conditions Prevailing during the Upper Palaeolithic.

Geist attributes two broad types of ecosystem to the 'ice ages' - tundra (arctic or alpine) and periglacial (Geist 1978:194). He associates the periglacial environment with extreme, cold conditions of the glacials, basing his discussion on areas in close proximity to glaciers themselves. Given the distance between the Massif Central and Pyrenean ice caps and our region of S.W.France, the relevance of Geist's periglacial ecosystem is sometimes less than clear. The tundra is, he believes, more appropriately defined as interstadial, but may, it is maintained here, also be associated with cold stages further away from the ice itself.

The vegetation cover typical of the tundra is low in fibre and high in protein, carbohydrates and minerals, being easily digested. Highly productive, localised patches of vegetation result in a complex mosaic of communities (Collinson 1977), especially close to water, the ground flora comprising lichens, herbs, bryophytes and prostrate shrubs (Geist 1978). The environment is open, broken only occasionally by stands of trees. The variety of tundra ecosystems which might be expected to have been associated with cold, glacial conditions is summarised in figure 3.4.

Figure 3.4 Arctic and Alpine Tundras.

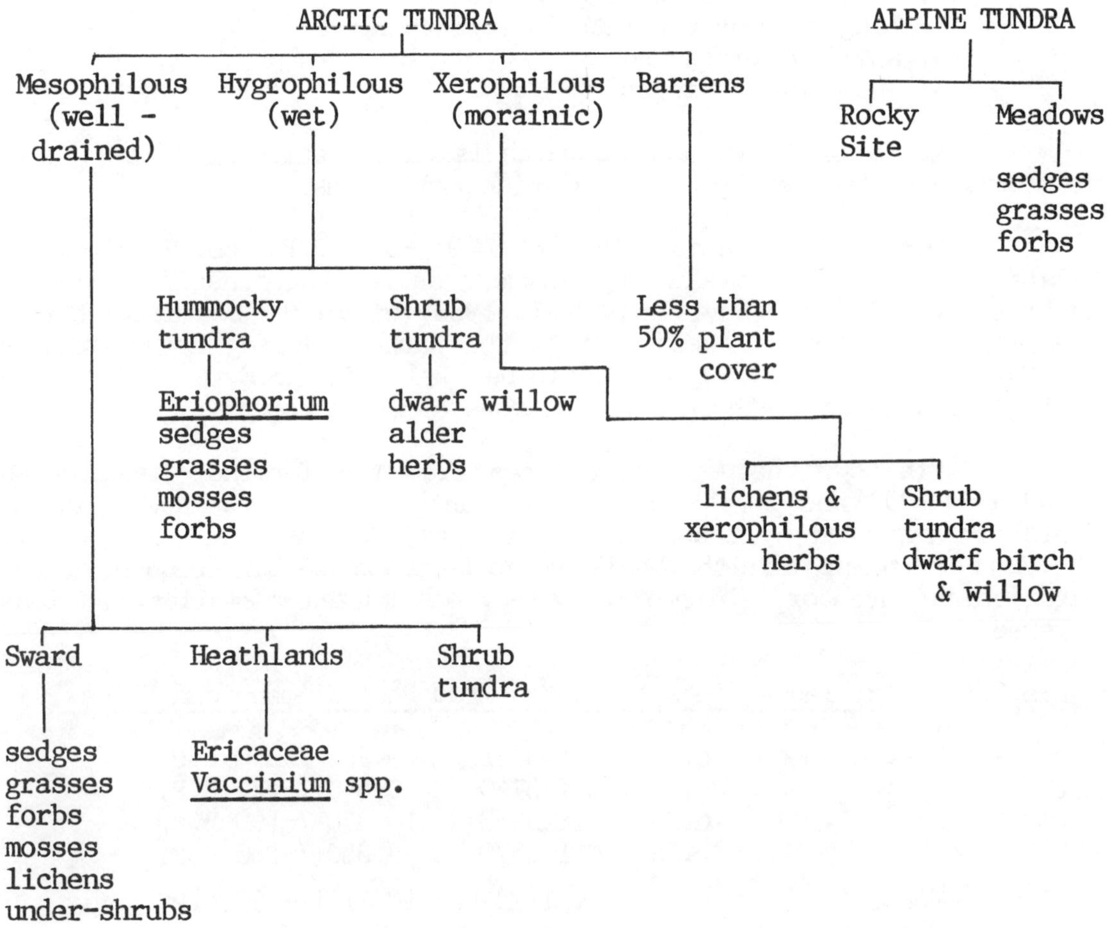

(Source: after Collinson 1977:180)

The steppe community, associated with cold, dry conditions of the glacials, is characterised by rich vegetation of xerophytic grasses and Compositae (Paquereau 1970). Cold meadowland of grasses and sedge is also attributed to glacial conditions, although a rise in humidity results in the greater tree cover (pine, birch, willow) typical of the boreal forest and taiga ecotone.

More usually associated with the warmer, more humid conditions of interstadials are communities of temperate deciduous woodland/forest (Collinson 1977), cool parkland (Paquereau 1970) and forest steppe (Zlotin & Khodashova 1980) with a strong birch and pine component. A wider variety of tree species occurs, while increased hydrophytes point to a rise in general humidity and growth of fenland communities, where annual biomass may reach as much as 1000 kg ha^{-1} (Collinson 1977).

For an in depth discussion of the species characteristic of each ecosystem type and the structure thereof, the reader is referred to Polunin & Walters (1985) "A Guide to the Vegetation of Britain and Europe".

Attention will now focus on the nature of the glacial and non-glacial environments reflected by specific examples taken from published palynological reports. No claim is made that all cold episodes are characterised by identical conditions; as with milder phases, the claim is one of general similarities which allow us to compare conditions over the course of the Upper Palaeolithic.

3.5.1 Stadial Environments.

Even during the glacial maximum, South West France, due to its latitudinal position, never experienced true tundra conditions, on account partly of its lack of long polar nights and the extreme obliquity of the sun's rays. However, evidence of periglacial conditions is regularly observed in much of our region, especially the Charente (in the north of our study-area), indicating that tundra-like conditions may have existed (see table 3.12 and fig. 3.5). The formation of a permafrost layer requires a mean annual air temperature no higher than -2°C, while the existence of fossil frost-crack features implies similar conditions. A maximum of -6°C is today recorded for ice-wedge casts (Washburn 1979), which lends support to the picture of cold, extreme conditions in Late Glacial Western Europe. Based upon the distribution of such periglacial features Maarleveld (1976, in Washburn 1979) identifies the -2°C isotherm in S.Europe, a boundary marking the southern permafrost limit which runs through the southern portion of our region.

During cold (stadial) episodes of the late Wurm in S.W.France we have a flora of low tree cover, pine, birch and willow making up the arboreal component. The herbaceous total is high, with a heliophilous group dominated by Compositae (eg: dry steppe at A. Pataud). Elsewhere, tundra communities are found, in association with those fossil permafrost features outlined above. The environment is thus a patchy one in which several types of landscape are to be found, ranging from expanses of steppe or tundra grassland and meadow to areas of more temperate woodland.

Table 3.12 Periglacial Features in S.W.France.

East of Angouleme	Lavaud	Cryogenic features. Grezes
	Mouthiers	Cryogenic features, soil polygons
West of Angouleme	Vindelle	Alternating thick/thin levels, representing running water under semi-permanent snow.
	Tillax-Mainxe	Ice wedge cast. Palaeosol
	Pt. Montbron	Talus, greze, involutions
Bordeaux		Ice wedge casts Fossil frost cracks

(Source: After Vigneaux (1975); Washburn (1979))

Features such as these (Vigneaux 1975) form when average annual air temperatures fall to at least -4°C; ice wedges require mean winter temperatures of -15° to -20°C before surface cracking commences and thus fossil cracks such as ice-wedge casts are acceptable proof of continuous or near-continuous permafrost (probably of glacial maximum age) (see fig. 3.5). We may therefore assume that much of our region represents a periglacial zone of semi-permanently or permanently frozen sub-soil (tundra) and can infer mean annual air temperatures of -8°C or less (Lowe & Walker 1984) with a high degree of winter freezing. Polygons form when mean annual air temperatures fall below -2°C to -10°C. Meanwhile, involution depth may shed light upon summer conditions since they develop in the active layer above the permafrost. Its depth is a function of summer temperatures (above 0°C) and hence melting.

Fig. 3.5

Map showing Periglacial Features in S.W.Framce.

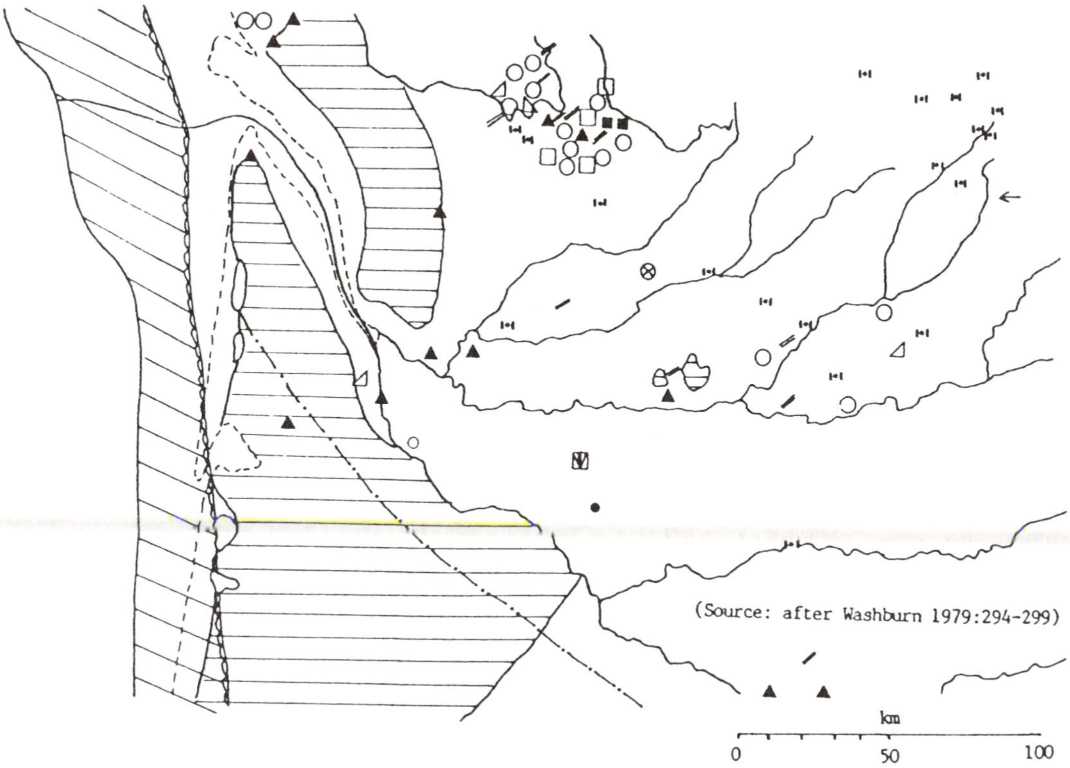

Key:

- ○ Garlands. ▲ Deflation & ventifact areas. ◿ Involutions.
- ○ Vertically-oriented stones. ⊗ Frostwedging. ☒ Periglacial fans.
- ⁄ Grezes litees. ı•ı Gravelly mudflows. □ Patterned ground.
- ■ Fossil frostcracks &/or ice wedge casts. ◺ Ice rafted erratics.
- ← Asymmetric valleys.
- Southern limit of continuous permafrost.
- —·— Southern limit of region with fossil frost fissures & frost cracks.
- ⟍⟍ Region beyond continuous permafrost (tree or shrub tundra ?)
- ═ Niveo-eolian sand.

Paquereau (as indicated above) identifies four environmental types (1970), two of which are characteristic of stadial conditions. These are summarised by Laville et. al. (1980:98). 'Cold Steppe' is characterised by cold, dry, open conditions of xerophytic steppe communities, with high Compositae totals. A.P. rarely exceeds 5%. 'Cold Meadowland' arises in conditions of increased humidity. A.P. is still weak due to low temperatures however, although some hardy, deciduous trees (birch and willow) occur with Scots pine. Ground flora is characterised by grasses and sedges, covering the open ground.

At Caminade Est levels G and F represent cold conditions of Aurignacian I (Paquereau 1978:) in which cold steppe and meadowland conditions might be expected to prevail. In G, we see low pine frequencies, an abundance of birch, and an important shrub layer with a marked Urticaceae component. Ground flora comprises Cichoreae and other steppe elements. Level F contains some juniper, indicating that conditions were becoming less extreme. This species is likely to be represented by a stunted form, due to sparse snow and cold frosts. Areas of woodland pockets, with scattered willow communitites and sedges, accompanied by shrub-layer vegetation, provides evidence of an environment in which reindeer herds are likely to have found suitable forage conditions.

Cold Aurignacian levels at Le Malpas comprise a mixture of canopy, shrub and ground flora (Paquereau 1978). Levels 7 and 9 present open vegetation of Arctic steppe-tundra character. Level 7 also sees an increase in hygrophiles, and suggests that cold meadowland conditions were prominent. The final level at Le Malpas (1) is also one of open country, but cold and dry: cold steppe. There are a few patches of pine, but the dominant feature is the shrub and herbaceous layers, albeit with the development of some Fenn wood or Carr.

During the Late Upper Palaeolithic, La Marche (Vienne), lying in a small sheltered valley, was, according to Leroi-Gourhan (1973) occupied between 13 500 B.C. and 11 500 B.C.. This south facing cave has yielded pollen indicating three identifiable vegetational components.

In front of the cave there was an open meadow where hygrophiles may have been associated with a stream (about 10 m from the site); however there is a marked absence of aquatic pollen. At the edge of the water Alnus sp. and Umbelliferae grew, surrounded by summer flowering pastures. Exposed plateau surfaces were covered by open communities and small pockets of tree and shrub growth, depending upon soil thickness and moisture (cold steppe ?). The third component is found on rocky valley slopes and the plateau surfaces: oak, pine, ash, box, ivy and filicales. In a situation in which we have three evident groups of species it is inevitable that pollen samples should take on a mixed appearance, including, as they do, both local and regional pollen.

Meanwhile, in Upper Magdalenian levels at Le Flageolet II, A.P. values reach 33% suggesting a relatively high tree cover. At the same time A.P. reaches only 5% at the Roc-aux-Sorciers, comprising pine and alder. Cichoriae totals 71% with filicales reaching 21%. The severe conditions experienced here are described by Guillien and Saint-Mathurin (1976) as monotonous, characterised by a thick, seasonal layer of snow and intense winter cryoclastism. The growing season is short, reducing vegetational productivity and almost precluding tree growth.

3.5.2 Non-Glacial (Interstadial) Conditions.

Paquereau (1970a) and Laville et. al. (1980) discuss two environmental types which may occur during mild episodes in the course of the Upper Palaeolithic. Cool Parkland is associated with mild, humid conditions. Open ranges have a more varied plant cover amongst which meadows, fenn communities and river-bank hygrophiles alternate. Although A.P. may still be relatively low (15% at La Ferrassie, Aurignacian II) forested areas do occur more often, comprising pine, birch, willow, hazel, alder and linden. Laville et. al. (1980) list three instances in which one might expect to find these conditions:

(a) brief, mild oscillations within a stadial.
(b) an interstadial of modest length and intensity.
(c) a transitional phase between two glacial environments and one fully temperate.

Temperate forest vegetation could be found in areas of humid and mild conditions; perhaps milder than today (Laville et. al. 1980). A.P. can exceed 60%, eg: at Roc-aux-Sorciers (Guillien and Saint-Mathurin 1976) and in heavily wooded areas oak, beech, elm, lime and maple are found in addition to species of the Cool Parkland type. We also have a thick forest ground flora, indicated by high frequencies of fern spores. In addition there are open meadows, fenns, swamps and marshes. Vegetational patterns are highly variable spatially, resulting in a mosaic of biogeographical associations.

In general we find, throughout the Perigord, trees in woods and light forests, considerable hazel copses, temperate undergrowth including ivy, blackthorn, hawthorn and, on calcareous soils, box and juniper. Associated with the latter may be the fenn communities typical of limestone regions (Donner 1975). Along streams and rivers willow and alder dominate, shadowing the reeds and rushes, Ranunculus sp. and floating or submerged aquatics found in pools and meres. Between forests and thickets we find heather, bilberry and other temperate grassland shrub and ground flora.

The Abri Pataud pollen samples show us that forests were, during the Aurignacian, much more widely distributed than is the case today, although for most samples A.P. remains below 50%, sometimes as low as 20%, comprising birch, pine, oak and alder, juniper, willow and hazel. Donner (1975:) suggests that park-like forests were prominent in sheltered valley areas while temperate but open, steppe-like vegetation predominated on valley interfluve and plateau surfaces.

Mild levels at the Aurignacian site of Le Malpas (Paquereau 1978) alternate with cold episodes throughout its sequence. The transition between levels 5 and 6 is marked by wetter, temperate, wooded conditions, with a hazel shrub layer of significance, despite the possibility of prolonged snow patches on north facing scree slopes near the site.

Levels 4 and 3 are separated by a brief, rich woodland phase, with increasing thermophiles. Level 2 sees the return of temperate conditions and a wooded environment. The open areas are covered by heather while increased precipitation results in marsh development. The period is notable for its lack of early and late spring frosts.

Level D_2 (Aurignacian II) at Caminade Est (Paquereau 1978) exhibits a constant pine cover. In the second half of this level we gain alder, hazel, elm and lime, with a consistent birch component. Birch forests are often associated with damp ground level floras. Sedges, reedmace and bulrush pollen are indicative of water edge plant communities and humid conditions. Thermophiles are also found, emphasising the warm, temperate nature of the level. There also exists a persistent shrub layer which, like other aspects of the vegetation, indicates that the period was not one of heavy frosts.

At Le Flageolet II, during the Upper Magdalenian, A.P. frequencies vary from 12 to 50 %. Levels 2 to 3c are representative of mild conditions set between two colder extremes.

Table 3.13 Reconstructed Vegetation at Le Flageolet II

Level	Vegetation
4	A.P. = 16% Mixed oak forest disappears. Shrub and ground flora - Compositae. Heliophiles (dominant)
3c	A.P.= 50% Pine & birch decline. Hazel dominates with mixed oak forest returning (oak, elm, lime, alder, willow, juniper, ivy, ash). Herbs include grasses Cyperaceae & hygrophiles.
3b	A.P.= 30% Pine, lime, elm, hazel. Herbs - humid.
3a	A.P. = 43% Hazel. pine, birch. Thermophiles and hygrophiles.
2	A.P. = 33% Pine dominates at base, birch at top (Birch forests).Alder, willow. Hazel appears, Compositae, grass etc.
1	A.P. = 12% Pine, birch. Some willow and alder. Compositae in herb layer. Grasses, heliophiles.

(Source: after Paquereau 1970b)

At Pont d'Ambon, Upper Magdalenian and Azilian levels present us with a mixture of vegetational types: light, open forest and open meadow-steppe tundra communities. The Final Magdalenian is associated with the end of a mild oscillation and declining hazel frequencies. Levels 3, 3a and 3b are dated to between 9 990 250 B.P. (Gif 3561) and 12 130+/-160 B.P. (Gif 3759), characterised by fluctuating temperatures and humidity. 3a is the most temperate and humid, with A.P. reaching

50%, comprising hazel, alder, oak, ash and ivy. Pine and birch are in decline. Level 2 (9 640+/-120 B.P. Gif 3740) is a cool but mild, humid phase, with an increase in forest cover. Reedmaces and waterlilies are found suggesting a stream or river bank community.

In general, mild conditions in Perigord are likely to have been more extensive than pollen samples suggest. Thermophilous species are likely to respond more rapidly to changing conditions than do their cold-tolerant counterparts, for warmth associated species exhibit steeper fall-off curves than do cold varieties. A single late spring frost can cause substantial damage to plant communities, retarding growth and even killing some species. In addition to this, cold phases show greater uniformity than do temperate equivalents - in terms of floral species and frequencies thereof.

During the course of the Upper Palaeolithic in the Massif Central pine and birch comprise the major tree species, supplemented by oak, hazel and alder. It is a region of greater microvariation (Lemee 1956) than the Perigord. During the Wurm III / IV interstadial temperatures rise (Protomagdalenian at Blot, Badegoulian at Cottier) (Debard & Moser 1976). Cool temperate conditions are identified by avifauna at Cottier (Mourer-Chauvire & Carbonnet 1976). However, rodents suggest colder conditions; only one forest species occurs, the majority being of mountainous nature. In the Cantal, the Oldest Dryas tree pollen are derived solely from a few refugia. The vegetation is steppic with grasses as the major component (Beaulieu et. al. 1982:). Pine provides the majority of the A.P. total, followed by birch. The Younger Dryas (10 750 240 B.P.) sees low birch totals, high N.A.P. (non-arboreal pollen) and moderate pine. In the Haute Loire, humidity appears to be relatively constant except for levels GC and GD at Blot which become drier. At Cottier bird species of C4 also suggest cold, dry conditions in a relatively rocky area. Wurm IV conditions are cold but dry in the east at Cottier, Rond du Barry and Blassac, becoming warmer and more humid at the former two. On the west coast the climate is oceanic with relatively cool summers and mild winters. We have an area of cool temperate forest of both deciduous and coniferous species, the growing season shortening as one progresses inland to more continental areas (Mellars 1985:274).

In general, there is little evidence to suggest that the influence of the Atlantic was reduced during the Upper Palaeolithic (Wilson 1975). The coastal end of the east (Massif Central)-west (coast) spectrum was relatively stable, being, in essence, similar to modern vegetational conditions despite lower temperatures.

In conclusion, one point should be emphasised when considering the environment of South West France. Climatic conditions provide the basis upon which topographic features work in order to produce a mosaic of microenvironments. Four groups emerge: river valleys, plateau tops, rock faces and valley slopes, and the coastal plain. In addition, once plant communities have become established, they too help to determine subsequent developmemts. Tundra conditions of the type described by Chernov (1985) do not occur in South West France. However, certain principles do emerge from his study which are of relevance. Tundra communities are diverse, ranging from the mixed vegetation typical of the forest tundra to semi-desert and forest steppe. On watersheds in the semi-desert areas shrubs and trees are known to thrive. It teaches us that we must not assume that glacials are treeless episodes and that interstadials are phases of thick forest cover. Variations and combinations occur and are duly reflected in the pollen spectra derived.

Chapter Four

MODERN ANIMAL ECOLOGY AND SELECTED HUNTING STRATEGIES.

This chapter concerns modern animal ecology and exploitation strategies employed by modern and ethnohistoric hunters. Its inclusion is justified on the grounds that an understanding of both chronological and spatial variation in the distribution of individual and groups of species depends on several principles amongst which are:-

1 - A knowledge of the conditions which determine the natural distribution of the species concerned.

2 - A working knowledge of exploitation strategies which, in part at least, are responsible for the state of the archaeozoological record.

Although questions are occasionally raised concerning the appropriateness of modern animal ecology to palaeolithic studies, it does provide a starting point - in the absence of a knowledge of one of the very things which we are, to some extent, attempting to reconstruct and explain.

The second variable is more problematic. Again the question is one of relevance. Can we justify imposing ethnographic, in the broadest sense, or historic hunting strategies on the palaeolithic hunter? The answer must be "no", but it does form a useful point of departure as, it is hoped, Chapter 9 (see below) shows. And while no attempt should be made to simply fit the archaeological data to the modern situation, it can be profitable to compare the two.

Following this consideration of ecological and ethological data, a brief discussion of hunting techniques will be provided, based upon ethnographic material available for selected species.

4.1 Ecology and Behaviour of Selected Species.

4.1.1 Reindeer.
(*Rangifer tarandus* L.)

Reindeer (or caribou) are divided into two broad groups: the barren-ground and the woodland varieties (Lawrence 1974:218). Darker than the former, the woodland species have slightly longer legs and shorter,flattened antlers. Barren-ground caribou bulls in winter prime, are a delicate brown, with a white ruff round the neck. Above the hooves, the lower limbs are circled with white, while similar colouring may be observed running from the ruff, over the shoulder and along the lower flanks (Lawrence 1974:220). The adult coat comprises thick, coarse hairs which are hollow, thereby increasing insulation. The nose is protected by long hair, allowing the animal to forage in the snow and thus it is protected from the cold (Clutton-Brock 1981). In the spring a shorter, smooth coat of a dark brown to grey colour quickly gives way to the slow growing winter cover. The reindeer is also well adapted anatomically to its environment. The feet are wide and splayed, the hooves being deeply divided and the accessory metapodia well developed, enabling it to walk on the snow and dig deep feeding craters. The feet are also suited to swimming (Ingold 1980).

The reindeer is unique among cervids in that both sexes have antlers, although at any one time only seventy percent of females may possess them (Banfield 1974). The antlers are different from those of other species, those of the female being markedly smaller than those of the male.

The smaller barren-ground caribou stands 85-120 cm. at the shoulder. Total body length reaches 220 cm., the tail being 10-15 cm. long. There are various estimates as to the average weight of the reindeer, examples of these being shown below.

Table 4.1 Mean Reindeer/Caribou Weights (kg)

Weight (kg)		Source
70 - 150		Bjarvall & Ullstrom (1986)
40 - 100		
50 - 150		Nelson (1973)
120		Ingold (1980)
85		
120		Spiess (1979)
75		
40	calf	
60	yearling	
120		Kelsall (1968)
70	1 - 2 y.	
85		
60	1 - 2 y.	

The weight of the individual varies according to the age and condition of the animal, subject to seasonal changes. Fat begins to accumulate in the male during autumn (usually at about the end of August). Winter fat is seldom extensive and is commonly lost during the rut (male) and in the course of spring migrations (female). The majority of the fat accumulates in the rump and saddle and may form a layer up to 7.5 cm. thick (Kelsall 1968). Fat is also deposited internally. Jacobi (1931) suggests that as much as 80 lbs of fat may be obtained from a 300 lb. mature male. Fat commonly exceeds 20 % of the total body weight but this is not invariably so. Figures provided by Spiess (1979) indicate a range of 2-10 % winter fat content. Bulls are at their leanest immediately after the rut, cows and juveniles by the end of the spring migrations.

The rut takes place during late September or early October. After a gestation period of 220-230 days calves (usually one) are born in the spring. The calving season is short; 90% of calves are born during two weeks between May and June (Bjarvall and Ullstrom 1986, Clutton-Brock 1981, Lawrence 1974). Calves, born in spring without a truly protective hair covering, do not possess an insulating layer of fat (Remmert 1980:41). Returning to the same calving grounds year after year (Spiess 1979), females are not prolific breeders, rarely producing more than six calves in the course of a lifetime, the expectancy of which may be approximately 15 years. Sexual maturity is reached by the third year and does may breed until the tenth. Malnourished deer produce relatively few calves (Bailey 1984:18).

Many cows come into oestrus together in the course of a seven day period. Thus it is impossible for dominant bulls to inseminate many females and repel rivals simultaneously. Wasteful competition amongst rutting males is thus avoided and domination assertion appears to be much less emphasised than in the case of other cervids.

The order of rank within a herd is largely dependent upon the possession of antlers. Those animals with the strongest antlers enjoy a dominant position especially in respect of food. Thus, while still carrying antlers, pregnant females have better access to forage during late winter and early spring (prior to the birth of calves) than do males and non-pregnant cows, the latter groups having lost their antlers immediately after the rut and at the end of the year respectively (Remmert 1980:94).

Immediately after giving birth the cow leaves the main herd with its young. The pair stay together so that the calf can learn to recognise its mother. On rejoining the herd, juveniles will be able to maintain contact with their dams until they begin to separate the following year. Summer herds generally comprise females, calves and yearlings, the bulls joining the herd just before the rut. However, geographical variations exist. The Norwegian Hardangervidda Herd (totalling up to 10,000 head and comprising groups of 100-500) coalesces in July, resulting in the total integration of both sexes and all age groups. Likewise, in Northern Scandinavia herds of more than 500 animals (Frenchen & Salomonsen 1955) present a mixed sex profile; a similar structure of 'herds' (1500-4000 head) was reported by Newhouse (1952).

The reindeer, tolerant of crowding, exhibit no form of territoriality, their social groups being fluid and open. The herd is organised by means of joint movement and defence, the behaviour of the individual depending upon its position with respect to others in the herd.

Herds generally comprise two groups of reindeer: the central body and the peripheral group. The latter consists of the vanguard, side and tail divisions. The first two are commonly more wary and restless than the better fed central and tail components. Recruitment to each group is by age and sex. Wild Norwegian reindeer exhibit a similar structure amongst mature females. In this case, we observe leader, lookout and defence groups which combine to co-ordinate movement and maintain the security of the herd.

In contrast to the Barren-ground caribou, the woodland reindeer lives in groups rarely exceeding 6-8 until the autumn (Lawrence 1974:220). While less gregarious than their tundra counterpart, woodland animals are individually more wary (Ingold 1980). As in the case of the caribou, Scandinavian woodland species exhibit patterns of seasonal dispersal and aggregation (Skogland 1983).

Population density figures vary seasonally reaching a peak at the time of aggregation and migration. The 'ecological density' (the number of animals in relation to both quantity and quality of habitat) is strongly influenced by changes in the absolute population size and the area over which the herds range. Figures provided for reindeer herd density fall in the range of 1 per square kilometre to 26 per square kilometre, although figures up to 75 per square kilometre in winter occupied spruce-lichen forest are quoted by Banfield (1974). In general, as table 4.2 reveals, population density among Norwegian herds appears to be lower than that in North America.

Table 4.2 Density Figures among Norwegian and American Reindeer/Caribou

Herd		Density (per km^2)
Norwegian		
Hallingskravet		2.0
Hardangervidda		2.5
Knutsho		1.0
Snohetta		3.5
N.American		
Manitoba (Jan')		4.5
(Apr')		26.3
St.Matthew Is.		18.12
Nelchina		1.15
Great Bear Lake	(1950/51)	2.5
	(1951/52)	1.8
	(1952/53)	0.8
	(1953/54)	0.0
	(1954/55)	0.5
	(1955/56)	3.7

(Source: after Kelsall 1968, Klein 1968 Skogland 1985, Spiess 1979)

Although not the only item in the reindeer diet (see below) the lichen supply is a critical factor in determining the total population and density which pasture will support on a year round basis. When optimal conditions of productivity prevail as much as 25-30 acres of continuous lichen are required per head of population (Ingold 1980). However, the consumption of arboreal lichens may reduce the land requirements and hence increase the potential population density figure. In areas which are predator-free, numbers rise phenomenally until a population crash occurs due to overgrazing and starvation. Overgrazed, trampled or burnt lichen areas require thirty or more undisturbed years to recover to medium height (Ingold 1980:20).

Caribou are eclectic feeders generally eating plants offering the highest available fat and protein content (Spiess 1979:31). During the spring, forage comprises growing leaves, shoots and buds of willow and herbaceous plants revealed by melting snow. Later on, newly growing grass and sedge shoots are consumed. During the summer, willow leaves, dwarf birch, grasses, sedges, blueberries, crowberries, mountain cranberries, horsetail and some lichen are taken (Lawrence 1974). Such plant food provides a relatively protein-rich intake. In autumn fungi are sought, providing a source of protein as the herbaceous and willow components are no longer sufficiently nutritious. Those caribou which remain outside forest cover rely on grass and sedge communities. It is only during the winter after snowfall has covered the remaining fungi and greenery, that the animal turns to lichen.

It has been suggested that a reindeer of 100 kgs weight requires 5000-6000 kcals day^{-1} during the summer in order to survive. In winter this falls to 3000-3500 kcals: there is a concomitant reduction in the basic metabolism rate which is associated with a pause in the growth of juveniles (Spiess 1979).

The ground and arboreal lichen upon which reindeer may be forced to subsist during part of the winter have a high acid content. The reindeer is adapted to such types of forage through the presence of certain intestinal bacteria adapted to a narrow acidity range. The digestive system of the species maintains a suitably narrow range of pH values, its flora being able to withstand lower alkalinity levels than usual. Physiological adaptation, therefore, enables the animal to exploit almost every type of terrestrial habitat (Larsen 1980). A climax lichen-bearing forest in close proximity to open areas of seasonally rich higher plant growth is the ideal (Spiess 1979). Thus the reindeer today is found in the circumpolar taiga belt adjacent to the Arctic tundra and mountain uplands in which alpine tundra communities are found, surrounded by climax forests. Kelsall (1968) identifies eight zones or habitat types which reindeer herds occupy:

1. Well drained higher heaths with thick lichens, grass, sedge, birch, willow.

2. Thin vegetal cover on crests and sides of eskers and beach ridges: grass and lichen.

3. Well drained outwash plain.

4. Medium drained: willow, dwarf birch, sedge.

5. Poorly drained: dwarf birch dominant, sedge, cotton grass, willow.

6. Poorly drained soil over bedrock: dwarf birch and crowberry etc.

7. Very poorly drained lowlands: swamp, marsh, sedges

8. Lakes, ponds and rivers.

Fig. 4.1

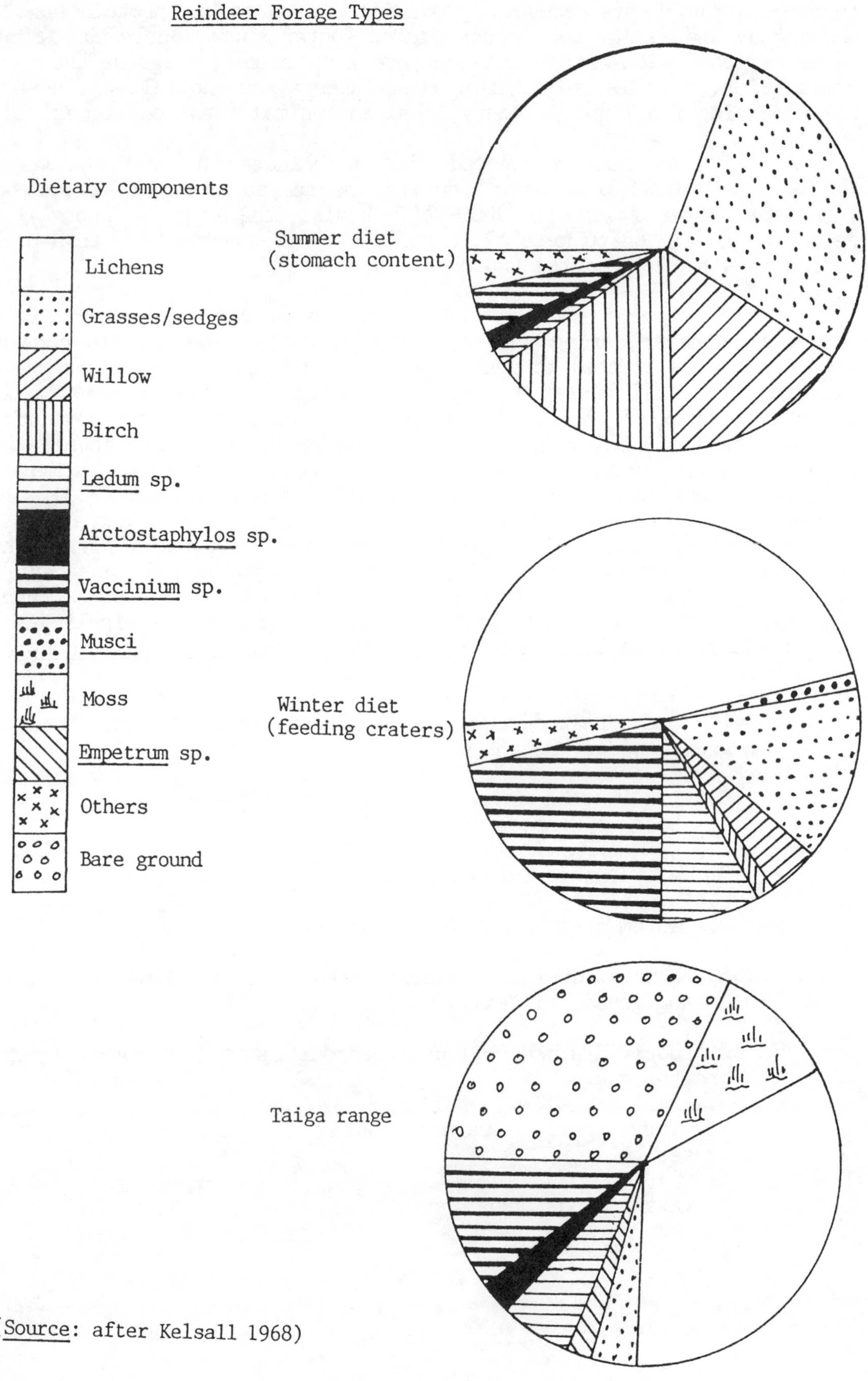

(Source: after Kelsall 1968)

The extent to which each zone is exploited may be estimated, in part, if we have some indication of the proportion of relevant plant species in the diet. This type of information is often obtained from the analysis of stomach contents. Kelsall (1968) summarises results of such analysis, the major components of the diet being lichen, grasses and sedges, willow and birch during the summer, lichen, Vaccinium sp., sedges and grasses during the winter (see fig. 4.1).

Traditionally thought of as being typical of flat, tundra (Remmert 1980) reindeer also inhabit treeless or open wooded, high country (Nelson 1973:112) in which habitats are selected in order to avoid areas where they are exposed to weather extremes and are vulnerable to predators (Crawley 1983); thus they may seek protection in the high, rugged terrain more commonly associated with the ibex and chamois. In regions of mountain heath, tundra and northern coniferous forest, snow cover is a crucial factor in reindeer survival.

Snow cover characteristics correlate with habitat suitability (Baker 1978) and associated herd movement (Skogland 1978). A maximum depth of 60 cm. is acceptable to the reindeer. Newly fallen deep snow is far from suitable, for the animal will sink deeply into it, hindering its movement and escape when necessary. In order to forage in such conditions, feeding craters are excavated for a depth of up to 2 ft. The area uncovered is small and hence, in some ways, the snow cover protects the lichen communities from overgrazing by the reindeer. In forests the snow may be deeper than on the wind-swept taiga (Clutton-Brock 1981) which, assuming a hard surface, enables the reindeer to browse on lichens hanging higher in the trees. Where snow has been blown away roots and runners may be consumed.

Moving toward tree cover in winter the reindeer is practising migratory behaviour in search of suitable forage and protection from gales, snow storms and temperatures as low as -50°C.

Thus, reindeer is a migratory species, making "directionally oriented, purposeful, and uninterrupted movements twice annually between distant winter and summer ranges" (Kelsall 1968:106). Migration may occupy up to four months of the year during which as many as 1200 miles may be covered (Chadwick 1979). The nomadic barren-ground caribou respond to fluctuating climatic and environmental conditions moving down slow gradients constantly in search of improved forage and travelling conditions. Four alternative responses are available to the reindeer:

1. They may seek a locality similar to that which they have left. This is the most commonly chosen option.

2. Some animals avoid the worst of the winter by migrating to areas in which they can survive winter conditions. The distance covered is not large but sufficient to enable herds to escape harsh conditions.

3. By moving a short distance, a seasonal change of biotope may secure the necessary habitat requirements.

4. Long distance migration may be undertaken.

In 1975 Leresche and Linderman listed several geographical principles which reindeer appear to follow when migrating. In general herds follow the contours of the local topography, using the line of least topographic resistance - when it coincides with the intended route (Burch 1972). They are more likely to cross hillsides than contours, for example in the North Brooks Range their east/west trail follows the 230 - 240 m. contours. When travelling through mountain passes, both valley sides and bottoms are used. When moving across hill country ridge lines, low passes and gentle slopes are used wherever possible, the herd spreading out in flat areas and moving together through narrower regions. They tend to course natural features for a considerable time before crossing them. The trails themselves form important features and influence animal movements, for herds follow the paths of preceding groups.

Causes of migration are believed to include changes in temperature, light and snow conditions, food availability and insect harassment (Remmert 1980). Autumn migrations commence in advance of the rigours of winter as temperatures fall and daylight hours shorten. Moving from the tundra to tree-lined regions migration is swift, following well used routes (Kelsall 1968). The caribou proceed more deeply into forest zones in years of high population levels unless movement is hindered by heavy, early snowfall or freezing conditions. In addition to the large scale regular movements, other less predictable and more unusual movements of small groups of animals separated from the main herd may be observed.

The length of the spring migration is determined by the distance between wintering and calving grounds. After calving, females and their young congregate, grazing until insect harassment forces them to seek alternative range. Blood sucking insects can only feed when temperature and wind conditions are favourable. On moving to the coast strong winds and lower temperatures may be encountered, so ridding the herds of the pests. Alternatively, a move to higher, cooler ground or more wooded country where mosquitoes are not so numerous achieves the same purpose.

Within two to three days of birth a calf can easily outpace a man as the herd moves twenty or more miles each day. Escape speeds of up to 50 m.p.h. are easily achieved while 25 m.p.h. may be maintained over extremely rough terrain. Herds of reindeer may return to favoured ranges year after year, suddenly neglecting these in favour of an area 50 - 100 miles away.

As Frenchen and Salomonsen (1959) point out, not all reindeer migrate. Some herds may be forced to remain in the tundra due to the fact that possible routes are too long and complicated to warrant following. This is commonly observed in areas in which strong winds blow the snow from the surface of the ground and reveal potential food sources (Chernov 1985).

Only the barren-land caribou undertake migrations of any distance. Forage availability in woodland communities means that it may not be necessary to travel long distances to secure winter feed. The woodland species may, however, undertake short seasonal migrations of a vertical nature (Ingold 1980).

Finally, reindeer are associated, in several ways, with a variety of other species. Predators include wolf, man, bear, lynx, wolverine and wild dog: the young may be taken by arctic fox, golden eagle, white owl and raven. Scavengers include the above, shrews, voles, marten, mink, ground squirrel, lemming, porcupine and birds such as crow and gull. The reindeer compete with goose, grouse, ptarmigan, moose, elk and muskox for pasture although the relationships are not antagonistic. By maintaining contact with herds of muskox, caribou seek protection against predation by wolves, by far the most successful of the species' 'natural' predators (see table 4.3). However, protection from human predation is practically impossible although the belief that it is not possible to miss the target when hunting reindeer is unfounded (Burch 1972, Spiess 1979).

Table 4.3 <u>Factors of Reindeer Calf mortality.</u>
(After Miller et. al. 1988)

Wolf predation	68.5%
Atelectasis	6.7%
Separation or abandonment	6.2%
Pneumonia	4.0%
Bear predation	0.4%

4.1.2 Red Deer.
<u>Cervus</u> <u>elaphus</u> L.

As in the case of the reindeer, the red deer may be divided into two broad ecological groups: the open (North American wapiti or elk) and woodland (European) varieties (Straus 1981). At least seventeen 'races' have been identified as a result of regional differences in size and appearance. In the present discussion general characteristics of the European red deer will be detailed, although, given the result of work described by Straus (1981), some attention will be given to the North American variety which is more catholic in its feeding habits.

Considerable variation in body weight of red deer individuals is recorded by several authors (Petersen 1966, Ahlen 1965, Leson 1978, Flerov 1952) as summarised in table 4.4. Seasonal variations in both weight and appearance occur. During the rutting season, the male possesses a layer of fat up to 2 cm thick under the skin: as much as 30 kg. of subcutaneous and internal fat may be obtained from one adult male (Flerov 1952). By the end of the brief rut total body weight may have fallen by up to 30 %.

Table 4.4 Red Deer Average Weights

	Weight (kg)		Source
European populations	500		Flerov 1952
	160 - 250	♂	Leson 1978
	90 - 130	♀	Leson 1978
S.W.England	180		After Harris & Duff 1970
Scotland	130		After Harris & Duff 1970
Scotland	56 - 132	♂	Corbet & Southern 1977
Scotland	38 - 77	♀	Corbet & Southern 1977
S.England	170		Corbet & Southern 1977
E.Europe (woodland)	255		Corbet & Southern 1977
	255	♂	Bjarvall & Ullstrom 1986
	120	♀	Bjarvall & Ullstrom 1986

All weights are those of mature adults.

The short summer coat is generally of a reddish-brown colour, the longer winter coat being dark grey-brown in colour. The moult begins in May and can be observed in various stages throughout June (Ryder 1977) while the calf remains spotted for the first two months only (Bjarvall and Ullstrom 1986). The male carries antlers for up to seven months of the year, time of shed varying from region to region, the old generally losing them before their juniors. In the Crimea shedding begins between late February and early March - in the Far East during March itself (Flerov 1952).

The rut, of approximately three weeks duration, occurs between the end of August and early October. After a gestation period of 235 days calves (rarely more than two, and usually one) are born in May or June. In any one year up to 92% of adult hinds bear young, of which 14% may be twins in non-hunted populations. In hunted populations this figure rises to 35%. In each case more male than female offspring are produced. In lowland woodland conditions the annual recruitment rate may be as high as 80-90% (Chaplin 1975). Females usually carry, the first time, in their third or fourth year while the male is sexually mature at the age of four.

There is a general lack of body fat during the animal's first year of life. Older stags (over 27 months) contain approximately three times as much fat as do juveniles; the bones of older animals have a substantially higher fatty marrow content. However, "while deer are usually thought of as being lean animals, this is really only true of the calf up to half its mature size, the adult stag after the rut, the hind at the end of the winter and the half-starved animal at any time. The late growing calf and the adult replenishing their fat reserves in summer are laying down quite a lot of fat" (Kay pers.comm.). Indeed calves reach approximately 100kg. in weight before laying down fat reserves: thereafter about 55% of empty body weight gain is due to accretion of fat (Kay 1985:207). Awareness of such carcase quality is of importance to the hunter.

Life expectancy varies geographically, dependent upon both natural and human factors. In managed conditions in the Scottish Highlands a life span of 15 years is typical, although by its thirteenth year the animal is past its prime (Harris and Duff 1970:30).

Only at the time of the rut are stags and hinds observed existing as a group. At this time, mid-September to mid-October, herds tend to be large, having one leader. In winter they live apart, the very old males living a solitary existence, stags occupying territories at a lower altitude than the hinds. However, in severe winter conditions male and female herds intermingle when their ranges overlap (Harris and Duff 1970).

In Europe, female winter herds (hinds/juveniles of both sexes) commonly comprise two to five/ten to fifteen head of deer. The Canadian wapiti may however be observed in herds of thousands prior to their winter migrations. The European deer is more sedentary, clinging to less snowy areas, dispersing in spring as snow melts, uncovering vegetation. Spring herds (totalling more than 50 individuals in many cases) comprise adult hinds, yearlings, juveniles of both sexes and males not yet fully grown. Ahlen (1965) provides detailed records of herd size observations in Scania, while emphasising the matriarchal nature of the species' social system.

Group size is not stable. Individuals may associate for a time with those of other groups; alternatively they may leave the herd to forage alone. Woodland species, however, rarely congregate in more than family groups (Corbet and Southern 1977). The size of group depends largely upon weather conditions and forage availability. Small groups shelter in narrow gullies and behind hillocks in high winds. When deep snow covers the ground large herds form in areas where grazing is clear.

As herd size fluctuates so too do population density figures, reaching a peak at the time of aggregation and migration (where relevant). Density is highest in 'optimal areas' (Ahlen 1965), commonly occupied by female herds, more dispersed at calving time and when food levels are low. Twelve acres of good pasture represent the maximum requirements of a single animal each year in open range conditions (Banfield 1974), providing 6-12 kg. of vegetation per day (Leson 1978). On Rhum, home ranges are well defined in terms of size but variable in shape: on average females require 400 ha. (4 km^2), males 800 ha. (8 km^2) per annum. The equivalent range requirements of woodland species are not known (Corbet and Southern 1977).

Like the reindeer, Cervus elaphus is an eclectic feeder, occupying both wood and moorland niches. The daily cycle incorporates periods of grazing and resting. Deer graze as long as dew remains on the ground surface; the animals then rest for two to three hours. Towards the evening grazing resumes and at sunset they leave forage areas for saltlicks, finally lying down to rest again (Flerov 1952).

This diurnal creature frequents forests in winter and more open park (Bordes 1984) and moorland in summer, taking to poorer terrain in an attempt to avoid predation. In general, woodland forage comprises grasses, moss and various arboreal fruits. Figure 4.2

Composition of Red Deer Diet in Southern Sweden (after Ahlen 1965).

Fig. 4.2

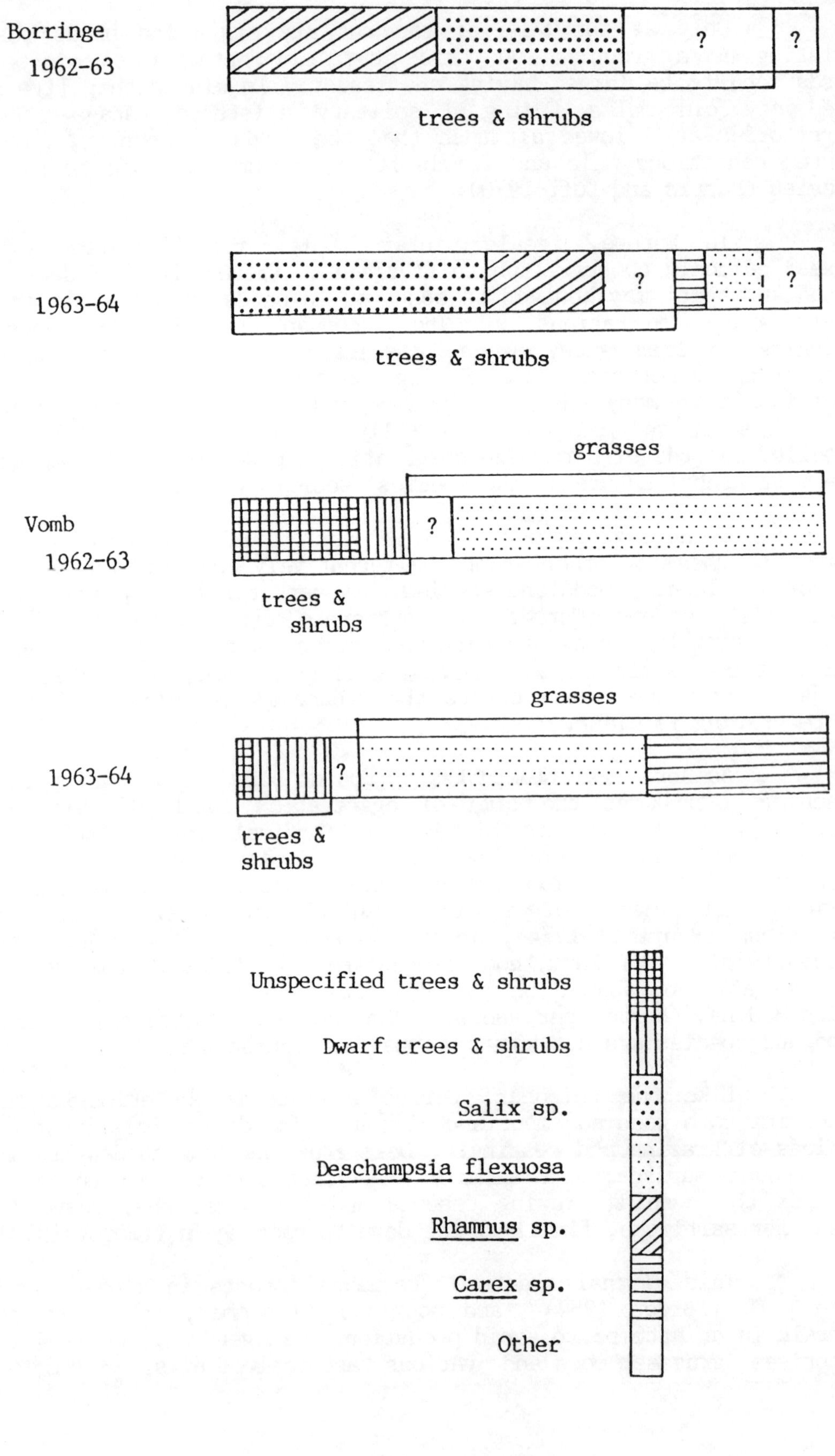

illustrates the annual variation in the make-up of the diet in Scania (Ahlen 1965). Upland forest or woodland areas with local open spaces are preferred. Deer may frequently be observed at the edge of wooded territory where they can browse and graze taking advantage of the low herbaceous vegetation in addition to deciduous and coniferous tree growth. Consuming leaves, shoots, bark or trees and bushes, grasses, fruits, berries, arboreal lichen, mushrooms, aquatics and algae (Flerov 1952), red deer exploit various plant communities. In Scania these include coniferous forests (including spruce, pine), deciduous woods (including beech and hazel with an important ground flora element - Deschampsia flexuosa), marsh scrub (including birch, willow, alder, sedges and grasses), open spaces, beechwoods and bogs (Ahlen 1965). On the Isle of Rhum Clutton-Brock et.al. (1982) list the following plant communities exploited by the red deer:

- Calluna heath
- Wet heath
- Blanket bog
- Nardus heath
- Schoerus Fen
- Mollinia flush
- Herb-rich heath
- Juncus marsh
- Agrostia / Festuca grassland.

Intermediate between browsers and coarse grazers (Kay et.al. 1984) the red deer increases its food intake from winter to spring. Such seasonal fluctuations in food intake are an adaptation to changes in forage conditions (Kay and Staines 1981). Late pregnancy, lactation, rapid calf growth, antler growth and replenishment of fat reserves (all of which are nutritionally demanding processes) occur in summer when climate is mild and good food abundant. As in the case of the reindeer and roe deer, seasonal changes in food intake result in body weight fluctuation of 20-30%.

Associated with a wide range of climatic conditions, the red deer is commonly considered to be a native of temperate regions. In Scania January mean temperatures approximate 0°C, July 16°C with an annual mean of 7°C. Precipitation totals about 600mm. per annum and lake surfaces freeze for up to three months (Ahlen 1965). The species is found at heights between sea level and 7500 feet.

The degree to which extensive seasonal migrations occur among red deer populations is debatable. North American Cervus sp. populations are known to congregate in herds of up to or exceeding 1000 head prior to long range movements. In Europe the animal is more sedentary. Extensive migrations in open regions are caused essentially by the idiosyncrasies of the nature of the snow cover distribution. In the summer season minimum requirements comprise sufficient food, reliable concealment facilities and avoidance of blood-sucking insects. Thus one finds hind groups in the taiga above the timberline. Stags are located in more restricted zones (the higher reaches of the forest zones), grazing on sub-peak glens among scattered trees. By autumn protection from cold winds and snow is required.

In the Caucasus (Flerov 1952) autumn migrations occur later than elsewhere. The deer congregate in valleys where there is little snow on the lower slopes of mountains in areas of relatively dense tree cover. Very few individuals remain in the mountains during the winter. Throughout their distribution, provided that there is little snow, deer will remain within their autumn territories. A snow depth of 40-50 cm. is critical (Flerov 1952), prompting movement to alternative ranges. In Scania deer survive the winter in the following snow depths (Ahlen 1965): 21.2 cm (Tynset), 37.7 cm (Opdat) and 31.4 cm (Storas).

While in the process of migrating (whether in vertical or horizontal fashion) the animals move in single file, forming distinct pathways across the landscape. In southern Sweden deer follow these paths in order to walk or run purposefully in certain directions (Ahlen 1965), attaining a speed up to 29 m.p.h. and negotiating watercourses with ease (Banfield 1974).

The red deer is associated with a variety of other herbivores and carnivores. However, the extent to which these associations are natural and thereby relevant to the present study remains questionable. The species commonly occurs with horse, bison, Bos sp., reindeer, roe deer, saiga and ibex. Predators include, in addition to man, the tiger, leopard, lynx, bear, wolverine, marten and mountain lion. The young are taken by eagle, bobcat and wolverine. However, it is the wolf which represents the major threat, selecting first the senile, young and weak. In addition to predation by carnivores the red deer is subject to the results of avalanche, drowning and starvation. Table 4.5 shows the proportion each of these represents in the death toll in Scania (Ahlen 1965).

Table 4.5 Causes of death amongst red deer

	n.	%
Hunting	109	85
Road/fence	5	4
Drowning	3	2
Starvation & other	12	9

(After Ahlen 1965)

4.1.3 Bovids.
Bison sp. Bos sp.

In the present discussion attention is concentrated upon Bison sp. due to the fact that most Upper Palaeolithic bovid material has been assigned to this taxon.

European and American bison belong to only one species, Bison bison L. of which there are four sub-species (Spiess 1979):

B. b. bison	American plains and woodland.
B. b. athabascae	American plains and woodland.
B. b. bonasus	European
B. b. caucasus	European

Meanwhile Degerbol and Iversen consider bison species in chronological terms (Spiess 1979:259): B.priscus prior to 10,000 B.P. and B.bonasus after this date. Broadly speaking Bison comprises two types, namely the woodland and open plain varieties.

The pelage is formed of long, coarse guard hairs and a matted undercoat, woolly in texture, the winter coat being substantially thicker than that of the summer season. The humped back, typical of the species is more marked in the male than the female, and non-existent in the yellow/red calf until a few months old (Bjarvall and Ullstrom 1986). Bovids are typified by the presence of permanent, keratinous horns, usually in both sexes, growing from the frontal bones (Petersen 1966:335). Growing from the sides of the heavy triangular skull, the short dark horns of the male have inward-turning tips. The female's horns are more slender, their tips pointing upwards.

Conflicting size estimates are provided by Banfield (1974) and Bjarvall and Ullstrom (1986). The latter are concerned with the European bison whereas Banfield (1974) is describing the herd animals of the North American continent. Size indications are provided in table 4.6. The weight and size of individuals varies with age and season. Bulls reach a maximum size at six years, cows at four. Laying down fat deposits during the summer for the winter will result in a seasonal weight increase, as in most species adapted to cold environments.

Table 4.6 Bovid Size Indications.

Variable	Source		
Body length	Banfield (1974)	♂	304-380 cm
		♀	213 cm
	Bjarvall & Ullstrom (1986)		250-270 cm
Shoulder height	Banfield (1974)	♂	167-182 cm
		♀	152 cm
	Bjarvall & Ullstrom (1986)		180-195 cm
Tail length	Banfield (1974)	♂	43-48 cm
		♀	45 cm
Weight	Banfield (1974)	♂	470 kg
			460-720 kg
		♀	420 kg
			360-460 kg
	Bjarvall & Ullstrom (1986)	♂	800-900 kg
		♀	500-600 kg
	Lawrence (1974)		1000 kg
	Arthur (1975)		350-400 kg

All figures relate to adult individuals.

The bison is sexually mature by the end of its second or third year. In the summer, the cows are joined by bulls having spent most of the year apart. The rut occurs in August and September (Roe

1970:95); bulls 'tend' cows in oestrus, ie: the male follows a female with his neck extended, head held level and upper lip curved (Banfield 1974). This may continue for several days until the polyoestrus cow is inseminated. Lactating individuals tend to conceive later than other cows and hence bear their calves later, after a gestation period of 270 - 300 days, but at some point between early March and the end of June (Arthur 1975:52). In this way cows are not lactating at the onset of the following season and may conceive at the beginning of the rut. Although twins are not unknown, each cow usually gives birth to one calf, perhaps twice in the course of three years.

The sense of smell is keen; calves follow their mothers for the first two or three weeks whilst recognition is by scent although they may nurse for three to four months (subject to forage conditions). A cow will reject calves which are not her own. Eyesight is also excellent and a moving object, for example a hunter or predator, may be spotted at a distance of half a mile.

The bison is regularly troubled by insects during the summer. Response to the blood suckers is one of wallowing (Banfield 1974). The animals roll in the dust, excavating wallows of considerable dimensions which may be used year after year. Thus relief from flies and mosquitoes may be obtained. In addition a symbiotic relationship with insectivorous birds exists: blackbirds and crows often perch on the relatively hairless back of the animal, feeding on the pests.

The European bison usually lives in groups of two to thirty individuals led by an old bull (Bjarvall and Ullstrom 1986). Very old individuals (life expectancy is up to 40 years) often live solitary lives as recluses, often failing to join the rutting herd. On the North American Great Plains however the bison is a gregarious herd animal. They travel together in bands of four to twenty head which have been suggested by some to be family groups (Banfield 1974). Bull groups comprise adult males and some old barren cows, while cow groups comprise females, juveniles and a small number of bulls (Arthur 1975:42). These bands coalesce seasonally, for example during the rut, forming large herds numbering several thousands, composed of both sexes. Although easily frightened the bison is a curious species. When attacked the animals mill around and may stampede, during which they have little control over their mass movement. Those behind the leaders may trample others should the direction change with little notice.

Leadership in large herds is usually found amongst the adult bulls which are likely to attack intruders with little or no warning. They are dangerous individuals and give ground very reluctantly (Banfield 1974), and thus a high risk factor is involved when hunting the species. In the case of smaller bands, an old female, usually the mother of most of the others in the group, has the role of leader, watchdog and general adviser (Lawrence 1974), signalling danger and leading the band towards new forage and grazing areas.

Few estimates of population density are available. Roe (1970) comments that historically herds were much larger than is the case today, and for this reason we may expect population density to have been somewhat higher, at least seasonally (at times of aggregation). In the Caucasus region of Europe Borowski et.al. (1967)

provide a maximum estimate of 12 - 15 head per one thousand hectares (10 km^2) with local concentrations reaching 40 - 50 per 1000 ha. where forage conditions are particularly good.

The bison varies in tastes and habitat: climatically it occupies regions ranging from the almost tropical to subarctic (Roe 1970:69). In North America it occupies grassy arid plains (Spiess 1979), meadows, aspen parklands, river valleys and even coniferous forests (Banfield 1974) while its European counterpart is at home in deciduous forest or mixed forest with a substantial deciduous element and scattered open glades (Bjarvall and Ullstrom 1986). McDonald (1981) lists eight vegetation types with which the bison may be associated, both now and during the Pleistocene:

Forest: closely spaced trees predominate, resulting in a closed canopy. The associated climate is relatively humid.

Woodland: trees and shrubs predominate, although the resultant canopy is broken.

Savanna: important trees and/or shrubs co-dominate with a strong herb component in a regime characterised by seasonal aridity.

Brushland: small shrubs form the major portion of the vegetation, the canopy being far from complete.

Parkland: herbaceous flora predominates, forming complete ground cover. Relatively large openings in forest and woodland areas typify this vegetation form.

Grassland: grasses predominate while tree/shrub cover is sparse. Short grasses characterise Steppe communities where ground cover is commonly complete. In Prairies tall grasses occur - adapted to more humid conditions.

Tundra: some shrub or dwarf trees occur in a cold environment of herbs, mosses and lichens

Desert: shrubs and herbs co-dominate, but ground cover is patchy.

McDonald (1981) goes on to describe "models of bison adaptation to different environments", in which three habitats are detailed, based upon the vegetation types considered: forest-woodland, savanna-wooded steppe and open grassland. The reader is referred to McDonald (1981) for details of each model.

Bison are less selective in their choice of forage than are many other ungulates in similar environments (Reynolds et. al. 1982:986). In a predominantly grass area such as the American High

Plains, Gramineae and sedges form the largest part of the diet (Speth 1983:118), shrubs and forbs comprising only a small part. As a result of browsing and grazing herbaceous material in bushy forests (Spiess 1979:259), up to 50 % of the diet may be browse and tree cover, although this usually forms under 10 % of the dry weight (Speth 1983). Even the forest bison prefers non-arboreal forage, selecting Rubus sp. and Festuca montana. Growing tips of willow, aspen and ash are chosen, supplemented, during the winter, by the thin bark of trees. The bison requires a high daily intake of food, ranging from 9 - 10 kg. dry plant food per day in winter to 10 - 15 kg. per day in summer (Borowski et. al. 1967), providing up to 5000 kcals. Most of this is consumed during the day, for bison are diurnal creatures. Foraging continues intermittently throughout the day from dawn to dusk, with rest periods around noon. However, on moonlit nights some foraging activity may take place, especially during the winter when the availability of food may be severely restricted.

In order to secure sufficient fodder to survive harsh winter conditions in North America, the Plains bison may undertake migrations of some distance although there is no reason to assume that these are of the same scale as those of the reindeer. Such migration may be directional or altitudinal. The agility and potential speed at which the animal moves is surprising (Arthur 1975). The average speed of 5 m.p.h. can rise to 32 m.p.h. when a frightened herd gallops to avoid danger. In general the movement is slow, the bovid walking in a plodding manner, moving in a single line along well-worn tracks or ploughing through the snow (Banfield 1974). Maximum speeds can be maintained for only a short period, while the animal is at ease in water, swimming with its head and shoulders above the surface.

Those herds which do migrate generally move south (by up to 200 miles) in order to find improved winter pasture, shelter and milder climatic conditions. From high mountain meadows they descend valleys in winter, seeking relief from heavy snowfall and cold temperatures. Solitary bulls, however, may winter in high mountain passes.

In November and May bison in the Wood Buffalo Park still migrate a considerable distance (150 miles) from wooded hills to river valleys (Arthur 1975:32). In Texas the search for cooler summer temperatures sends the herds northwards in spring. The extent to which such movements are determined by temperature conditions is still a matter for debate (Roe 1970:73). Large, windswept open plains offer relief from insect harassment, while snow conditions and the availability of forage of sufficient quality can also be invoked as causal factors. The aggregation of bands to form large herds necessitates slow, constant movement in order to locate food, the alternative being severe pasture depletion and destruction.

In general the ungulate species of the North American Plains are local in their habitats: not all bison migrate in a regular fashion. Indeed, some authors look upon such migration as the exception rather than the rule (Roe 1970) and favour the concept of large winter herds, high population density and an element of sedentism. Roe (1970) discusses two types of 'irregular migration' which he considers to be more typical of this 'highly erratic species':

1. Regular movements not occurring in the usual directions at the expected times,

2. In any direction at any time of year.

He emphasises that uniformity in the wanderings of bison is not to be assumed and that a pattern of generally regular migrations is not to be supposed (Roe 1970:594).

Finally, four non-predatory agencies may be identified which are destructive to bison populations and their habitat. These can act as stimuli to the permanent or temporary geographic displacement of populations or the immediate destruction of the animals concerned.

Not only can fire destroy the habitat of the bison, but it can also cause loss of sight in individuals and other physical harm. Physical harm may also be caused by diseases running rampant through the herds, for example the foot and mouth disease of domestic cattle in the U.K. during the 1960's and 1970's. Although the species can swim across water courses, a swift swollen river is likely to take its toll of young and old alike (Roe 1970:178) especially in cases in which river banks are not secure. Increased run-off from melting snow may cause many a fatal accident.

Snow conditions have, in the past, been invoked as a cause of the ultimate destruction of much of the bison population of N. America. This is highly questionable. The animal is well adapted to cold conditions and is often to be seen standing facing a snow storm for a considerable time. After the storm the bison group disperses somewhat and moves away to seek food which has been covered by snow. Excavating feeding areas, the bison continues to graze on grass communities below the snow surface. However, where snow drifts into ravines to great depth bison may not be able to negotiate conditions sufficiently quickly to avoid predation (by man and carnivores) (Arthur 1975:38). It was in fact the increase in human predation during the eighteenth and nineteenth centuries (with the use of firearms) which more probably caused the eventual decline of the American buffalo.

4.1.4 Horse.
Equus caballus L.

Details of the behaviour and ecology of the wild horse are few and far between due partly to the fact that there are hardly any remaining other than those bred in captivity (Spiess 1979). For this reason, the discussion of the horse will necessarily be somewhat briefer than that of the other major species.

The Przewalskii horse is commonly considered to be the species most similar to the Pleistocene horse and hence attention will be concentrated upon this rather than the feral populations of N. America and elsewhere.

The adult stands about 1-1.2m at the shoulder, is 1.8-2.0m long and has a tail of approximately 90cm length. Walker (1968:1340) estimates that the average weight of a stallion is approximately 350 kg., while that of the mare is 250 kg.. A yearling weighs in the region of 150kg while a new born foal weighs 15-20 kg.

The adult summer coat is short and smooth. The back and sides of the animal are a reddish-brown, becoming progressively paler towards the belly during the summer. During the winter the horse is different in appearance: the hair is longer and lighter in colour. It should be noted that colour variations have been recorded differing in various habitats (Bokonyi 1974:50). The flat steppe Przewalskii of the Soviet Union is paler than its mountain counterpart.

In general, the horse is a massively built, powerful animal. The head is large but short in relation to the body. With a powerful jaw, well-developed masticatory muscles (Bokonyi 1974:48) and large teeth, the stallion is capable of inflicting severe damage by biting.

Females come into heat several times a year, and oestrus continues for a few months if the mare is not bred (Walker 1968:1342). After a gestation period of 11-12 months the foal is born in mid-April to mid-July. A few days beforehand the mare retires, eventually giving birth early morning or at night (Bokonyi 1974:62). In general one foal is born, twin births usually being abortive. From the age of 3-5 years, reaching sexual maturity, the female may bear a foal alternate years until after her twentieth year, the reproduction rate varying from 20 to 40 foals per 100 adult mares, dependent upon forage conditions (Slade & Godfrey 1982:1090). Life expectancy lies between twenty five and thirty years.

Although formerly the horse congregated in large herds, i.e. before its extermination by man, the number of individuals in herds (often based upon the family unit) generally varies from 3 to more than 100. This group includes young and adults of both sexes, amongst which the leader is usually a stallion, but may be a mare. Bokonyi (1974:20) describes a group comprising one stallion, seven mares and no juveniles. In the majority of cases mares keep foals away from stallions thus protecting the young. In cases in which a stallion leads the band in terms of rank, a 'leading' mare will follow the male but will be superseded from time to time. There is, however, an almost reverse foraging order: mares with foals eat first followed by those without young and juveniles (up to three years). When they have finished the stallion arrives: if, however, he is impatient the others may be routed but may return once he has started.

Although no details of potential population density have been recovered, Spiess (1979) states that a range of 200 - 250 acres per group is necessary, to include grazing, shelter, water and shade. It appears to be the distribution of drinking water which determines that of the horse (Walker 1968:1340). In general equids inhabit grassy and shrubby landscape avoiding soft, marshy ground and deep snow. The Przewalskii horse is to be found in areas of sparse, low quality vegetation unsuitable for many artiodactyls. The saline steppe of vast sand dunes and rocky dry lands is the home of the Asiatic Wild Horse described by Mohr (1971:64). The saxual (Haloxylon ammodendron) characterises the landscape, standing to a height of six metres. The wood of this shrub-like plant is hard and heavy and the bark is juicy, serving as a reservoir for water. Local herbaceous ground flora grasses, sedges and various root tubers are also included in the diet

(Bokonyi 1974) which is one of high fibre and low protein intake, for although primarily grazers, the horse can survive on a wide variety of plant foods and is known to take such forage as Artemisia sp. and Juniperus sp..

The horse is well adapted to running, achieving considerable speed: such speed is generally reserved for escape from danger. Przewalskii horses cover large distances in their annual movements, for example leaving the Djungharia Plain in autumn for the mountains of S. West Mongolia. The primary stimuli for migration are the changing nature and availability of forage and water and a need to seek shelter during the winter.

Finally, a note should be made of the horse population of the New Forest (Hampshire, U.K.) where we find 2500 ponies in an area of approximately 296 km^2 (density: 8.44 p.s.k.); the size of this area is known to change on a regular basis (Putman pers. comm.). Broken deciduous woodland, grassland and both marsh and heathlands form the major habitat (Putman 1986). Much food is consumed at night, the animal being largely nocturnal. Alternatives and supplements to the diet in bad winter conditions include bracken, gorse, bramble, oak and beech shoots (Spiess 1979). Here we find family groups of mares and immature offspring, female relatives, some unrelated females and an associated stallion.

The bands are unable to migrate to alternative, improved winter pastures and in severe winter conditions depend upon human assistance for their survival.

In addition to the 'horse', the extinct European ass (Equus hydruntinus) is encountered in South West France during the Upper Palaeolithic. This small herd, grassland species (Spiess 1979:258) appears to have been characteristic of both the tundra and the loess steppe south of the European ice-sheets (Groves 1974:119), its prime habitat being open plains, although wooded steppe may have been chosen as a source of winter protection from extreme temperatures, strong winds and deep snow, for as Thevenin (1943) suggests, the species does not appear to be well adapted to the extreme conditions of the late Pleistocene stadials.

4.1.5 Roe Deer.
Capreolus capreolus L.

Standing approximately 60-70cm. at the shoulder the adult weighs 18-20 kg., the male usually being 2-3 kg. heavier than the female. With densities ranging from 8-20 p.s.km. territory size varies from 0.01 to 0.1 km^2 , its size depending upon the behaviour of the dominant buck and habitat, ie. the quantity of cover and food (Freethy 1983).

The rut occurs between mid-July and mid-August. Unlike most cervids, herds do not form during the rut: the lone buck pursues does in territories adjacent to his own. Those does not fertilised enter a second rutting period in November/December and all fawns are dropped, at approximately the same time in July or November.

It is in autumn that one encounters small bands generally comprising a family group of buck, doe and yearling twins. Group size may rise to twelve during the winter, although these split up in early spring (Leson 1978:57). Despite a high potential annual increase in the population total, high mortality rates are common, ensuring that the carrying capacity of the area is not exceeded. Thirty percent of fawns may die, while infections, caused by pulmonary and intestinal parasites, take their toll, as do stress and variations in cold, humidity and forage conditions.

The major criterion determining the distribution of the roe deer is the availability of undergrowth, close to open grassy areas or forest clearings (Taylor 1982). Protection from strong winds is sought in river valleys (Geist 1978) and in areas in which reed beds exist alongside patches of forest the roe deer may be observed. Marshes and other water sources also attract individuals and small groups. In addition, and of particular significance in the limestone areas of South West France, there are some places, such as the above and chalk downland, where roe deer may survive at a considerable distance from water sources. During the Pleistocene the roe deer would probably have done quite well in the Perigord, given river valleys, its southern latitude and hence abundant sunshine, thin, soft snow blankets and vegetation comprising a mixture of meadows, steppe, dwarf birch, willow flats and taiga.

The animal is well adapted physiologically to changing environmental conditions. Consuming very little energy between September and April, the roe deer may stop eating altogether in deep snow conditions of the winter. When new, fresh, easily digested shoots are available in spring the animal forages voraciously. Forage generally includes trees and shrubs, grass, berries, fungi and acorns, for the species is to be observed in both deciduous and coniferous forest environments, provided that a shrub layer a sufficient ground flora is available.

4.1.6 Saiga
<u>Saiga tatarica</u> L.

The saiga is a hypsodont bovid (Janis 1984) of 40 - 45 kg. average weight (Eltringham 1984:126). Herd size ranges from 40 to 120 during the summer and fall, the rut occurring between September and October. During the spring however, groups may be as small as 2 to 6 individuals. Fawns (often twins) are dropped in late May or early June by females as young as 18 months. Calf survival rate is usually high.

The saiga is a dry steppic animal preferring habitats in which there is not a heavy or consistent snow cover. In the fall in Central Europe large herds migrate southwards to warmer, grassy valleys to avoid bitter cold and snow (Walker 1975:1465), returning northwards in spring to rich, short-lived summer grazing. This low level feeding animal consumes almost any vegetation available, although it prefers grasses and shrubs. The distribution of the species is determined largely by the nature of the relief: the saiga will only inhabit those areas which furnish flat, open terrain, such as the Pleistocene Gironde might once have been.

4.1.7 Ibex.
<u>Capra ibex</u> L.

Like the saiga, the ibex is a hypsodont ungulate (Janis 1984), weighing between 40 and 125 kilograms (80-125kg adult male, 40-55kg adult female) (Leson 1978, Bjarvall & Ullstrom 1986).

The rut normally occurs in mid to late November and early December or December and early January. Females usually give birth every two years. Adult males and females live in separate herds during the summer but aggregate for the rut and winter, the males being more solitary during the autumn prior to the rut.

The ibex is a montane animal, often living at 2000m above sea-level while grazing on short-lasting communities of dry, abrasive plants with high-standing crops (Francisci et. al. 1985:124). Poor quality forage utilisation (including grass, forbs and branches [Spiess 1979]) ensures that the animal can survive in seemingly inhospitable areas, given home ranges of 1.29 to 51.8 km^2.

Adapted to rock faces and steep slopes, the ibex is unable to cope with snow and icy surfaces due to the form of the hoof. Winter mortality can decimate populations, their survival depending upon the avoidance of adverse conditions: in 1963 518 bodies were discovered in one valley in the Val d'Aoste, each a victim of the cold and avalanches (Leson 1978:91), a phenomenon to which males are more prone than are females. The scavenging potential was considerable.

Still surviving in the Pyrenean region of France, these animals descend, during the winter, to the coastal plain and to mountain valley bottoms (Spiess 1979, Stuart 1982) and are seen below the treeline in spring (Bjarvall & Ullstrom 1986). The migrations are vertical rather than horizontal within the Pyrenees (and Alps) although descent to the more northerly Atlantic coastal plain is possible.

A picture of two movements each year is a gross simplification however, there being five main periods during the year when migration is possible:

1. Late September and early October, when rams and ewes move to winter ranges (pre-rut home ranges in the case of males).

2. Last week of October and first week of November, as rams move to their rutting grounds.

3. Last half of December and first half of January, when rams leave the rutting grounds.

4. Late March and April, at which point both rams and ewes move to late winter/early spring ranges.

5. Late May, June and early July. Females now move to lambing grounds. Then rams and barren females, as well as juveniles move to summer ranges.

4.1.8 Chamois.
Rupicapra rupicapra L.

The chamois, the Wurm and Holocene range of which is shown in figure 4.3 (Masini 1985), is essentially a mid-mountain species (Leson 1978), weighing between 20 and 50 kilograms. Mixed herds are commonly observed although males can live a solitary life, joining others at the time of the rut (Leson 1978). The mixed herd is led by adult or old females, the mean size of ranges varying seasonally (Hamr 1985), from, for example, 60ha (winter) to 74ha (summer/autumn). Distances travelled between such ranges are not large however, the maximum reported by Hamr (1985) being 2.70km.

In addition to mountainous areas, the chamois is adapted to steep cliffs at lower altitudes. As a result, two ecotypes have been identified (Spiess 1979). One of these inhabits cliffs, sheer rock and steep valley slopes in wooded and forested regions, while the other seems to prefer more exposed mountain territory. The latter is more commonly observed with the ibex.

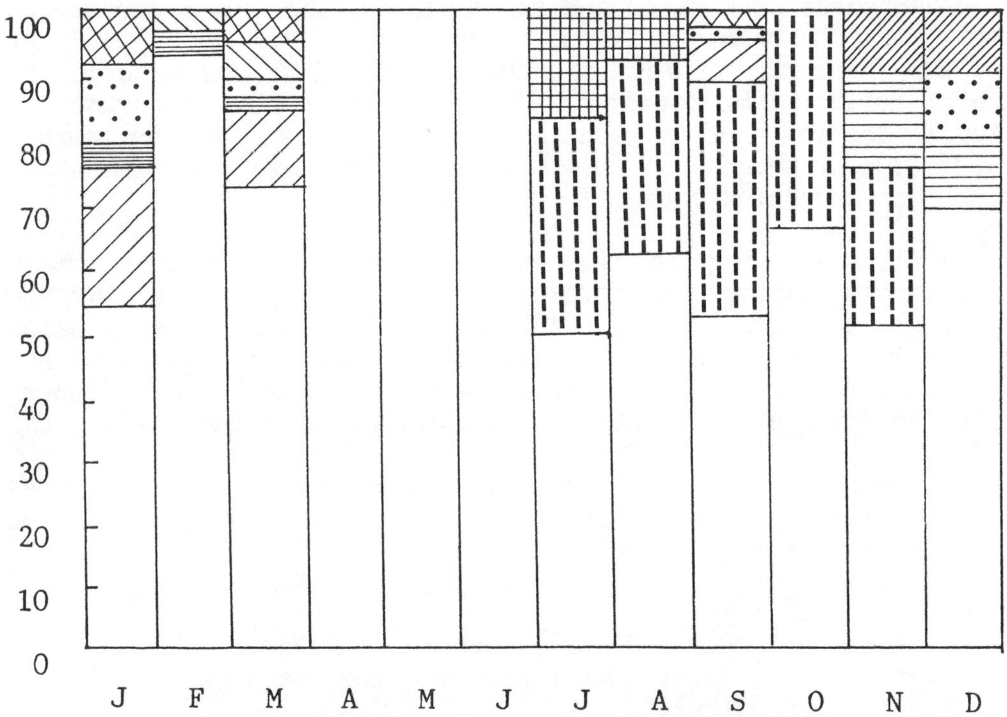

Figure 4.4 Monthly Diet of the Chamois based on Field Observations (after Perle & Hamr 1985, table 8.2)

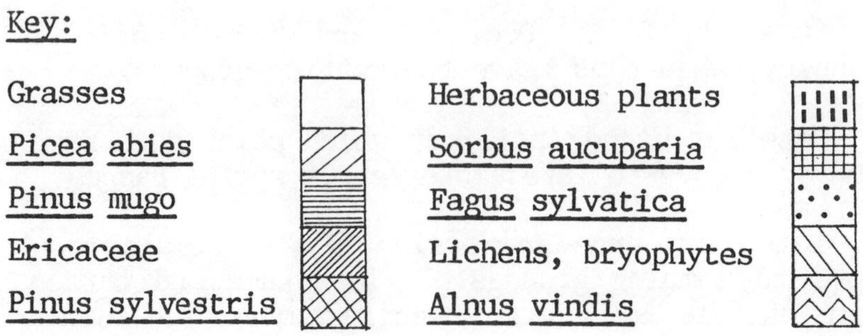

Fig. 4.3

Map showing the Pleistocene and Holocene Range of Chamois in Western Europe.

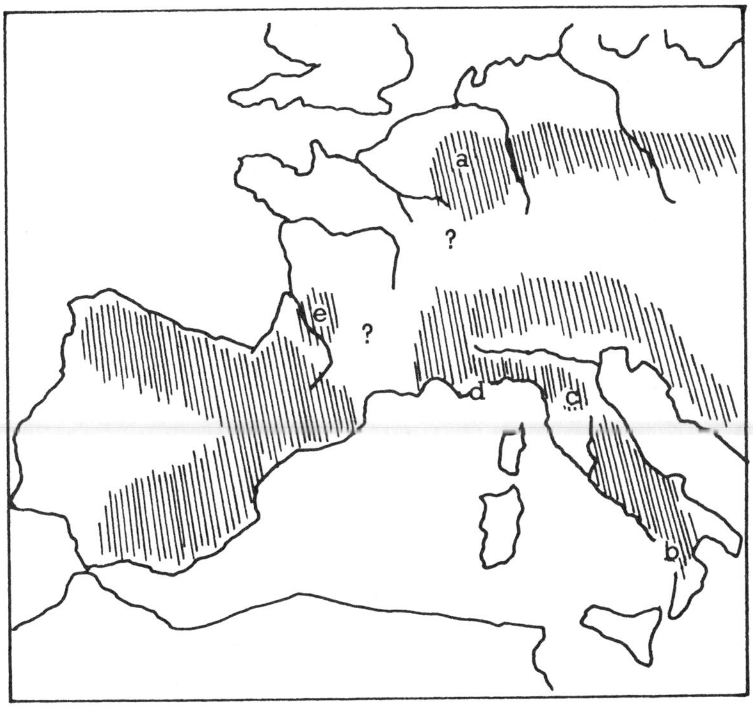

a Trou de Nutons - Wurm IV

b Papasidero - Wurm IV

c M. Sumbra - Holocene

d Grotta dei Fanciulli - Wurm III

e Combe Cullier - Wurm IV

(Source: After Masini 1985:42)

Clover, plantains and forbs provide much of the summer diet of the chamois, in addition to grasses (see fig. 4.4). Their seasonal movements have been likened to those of the reindeer as they descend in search of winter and autumn forage in forests, such as fungi (Spiess 1979:263). As in the case of the reindeer it is thought to be the snow cover which prompts these migrations, a variety of slopes being occupied, with preference being given to those which are east and south-east facing. Extremes of relief are exploited, ranging from the very flat to the very steep and rocky, selection depending upon the extent of the snow. When snow is thick steep slopes are chosen, open terrain being preferred, although some tree cover is sought for protection from strong winds and energy loss due to radiation (von Elsner-Schack 1985:76). During snow-free periods flatter land is chosen, grassland again being the most commonly exploited vegetation type.

4.1.9 Boar.
<u>Sus scrofa</u> L.

Although it would appear to have been relatively rare during the Upper Palaeolithic in S.W.France, the boar today provides a major game species in the region. Weighing 120-200kg. (Leson 1978:84), average densities may be as low as from one individual per 20-50 km^2, for each animal requires a very large territory.

The timing of the rut is variable depending upon such factors as the intensity of insolation, temperature and food supply. It usually occurs between November and January, but may take place in late October. The male is a solitary animal which, once alarmed, may run for up to 60km. in the course of one night (Leson 1978) crossing water without any difficulty. The summer herd (totalling from 6 to 50 individuals) is largely composed of adult females and their young. The younger males (up to 5 years) form small bands, which are chased off by a dominant male at the time of the rut.

The boar requires a large forage intake which includes acorns, nuts, fruits, roots, fungi, rhizomes, legumes and carrion. It is associated with deciduous woodland (Stuart 1982), especially the broadleaf forest, being most active during the evening and early morning (Walker 1975).

An injured boar can be a very dangerous individual. Once wounded or chased for a long time, it will head for water as quickly as possible or lie belly to ground as a domestic dog (Leson 1978:88), hoping to avoid eventual capture. The risk of injury to the hunter and risk of failure to ensure capture of the prey are such that the species is unlikely to provide a major component of the diet of predators forced to stalk or trap the animal on foot.

4.1.10 Mammoth and Woolly Rhinoceros.
<u>Mammuthus primigenius</u> Bl.
<u>Coelodonta antiquitatis</u> Bl.

Despite common belief, the mammoth was no larger than the modern day elephant, standing about 10ft. at the shoulder. It was a species well adapted to steppic conditions (Kurten 1968), feeding on tough, poor quality siliceous grass. Stomach content analysis from fossil finds have revealed a tundra and arctic vegetation, although a mixed grassland-forest biome of low quality grass is also often suggested (Soffer 1985; Guthrie 1982). However, these carcases are discovered in areas which today have a substantially colder climate than has S.W.France; it is possible that Pleistocene conditions in France and Siberia were equally dissimilar. One must not assume that these results automatically reflect conditions in the Perigord. Milder climate may well explain the relative paucity of available mammoth data (reflecting a lower population density).

It is worth noting that during earlier episodes of the Pleistocene the mammoth was associated with boreal forests (during the Chelford Interstadial) and temperate deciduous and coniferous forests (during the Ipswichian Interstadial) (Stuart 1982). Sutcliffe (1985) suggests that the mammoth was an animal of broad ecological range, eating plants selectively and moving to forested regions during winter. Tree, shrub and ground flora were taken as forage.

The Pleistocene rhinoceros was somewhat smaller than its modern counterpart. Living in forested areas, its forage included low-growing grasses and other herbaceous elements (Stuart 1982). In terms of associated ecological conditions the woolly rhinoceros was a catholic species. In Northern Spain associated pollen deposits indicate that the climate was relatively mild with extensive grasslands and some broad-leaved trees (Kurten 1968). Some form of wooded steppe is suggested here as a possible ecotone. Like the mammoth, the woolly rhinoceros merely shows how little we know about Pleistocene, as opposed to modern animal ecology.

4.1.11 Lagomorphs
Rabbit *Oryctolagus cuniculus* L.
Brown Hare *Lepus europaeus* L.
Mountain Hare *Lepus timidus* L.

The adult brown hare ranges from 3.5 to an absolute maximum of 9.8 kg in weight, in general retaining the same colouring throughout the year. After a gestation period of 40 - 46 days, the first litter of the year may be born as early as February, two further litters following during the year. Litter size varies from 1 to 5 young but average 3, the first usually being the smallest. Highest population densities (2 females per square kilometre (Bjarvall & Ullstrom 1986:60)) are found in open country with a rich bush component. Preferring such habitats, the brown hare avoids dense woodland pockets, selecting grasses, buds and bark as food, although during winter months it is known to dig for green plants under the snow (Bjarvall & Ullstrom 1986:60).

The mountain hare is smaller than the brown variety, averaging 2 to 5.8 kg in weight. Regional variations in coat colour occur seasonally; for example, during the winter, in the north of its range it is white while to the south the coat is greyer. Such colour changes are governed largely by climatic (light and snowcover) factors. Females come into season in February in the south, in late March / early April to the north, and young are born after a gestation period of approximately 50 days. Up to three litters in the south and two in the north are produced each season, commonly consisting of 2-5 individuals, occasionally rising to 8. The mountain hare is found mainly in woodland environments (occurring in both birch and beech woods) and mountain grassland (Corbet & Ovenden 1980:145), the primary determinant of distribution being the continuity of suitable cover. The species prefers areas in which deciduous trees form an important part, selecting deep, damp valleys around the deciduous borders of the forest edge and shrubs along watercourses (Bjarvall & Ullstrom 1986:63). However, where the brown hare is absent, Lepus timidus may occupy open terrain. Bark, twigs and shoots of sallow, willow, aspen, rowan, birch, bilberry and heather make up the bulk of the mountain hare's diet, the animal being more active during the night. Up to 4 km may be covered during the course of a single night, the hare spending daylight hours below ground in dens which usually provide a strategic view of its surroundings.

The rabbit is the smallest of the Leporidae, averaging only 1.3-2.5 kg in weight. In general the female produces 3-6 litters per annum, each comprising 3-12 young. The distribution of the rabbit is governed by the nature of the soil; sand, or another easy-to-dig substrate is preferred. Open meadows and grassland, scrub and woodland, dunes and grassy cliffs are all inhabited, the species generally avoiding coniferous woodland and heather covered moors (Corbet & Ovenden 1980:143), although coniferous buds and bark are included in the diet, along with most deciduous trees and grasses. Like the mountain hare, the rabbit is essentially nocturnal in its behaviour (Bjarvall & Ullstrom 1986:67).

4.2 Hunting Strategies.

Various hunting strategies may be adopted in the exploitation of herbivore faunas, the suitability of each depending upon several factors including the geography and ecology of the surrounding environment (i.e. density and completeness of ground flora, tree/shrub cover), animal behaviour (social ungulates which can maintain high speeds can be herded over cliff-faces, while solitary animals are more usually stalked), the knowledge of the hunter concerning the potential resource base, the hunter's requirements and available technology, and the social organisation and size of the human population and hence labour force (e.g. the number of hunters must be of sufficient size to successfully attack either a very large, dangerous individual or a temporary overabundance of smaller-bodied game).

Geist (1978) discusses various adaptive hunting strategies, such as jump hunting and close confrontation hunting. Three categories of hunting strategy are described, examples of which are discussed here in relation to species typically exploited in these ways. His categories are:

1. **Opportunistic Hunting**, in which prey is taken as and when encountered and required. Preparation and forward planning are not involved (Geist 1978:251).

2. **Systematic Hunting**, including such techniques as stalking and coursing. The hunter must be able to locate the prey, approach unnoticed (the element of surprise) and dispatch the animal swiftly. Some element of planning is involved.

3. **Co-operative Hunting**, adopted when it is not possible to creep up on prey unnoticed, e.g. in dry, short-grass steppe. Co-operative hunting involves the co-ordination of several individuals. All co-operating parties must share in the kill, and therefore either a few large-bodied or many smaller animals must be taken. Horse, reindeer and bison can be lured to their death at a jump or river crossing while co-operative confrontation hunting (either close or distance) is commonly involved in the taking of larger and more dangerous species, e.g. the individual mammoth or isolated, dangerous, social bovid (Geist 1978).

In the following discussion various strategies described in the available ethnographic literature as being commonly employed in the hunting of reindeer, bovids, red deer and horse will be considered. In addition some possible means of exploiting mammoth are briefly considered. Although the data are limited in terms of prey species and hunter-gatherer community, it is hoped that some insight may be gained as regards the possible means of dispatching prey.

4.2.1 Reindeer.

Co-operative hunting techniques are more commonly employed in the taking of reindeer than are systematic hunting techniques.

Mass culls occur where the human population depends upon the species in question to a greater extent than any other taxon encountered in the surrounding environment at the time concerned. Thus mass killing may occur on a seasonal basis. The existence of a relatively large human population and labour force is essential: firstly, sufficient manpower must be available to allow relevant techniques to be put into operation and secondly, sufficient demand for the yield must be maintained, unless caching is practised on a large scale. In addition, in order for mass killing to be successful, hunters must have a good knowledge and high expectation of moderate or large herd availability (Spiess 1979). A culling strategy employed at a distance from the herd will yield little return for the energy output involved, and therefore areas selected will be those of high prey density, possibly along proven migration routes.

The majority of mass culls take the form of either mass traps (corrals, pounds), or drives into water, followed, in the case of the reindeer, by dispatch by hunters in kayaks or canoes awaiting the herd's arrival. Central Arctic Eskimo caribou hunting entails driving animals into inland or coastal waters and spearing them from canoes. The caribou may also be shot at from behind mounds of stones from which the herds have been watched (Spiess 1979:104).

The Copper Eskimo construct drive-fences of stones, snow blocks and moss, to the leeward side of grazing herds. Hunters wait, concealed, at the end of the fence, while their companions (women, children and dogs) drive the caribou towards them. Such small scale drives require a workforce which can be provided by two or three extended families.

The Upper Tanana Indians also employ fences, of two varieties:

1. Two converging lines of timber are set up, the apex sometimes being furnished with snares and traps into which herds are driven.

2. Straight fences, running for a considerable distance, may be broken up by snares at various intervals.

Large corrals were operated by 50-400 Chipewyan Indians during the fall and winter in the eighteenth century. These were to be found at the edge of areas of woodland, or a short distance from the southern edge of the forest-tundra boundary zone (Smith 1975, Spiess 1979). The intention was to exploit the migration of the herds from summer to winter pastures.

Up to 400 head of reindeer were killed at a time by the Kutchin of N. Alaska using such pounds and corrals, especially during the autumn and early winter when the corral-based hunt formed the basis of the economy. Throughout the winter, however, snaring becomes important.

Winter stalking, (an example of systematic hunting) which may be an individual pursuit, is a selective hunting strategy through which it is possible to choose those animals to be tracked. Both stalking and snaring result in the taking of a small number of prey and thus are not commonly practised by large scale human aggregations; they are likely to be employed at times of population dispersal. Winter big-game hunting of the Caribou Eskimos of the Alaskan Interior is likewise based upon snaring and stalking along game trails.

The Nunamiut practise individual or small group hunting during the summer. Binford (1978:169) refers to this as an 'encounter' strategy, whereby large tracts of land are crossed in search of game. Employing this strategy during the spring, some caribou may be killed, supplementing the dwindling supply remaining after the winter. At the same time, information is gathered regarding herd movement and location, so as to enable hunters to plan their 'intercept strategy', which is adopted as herds follow their migration paths. It is through the fall that the Nunamiut hunting effort increases and greatest yields are obtained (Binford 1978:416).

A large amount of forward planning and organisation is involved in the annual subsistence strategy of the Nunamiut. Seasonal special purpose sites are established at which specific tool-kits and 'gear' types are to be found (Binford 1980). Caches are marked by piles of caribou antlers and may be exploited as and when the need arises and proximity permits. Lookout posts are established from which herd movement is monitored. In addition kill sites, butchery sites and base camps are all important components of the Nunamiut 'logistically' organised strategy in which the settlement reflects the management of human population groups and the available resources.

The choice of individual strategies employed by hunters largely depends upon the abundance of prey, the topography of the region and the number of active hunters. The choice between mass traps, corrals and drives into water depends upon the availability of suitable watercraft and the topography. There is, for example, less energy investment in the construction of fences and decoys leading caribou to water than there is in the setting up of traps and corrals elsewhere (Spiess 1979). However, kayak based hunting in the Alaskan Interior is relatively inefficient unless the reindeer is migrating in large numbers, passing the hunters frequently in the same direction.

Mass killing of migrating herds will result in a mixed age and sex profile. Despatch, by hunters, of herds which have been driven into corrals and pounds is unlikely to be selective. Stalking, running down and decoying, however, may be age and sex selective (Spiess 1979), as the hunter is able to choose the individual taken. Alternatively he may elect to look elsewhere for more rewarding prey.

4.2.2 Bison.

Bison-hunting amongst the Blackfoot Indians, which varies seasonally, provides a useful model of N. American Plains hunting which may be of relevance to our study of South West France.

During the summer months, as bison move into flat country forming large herds, human population aggregation occurs as hunters congregate on tribal camps (Daryll Forde 1934), resulting in the practice of a form of co-operative hunting. The general aim is to surround the bison in the open: in order to restrict the movement of the herd, grass may be fired on the windward side, allowing the channelling of animals in a selected direction along a route where hunters lie in wait.

During the winter bison may be driven into pounds which are large enough to hold more than 100 individuals. Such pounds are commonly placed at the foot of steep cliffs, their walls built of earth, rocks, logs and brushwood. Although high and closely woven, they are not particularly strong, for unless able to see a way through, the bison rarely charge the walls. At the top of the cliff two lines of mounds of brush or rock are laid out for perhaps, a quarter of a mile, narrowing the passage towards the edge of the cliff. At the entrance to the channel the piles may be a few yards apart, whereas at the bluff end these are larger and closer together. For a short, final stretch they may take on the form of a continuous fence. The bison are encouraged to move between the converging lines eventually to stampede over the cliff face. The injured animals, having fallen into the pound below, are dispatched by the waiting hunters. Chosen sites are refurbished and used in subsequent years. Such corral trapping is reported by Speth (1983).

When snow deposits are deep, hunting parties may be able to drive individuals into drifts, hampering the bison's movement, enabling trackers to reach and kill the prey.

Although the majority of winter kills are based upon the use of corrals and pounds, when a fairly large hunting party encounters a relatively small herd on a still day, it may be possible to surround and close in on the prey (Daryll Forde 1934). Such co-operative ambushing may occur whilst animals are grazing along water courses (Speth 1983). The response of the herd is to run around in a circular movement until exhausted. At this point the animals are killed or wounded by arrows and despatched at close quarters.

4.2.3 Red Deer.

A species such as the red deer, which is commonly observed in small groups or as individuals, does not lend itself to mass drives or corralling, and thus a 'systematic' approach is adopted. Instead, the hunter must depend upon stalking and coursing, each of which yields relatively low return for the energy expended.

In Britain, stag harbouring (coursing) depends on prey availability and the harbourer's ability to put this information to use. It may be assumed that this was the case during the Upper Palaeolithic. Goss (1933) recalls incidences in Somerset where some woods were found to be frequented only by hinds; others were preferred by stags. Harbouring comprises early morning reconnaissance trips, the purpose of which is to locate suitable prey. The information gained is then given to the Master of the hunt enabling the band to lift and follow the stag. Stag hunting in S.W.England occurs between early November and late February, after the rut, while males continue to carry antlers. In cases in which the meat yielded by the animal represents the primary target, one would expect the timing of the hunt to vary according to the nutritional quality of the meat. Sexual differences are observed in this variable, the quality of the meat of the male reaching a peak just prior to the rut. The meat is lean after the rut. The female, on the other hand, remains in good condition until the end of the winter (Kay pers. comm.).

Deer stalking in Scotland is a solitary or small group activity. It cannot, successfully, be otherwise. A large group or band of hunters is likely to distract the prey - due to noise, movement, scent etc. (Geist 1978:274). Having located the prey, the hunter must approach stealthily downwind prior to dispatching his target. Stalking is a time consuming activity during which concealment may be necessary. Thus the positioning of the hunter is of importance if the expedition is to be a success. The adoption of a similar strategy will be necessary in the exploitation of other small-group or solitary, woodland species.

4.2.4 Horse.

Travelling in herds of a considerable size, the gregarious nature of the horse and its potential speed can be exploited through co-operative hunting (Geist 1978; Reiger 1980:124). By lighting fires behind the animals, the hunter can drive herds towards precipices and similar topographic features over which the horse falls. Along the path of the herd, beaters and obstructions prevent the animals from changing route. Such a technique is postulated to account for the magma at Solutre, near Lyon. The method is similar to that often employed in reindeer exploitation, but the horse's speed and the chaos of the stampede make the horse particularly suitable for this strategy.

Coursing and stalking, as described by Levine (1979) appear to be less appropriate for the horse than is the drive when the latter is possible. Topographic conditions have, however, to be suitable, and determine the strategy adopted to a large extent.

4.2.5 Mammoth and Woolly Rhinoceros.

Species such as the mammoth and woolly rhinoceros, the largest known of the range of resources available, were, in all probability, chased to drifts of deep snow or into muddy swamps where they mired down and could be dispatched (Reiger 1980:124). Deadfalls, traps in which the animal was killed or disabled after springing the trigger, may also have been employed, as may pitfalls and other holes dug into the ground and covered with a layer of vegetation. A labour force sufficiently large to drive or beat the animals towards the trap is commonly assumed when discussing the significance of potential cliff-fall sites such as is encountered at Solutre with horse: nothing more complex is required in the case of Pleistocene megafauna.

Chapter Five

SEASONALITY: AN HISTORICAL PERSPECTIVE.

The question of seasonality is one of importance when considering faunal resource representation and exploitation. The presence of species in an area depends primarily upon whether or not conditions are suitable for it, so that seasonal changes in such environmental conditions may result in concomitant fluctuations in animal populations as herds migrate in order to exploit alternative resources elsewhere. Furthermore, if we have reason to assume that faunal assemblages are the result of human predation we may also be able to draw tentative conclusions regarding the season of occupation of sites. We should not, however, assume that the seasonality of a single site reflects the season of occupation of an entire region, as the occupation of different sites at different times of the year in a hunting territory has been shown to be an integral characteristic of the subsistence-settlement system (Binford 1978).

5.1 Early Studies.

The question of seasonality has reared its head periodically since Lartet and Christy (1877) raised the subject in their account of the palaeolithic material of the Perigord in Reliquiae Aquitanicae: being contributions to the archaeology and palaeontology of Perigord and the adjoining provinces of Southern France. Since the mid-1950's however, it has received near-constant attention from several authors working with different material in different ways. In the present discussion, only those studies which relate to South West France will be considered, based, as they are, primarily on reindeer although more recently, fish material has also formed a focus of attention.

The majority of research undertaken prior to approximately 1950 was based upon reindeer antler material, although dental studies have gained in importance only in the latter half of the twentieth century. Lartet and Christy (1877) raise the question of seasonality in their discussion of the Palaeolithic of Aquitaine, based upon the then-available ecological and ethological literature (Austen 1877; Anderson 1877).

In reindeer both sexes carry antlers during some part of the year. These are regrown during the summer, having been shed by adult males during the winter and by steers, barren females and yearlings in spring. Pregnant cows carry their antlers through the winter, until parturition. Velvet covers the antler during the summer, but is scraped off prior to the rut (November). By determining the sex of the antlers (based upon distinctive morphological criteria) and the age of the individual at death (dependant upon developmental stage and whether or not the antler is cast) the seasonality of the deposit may be determined with a varying degree of certainty.

Lartet and Christy (1877) prefer the concept of annual reindeer migration (caused by insect harassment) to the notion of year-round occupation of the Perigord. In order to explain the presence of antlers indicating otherwise, they suggest that reindeer were killed elsewhere throughout the year by humans, their antlers being brought to the area by migrating hunters.

In 1920, Saint-Perier discussed the regular annual migrations of Magdalenian human populations in the Pyrenean region, largely basing his argument upon the available reindeer and red deer antler. He concludes, for example, that the Lespugue sites yield evidence of winter occupation, the hunters being out of the area between July and November. He suggests that the summer was spent on the western seaboard.

Winter occupation of the present coastal region of the Gironde is hypothesised by Lacorre (1939), for at Grotte des Fees (Marcamps) reindeer antler is considered to represent this season. Ten years later Malvesin-Fabre (1948) describes the fauna of the Gironde, maintaining that reindeer moved to the Perigord-Gironde in the autumn, in order to over-winter in the region.

In 1953 Lacorre once again considers the information yielded by antlers, on this occasion from the site of Badegoule. He identifies both forest and tundra varieties, based upon antler morphology and proceeds to discuss their seasonal distribution. He maintains that the tundra reindeer was to be found in the area during the winter, based upon the fact that a large quantity of adult male, cast antlers was present. Meanwhile the forest reindeer predominated from May until the winter migrations began, when they left the area. This hypothesis is based upon the age distribution of the antler remains recovered, the majority being juveniles of 1 to 6 months of age.

5.2 Essai sur le Renne: 1954-1975

Much of the analysis reported between 1954 and the late 1970's was undertaken by Bouchud (1954a, 1954b, 1954c; 1956; 1966; 1975), employing modern Scandinavian populations as his comparative material. He set out initially to consider the Palaeolithic faunas of France from primarily climatic and evolutionary perspectives, conducting a large amount of biometric analysis. However, the material with which he was primarily concerned comprised two forms of data relating to seasonality studies: reindeer antler development and dental material (Bouchud 1954c; 1966).

Bouchud considers uncast antlers (bois de massacre), divides them according to sex, and sets about ageing them in terms of monthly development. Tables 5.1 and 5.2 summarise the ageing criteria adopted, antler size and the form of the pedicle being his major concern.

In addition to the assumptions made by earlier analysts, namely that both sexes have antlers, adult males shed their antlers after the rut, adult females in early April, and juveniles in late May or early June and finally that velvet is shed in mid-September by males over 3 years of age and by females and young in October, Bouchud (1954b, 1956, 1966) assumes that antler growth is the same in both

Table 5.1 Annual Antler Development of Reindeer according to Bouchud (1966)

Months	Adult Males	Adult Females
January	Porous pedicle. Signs of shed, especially 2-3 years.	
February	Pedicle becoming porous New growth minimal	Pedicle & antlers very compact.
March	Beginning of growth	
April	Rapid growth ca. 15cm	Pedicle very porous. Signs of shed: adult
May		Signs of shed: young
June	Rate of growth ca. 2cm per day (max.). Development of compact bone	Signs of late shed: young Start of new growth.
July		More rapid growth
August	Growth, slowed by development of compact bone, stops.	More rapid growth
September	Hardening of antler. Velvet shed mid-month.	Very rapid growth. Development of compact bone.
October	Antler pedicle very compact.	Rapid growth, slowing down & stopping. Hardening of tips. Shed of velvet.
November	Rut. Shedding "adult"	
December	2-3 years shed, end of December.	Antler & pedicle very compact

(Source: after Bouchud 1966:99)

Table 5.2 Annual Antler Development of Reindeer Fawns according to Bouchud (1966)

Months	
January	
February	
March	First year antler & pedicle very compact.
April	
May	
June	Porous pedicle. Shed early summer
July	Growth of first antler
August	More rapid growth
September	Maximum growth
October	Growth slows & stops. Compact bone forms Velvet shed
November	
December	First antler & pedicle very compact

(Source: after Bouchud 1966:99)

sexes, although shed time is acknowledged to vary to a limited extent. In both sexes, antler size increases with age, the animal being at its prime at 6 - 8 years. After this however, antlers begin to get smaller and less complex. Thus we are warned that the use of reindeer antlers is less reliable, in ageing studies, than that of red deer (Bouchud 1966:69), for growth may be seen as a function of soil types, food and general health.

Bouchud (1954b) prefers to study the available dental material. He claims that such a study is more precise than that based upon antlers for the latter are more often employed as raw material, used in diverse ways. In this way he explains sexual differences in antler representation; for male antlers are seen as a much more desirable form of raw material than the female equivalent due to their more robust physical structure. The lack of shed female antlers is thus seen as a result of human choice and hunting strategy rather than the absence of herds from the area at the relevant time (1966:100). He is able to employ both complete and fragmentary mandibles and (he believes) isolated teeth. Permanent and milk dentition are separated and, in each group, M1, M2 and M3 etc are aged by comparing the isolated teeth with those in mandibles of known age. He claims an upper limit of 26 months (Bouchud 1954b:80) (see table 5.3, based upon comparative material from Uppsala University and the Stockholm Naturhistorika Riksmuseet).

Despite the fact that Bouchud (1954b) recognises that births are not simultaneous, he assumes that all the young are born in June. Thus an individual of three months of age died in August. Bouchud repeats his basic assumptions regarding simultaneous June births and the timing of the rut, and adds to these the following:

1. Attrition occurs at a constant rate.

2. Full eruption takes 27 months. There is no difference in wear and eruption between tundra and forest species.

3. Sample size exceeds 100 lower teeth. Lower dentition is preferred as such teeth have more regular wear and simpler roots.

Based upon these assumptions and the comparative material made available to him, Bouchud (1954a, 1954b) claims that it is possible to age individuals up to 2 years of age to within one month. It is upon such results that seasonality implications are based. For example, at Les Rois (Charente) during Aurignacian I, reindeer were, he claims, killed on a year round basis. A similar pattern is seen during Perigordian IV at Roque Saint Christophe and Bourdeilles (Dordogne). Such results are obtained from the majority of assemblages studied, a notable exception being that of Cottier (Haute-Loire) where, during the Early Magdalenian, reindeer appear to have been taken between May and September. Finally Bouchud (1954b:84) concludes that the reindeer did not make long-distance migrations during the (Middle and) Upper Palaeolithic, being able to satisfy their requirements within easy reach of their presumed home range. He

Table 5.3 Dental Eruption in *Rangifer tarandus*

	Scandinavia	Siberia	Alaska	South West France (Palaeolithic)
BIRTH	Incisors & canines present. Deciduous molars at 3rd week	Incisors & canines present	Incisors & canines erupted. Premolars erupt a few weeks later.	Incisors & canines present. Deciduous molars at 2nd-3rd week. Use of m1 by 12 mths, m3 by 36mth
MOLARS	M1 at wk 10, used from 4th mth. M2 at c.10 mths, used c.15/16mths: M3 at 19/20mths. Used after 24mths	M1 erupts at 3½mth used in 5th mth M2 erupts at 1yr used c.15mths. M3 at 29mths, used c.30mths	M1 visible under skin at birth M2 used end of 1st yr. M3 erupts end of 2nd yr	M1 erupts 3rd month. M2 erupts 13th mth M3 erupts 24th mth used at 30 mths
INCISORS & CANINES	Deciduous replaced 12-14mths. Upper canine at 15mths. Used c.16mths.	I1 replaced at 10 mths. I2 & I3 at 12 mths. I1,2,3 used at 17 mths (Sokolov (1937) Gryuner (1931)) = Replaced at 6mths. Used at 10mths. Canines replaced at 10-15mths.	Replaced c.10-12mth	?
PREMOLARS	Deciduous replaced c.27mth. P4 used c.28mths. P3 used c.30mths.	Replaced 29mths used at 36mths.	P3&4 replaced end of 2nd season. P2 erupn ends 36mths.	Erupt simultaneously at c.27mths. Used at 30mths.

(Source: After Bouchud 1966:103-108)

discusses the possibility of domestication and the existence of two forms of reindeer as possible explanations of the permanence of the species. His argument supports the latter. The reindeer is not considered to be sedentary, but undertaking small-scale, limited migrations which may have been followed by groups of human hunters. Bouchud maintains that the absence of reindeer from the area did not preclude the continuous presence of human hunters in the region. Despite pointing out that the reindeer was to be found in the area only during spring and autumn, Bouchud (1966) concludes that the Vezere valley sites were occupied throughout the year. He maintains that we must infer movement of hunters to reindeer summer and winter pastures in order to explain the apparent year-round data (Sturdy 1975). Bouchud allows for movement to the coast, visualising two possible mechanisms:

1. 1 or 2 years absence from the site in question. This would not, however, be visible as a sterile level.

2. Coincidental absence from the site. Such absence causes the (age) distribution of teeth to appear even throughout the year.

Between 1954 and 1975 we see the appearance of a number of publications either refuting Bouchud's results or simply adopting his methodology.

Lacorre (1956) criticises Bouchud's conclusions, results and his interpretations thereof, as opposed to his basic assumptions and methodology.

Four problems of Bouchud's research are identified by Lacorre (1956) who maintains that reindeer herds migrated through the Les Eyzies area twice a year, undertaking longer-distance movements than those proposed by Bouchud (1966). He points out that Bouchud is forced to assume that the reindeer has become migratory (ie: on the Canadian scale) only in its recent Arctic history, and questions his reasoning on the basis that no other extant species has undergone such a fundamental behavioural change.

Lacorre (1956) questions the extent to which fossilization can, as Bouchud (1954b) claims, be held responsible for the presence or absence of certain classes of data. Fossilization conditions in the limestone deposits of South West France are good and 'complete' skeletons of microfauna are often uncovered. Thus Bouchud is wrong to assume that the lack of material can be simply explained by preservation, as Lacorre's example of La Gravette mustelid data shows.

Bouchud also employs human choice in hunting strategies as a means of explaining the lack of some data. Lacorre (1956) does not believe that hunters could easily distinguish between the sexes and age groups, especially when encountering large, moving herds. Furthermore, the available technology was, he believes, such that not every hunted reindeer was dispatched. Thus there may have been a population of reindeer, wounded and temporarily incapacitated, which may have remained in the region all year.

Finally, Lacorre (1956) criticises Bouchud's (1954b) assumption that dental attrition and eruption form a continuum on which to base his year-round conclusions. He points out that wear varies seasonally, according to dietary change. Abrasion is thus highly variable. Furthermore he maintains that since we cannot distinguish between forest and tundra subspecies by dentition, it is inadvisable to attempt to age individuals of both sorts by wear due to the fact that their respective diets are different.

While discussing the Aurignacian at La Quina, Guillien and Henri-Martin (1968) consider the seasonality of occupation based on dental material. Summer occupation is proposed. However the paper is today of interest due to the methodology and problems discussed rather than the results themselves, which may be questioned in the light of criticisms of much of the earlier seasonality research (cf. Binford 1973).

The methods employed are reminiscent of those advocated by Bouchud (1966), based upon dental eruption sequences and the speed of abrasion. Guillien and Henri-Martin (1968:337) assume that development is rapid and uniform up to 18 months after May-June births, at which point puberty is assumed to set in.

Tooth eruption and wear are employed as means of ageing reindeer until the appearance of M_3(third lower molar), for prior to this, the developmental sequence derived from studies of modern reindeer/caribou populations is considered to be fairly reliable. However, the authors isolate an important problem which should here be briefly considered given the preponderance of dental studies in recent years.

Mammalogists commonly refer to the eruption of the tooth through the gum, and it is upon such data that our comparative frameworks have been largely based. However, archaeozoologists are not in a position to do this, and refer instead, to the eruption ('sortie') of the tooth from the bone. Guillien and Henri-Martin (1968:338) point out that in modern Lapp reindeer there is a time lag of as much as 7 to 8 months between emergence through the bone and piercing the gum. A Pleistocene reindeer which is considered to be a year old may, therefore, be only 4 months of age.

Having considered tooth eruption the authors then proceed to discuss attrition or wear. They assume that abrasion progresses swiftly from the distal to proximal facet of the tooth, and that it appears successively on distal (a), medial (b) and proximal (c) faces. Therefore m_4b refers to a fourth deciduous molar, worn exclusively on the medial and distal sides.

During the first and third six months, development is very rapid (Guillien and Henri-Martin 1968:339). Calcification, eruption and abrasion are all apparent while during the second six months (after the eruption of M_2) growth is much slower and the stages in development less distinct, finishing at M_2a (the occlusion of M_2). As a result, teeth derived from individuals of 1 to 6 months and 13 to 18 months of age are more readily recognised by Guillien & Henri-Martin (1968) than are those of age 7 to 12 months (stage M_1c). The analyst who is unaware of this difference will not attach particular significance to the small number or absence of identified M_1c, and will assume the length of the phase to be the same as that of M_1b and M_2a, whereas it may perhaps be ten times longer. Guillien and Henri-Martin (1968:340) claim that the informed analyst will immediately conclude that reindeer and man ceased to co-inhabit the area during the winter months, as they do at La Quina.

1970 sees the publication, by Delpech, of the Flageolet II (Bezenac, Dordogne) material in which she identifies occupation during March, April, June, September and December, having adopted Bouchud's methodology concerning antler and dental material.

In 1973 Binford presents the results of his own analysis of both modern ecological and biological data available from Siberian reindeer populations and archaeological material. He points out that a birth cycle is clearly seen, in which the majority of young are born in late May and early June. Births are not, however, simultaneous, and annual variation in the timing thereof does occur.

By assuming that complete maturation of premolars takes 27 months, Bouchud (1966), we are told by Binford (1973), is overestimating the time factor involved. Binford points out that among modern Alaskan caribou eruption may be completed by 22 months, although the process averages 24 months. He provides an upper limit of 26 months and therefore Bouchud reduces a period of 4 months to a single month. (Similar problems are encountered in his statistical manipulation of antler data, through which six antlers representing October to March are equated with a single month, despite the fact that they might all have been killed at the same time (Sturdy 1975)). Binford (1973) also maintains that Bouchud's estimates of molar eruption are correct only for M_2. With errors in his estimates for M_1 and M_3 Bouchud is in a position to conclude that three molars derived from one individual belong to animals killed during different months. Binford (1973:239) provides the following figures for M_1 and M_3 eruption:

M_1 = 4+/-2 months
M_3 = 25+/-6 months

By employing isolated teeth Bouchud's results become highly problematic.

The rate of attrition is dependent upon various factors including both the individual's diet and the nature of the soil. Seasonal variations in diet result in differing forage intake and thus wear patterns. Attrition is highest when sedges are consumed, and, among the Canadian / Alaskan caribou, reaches a minimum during the winter, as do tooth and mandibular growth. Thus Bouchud (1966) is wrong in assuming that attrition occurs at a constant rate and we can conclude that his estimates based on his assumption of constant attrition will be incorrect.

The highlighting of weaknesses in Bouchud's approach (be they analytical or statistical) leads us to question his results and those subsequently produced by other analysts following his lead. Alternative techniques and associated conclusions must be sought.

5.3 The Abri Pataud: 1975-1979

In 1975 Movius published an important volume concerning excavations of the Abri Pataud (Dordogne), in which Bouchud (1975), employing the methodology developed earlier, presented a detailed study of the fauna, concentrating upon the recovered reindeer material upon which he based his claim for year-round occupation. 1975 also sees the appearance of Delpech's Doctorat d'Etat thesis in which she adopts Bouchud's methodology, despite disputing his claim for the existence of several subspecies of reindeer. However, seasonality issues appear to be of minimal importance in the thesis, receiving, as they do, scant attention. Bouchud later qualifies his conclusions concerning the Abri Pataud (Gordon 1980). He maintains that the shelter itself was occupied during the summer, while during the winter, hunters camped outside the site. Meanwhile 1977 also sees the appearance of a paper in which Bahn adds his support to the growing body of criticism of Bouchud's (1966) methods and conclusions.

In 1979 Spiess also presents a study of the Abri Pataud reindeer material. He provides evidence for late autumn, winter and early spring occupation, in the form of antler, foetal long bone and dental data and finds no evidence for late spring, summer or early autumn occupation.

When considering the various types of data, Spiess (1979) makes two basic assumptions which remain constant throughout his analysis. He assumes firstly that calving occurred in mid-May (in contrast to Bouchud's (1966) June assumption) and secondly that the rut took place around 15 October. He discusses the use of antler development in seasonality studies, pointing out, in particular the interpretational problems involved in Bouchud's approach. Even if we were in a position to determine the time of shed, the existence of cast antlers does not necessarily imply that the hunter was in the region at the time of shed, for such antlers may be collected at any time of year. Only uncast antlers may be used, Spiess (1979) claims, in the determination of the season of occupation of a site, assuming that the individual did not die a 'natural' death. However further problems arise, for both antler growth and shedding are determined by changes in hormone levels and vary considerably in animals in poor physical condition. Spiess (1979:95) casts doubt upon Bouchud's ability to measure the base of an unshed antler and thereby determine the age and sex of the individual in such a way as to allow him to determine the season of death.

The use of mandibles as a means of determining seasonality is summarised. Based upon tooth eruption and the nature of wear on crown surfaces, Spiess (1979:70) claims that the best that can be achieved is an estimate of age in terms of multi-year categories, once the reindeer is more than 2 years of age. Only up to this age can seasonality be gauged.

When considering Bouchud's (1966) analyses Spiess (1979:71) rightly points out that there is a great deal of variation to be found within a reindeer population in terms of age at eruption and progress of wear. By comparing Bouchud's results with similar work concerning the Kaminuriak caribou in Canada, Spiess (1979) shows that each of the former's ageing criteria covers a wider time span than was initially claimed by Bouchud (1966). By employing artificially narrow ranges, year-round occupation (ie: kills over a wide range of time) is an almost inevitable conclusion.

Spiess (1979) also considers the use of foetal long bone length in the determination of seasonality. In order to do this he assumes that:

1. Foetuses can be aged to within one month by employing long bone measurements.

2. Calving and rutting times are known

3. The 17 foetal long bones which are considered do in fact belong to R. tarandus.

Given that foetal growth is rapid, long bone length should provide a good indication of the age at death, and provided that we can assume that we have correct dates for both rut and calving, the stages of foetal growth observed should indicate the season of death. In addition to questioning the reliability of the dates provided for the rut and calving, it is important to consider whether the size of the sample employed in the analysis is sufficiently large to be considered representative.

Finally, two techniques of tooth cementum analysis are considered, that developed by Bourque et. al. (1978) and that involving decalcification.

After mounting tooth root fragments in resin. grinding and polishing them to a flat surface, microscopic observations are made. A dark stained line occurs in the cementum between December and April in modern Arctic/Subarctic caribou populations, and Spiess (1979) assumes this to be the case in South West France during the Upper Palaeolithic. The annual layers form translucent and opaque bands which are examined under high intensity light. Although the contrast between the bands is poor in terms of visibility, the advantage of the technique is substantial; it provides a permanent record, in that the technique is not wholly destructive.

An alternative technique is that involving the decalcification of the tooth in question in acid, its cutting, sectioning, staining, mounting and microscopic examination. Problems arise however, due to the fact that the deposition of the dentine/cementum is a seasonally related phenomenon which also varies from year to year and decreases with age. Nevertheless, estimations of seasonality are made dependent upon the 'colour' and thickness of the outermost layer of cementum and/or inner layer or dentine. If the deposit is dark, then the animal died between early December and late April, if light and very thin, a few months later; if light, but of thickness equalling that of earlier layers death occurred just prior to the following winter.

Decalcification of the Abri Pataud material was not particularly successful. Only 10 of 171 tooth root fragments could be read and results obtained. Spiess (1979) concludes that the techniques should be employed only when in possession of large, well-preserved (ethnohistoric) samples. On the other hand, the grinding technique is suitable for small samples (thereby valuable) which are very old and/or poorly preserved.

Spiess (1979) assumes that similar lines occur in other ungulate species (namely horse, bovids and red deer) although the monthly span may vary from one species to another. Seasonality based upon red deer is now being systematically studied (Gordon pers. comm.). Only a small number of teeth were sectioned by Spiess the results being summarised in table 5.4 by stratigraphic division. On the basis of these results, Spiess presents the hypothesis that the Abri Pataud represents a late autumn and winter base for the hunting of all the available ungulate species (Spiess 1979:190). This would appear to preclude the possibility that different species were exploited at different times of the year.

Table 5.4 Summary of Seasonality by Stratigraphic Division at the Abri Pataud. (After Spiess 1979:195)

Level 2	
Reindeer	Tooth eruption: March/April. Foetal long bone: Dec. & Jan.. Tooth sectioning: December to April
Salmon	July to early November.
Level 3: main occupation	
Reindeer	Tooth eruption: March or April Foetal long bones: February
Salmon	September to early November
Level 4: 0 - 2 living floor	
Reindeer	Tooth eruption: November to mid-March
Level 4: Other than 0 - 2 living floor	
Reindeer	Tooth eruption: Late September / October
Red deer	Tooth sectioning: October to March
Level 5:	
Reindeer	Antler: November. Tooth eruption: Sept.-March. Foetal long bones: Feb. & Jan. Tooth sectioning: December to April
Red deer	Tooth sectioning: Winter
Level 6:	
Reindeer	Foetal long bones: November to December
Level 11:	
Reindeer	Foetal long bones: January
Horse	Tooth sectioning: Winter
Level 14	
Reindeer	Tooth sectioning: December to April

To summarise, Spiess (1979) concludes that levels 2, 3 and 5 of the Abri Pataud were occupied from October to March. Level 4 was occupied during this time, but lacks evidence for January and February.

Finally, Spiess (1979) discusses the implications of seasonality and associated population studies at the Abri Pataud. He concludes that hunting techniques employed were not age or sex selective, and that the 'total population' was available for exploitation in the area around Les Eyzies - one which was good rutting territory.

5.4 Studies of Magdalenian Seasonality: 1982-1986

Since 1980 the Magdalenian of France has received a great deal of attention. This is a result of the work of two specialists, considering different types of material. Le Gall (1982, 1984) has based his reconstruction of seasonality on fish remains from selected sites in South West France, while Gordon (1982a, 1982b, 1982c; 1986a, 1986b) concentrates on reindeer dental material from French Magdalenian assemblages, having developed the adopted methodology in a study of Canadian Barrenland caribou.

The research reported by Le Gall was undertaken as a doctoral thesis (1982) at the University of Bordeaux I (Institut du Quaternaire) and was finally published in 1984. In this he considers osteology, palaeoecology and ethnology - although in the present discussion only the seasonality aspects will be discussed.

Le Gall's (1982, 1984) discussion is based primarily on the study of the occurrence of seasonal annuli on fish vertebrae, allowing him to estimate the time of year at which the fish died. The study of seasonal annuli in fish vertebrae is, in principle, similar to that of reindeer cementum analysis. Wide/light and narrow/dark rings are believed to equate with alternating periods of activity and passivity in growth. Such growth ceases completely during the coldest period of the year, temperature being the primary determining factor. The season of death is determined by the identification of the ring forming, comparing it with the one most recently developed. However, it is difficult to identify winter kills based upon this principle, and the result is rarely as specific as those relating to early / late spring, summer, or autumn. Finally, problems arise when employing comparative material, for seasonal rings are more easily seen in modern than fossil material.

Three case studies are presented, based on material from the sites of Pont d'Ambon, Grotte des Eglises and La Gare de Conduche. In the present discussion attention will focus upon the first of these, brief summaries of the results from the others being provided.

Table 5.5 shows the frequency of each of the taxonomic groups considered by Le Gall from each level at Pont d'Ambon. Several observations are made by Le Gall (1984:141), including the apparent over-abundance of carp, the constant presence of pike, the under-representation of salmon and the abrupt decline in eel frequencies in level 3a. The material employed by Le Gall is well preserved. It was obtained by careful wet-sieving of sediment through a 1mm mesh. A total of 4412 fish bones were used. He begins his discussion by rejecting the suggestions that the fish represent the result of rapacious bird activity and that periodic flooding of the site by the Dronne could be responsible for the deposits. With a density of as much as 1400 bones per $1m^2$ in one level, these are not considered as important agencies of accumulation.

Table 5.5 Frequencies of Taxonomic Groups at Pont d'Ambon.

	%				
	c.4	c.3b	c.3a	c.3	c.2
Carp	70.75	65.82	86.68	86.32	68.28
Eel	24.90	28.81	8.61	11.13	16.55
Pike	3.30	4.92	4.25	2.39	13.79
Salmo sp.	1.05	0.45	0.46	0.16	1.38

(Source: after Le Gall 1984:141)

Le Gall's general conclusion is one of all year round fishing, based upon the results presented in table 5.6. In level 4 the carp (chub and dace) are taken all year round, although the chub is exploited mainly in autumn and early winter and dace in spring and autumn. The foragers may be taking advantage of the dace's February to May spawning period while the fish are closer to the surface than during late autumn and early winter. At this time temperatures fall and both species seek food in deeper water. Similarly, the spring spawning of pike takes place in shallow water close to the river bank so that the spring pike catch may be intentionally exploiting this fact.

In level 3b the bleak (or ablet) makes its first appearance. Once again year round exploitation is proposed, although it is pointed out that summer fish are larger and thus perhaps more desirable in the case of the dace and chub. However, spring and autumn provide the main eel catches, while the pike is taken in autumn or early winter and spring. Fishing at the height of winter is rejected on the grounds that there is the possibility that the water surface was frozen for a short period of time. It is noticeable that there is a lack of midwinter fishing.

Apparent seasonal selection is repeated throughout the levels at Pont d'Ambon, although in level 3a sup. the preference is markedly less pronounced. Throughout, each fish species appears to have been taken at its most favourable time - a fact employed to reinforce Le Gall's conclusions. He concludes that the site was used only sporadically (although the episodes were distributed throughout the year) - presumably in order to exploit the fish at the most profitable time.

At the magdalenian site of Grotte des Eglises (Ussat, Ariege) dated to 11800+/-500 B.P. (Gif 1434) and 12900+/-220 B.P. (Gif 3923)

Table 5.6 Seasonality Indications from Pont D'Ambon.

Species	couche 4 carre 17	couche 3b carre J7	c.3a inf. carre J5	c.3a sup. carre J5	couche 3 carre K6	couche 2 carre K5	couche 2 carre H8	c.2 sup. carre H8 -G8
Salmon/ trout	autumn/ winter							
Pike	winter/ spring	autumn/ winter spring	autumn/ winter spring	spring	autumn/ winter		spring	summer autumn/ winter
Eel	summer/ autumn	spring summer autumn	autumn spring		summer autumn	spring	autumn	spring summer autumn
Carp	autumn/ winter	spring summer autumn	spring summer autumn/ winter	summer autumn	spring autumn/ winter	autumn/ winter spring		

(Source: After Le Gall 1984)

for levels 8 and 8b respectively, Le Gall (1984:176) concludes that most fish were taken at the end of autumn. This supports Delpech's claim that ibex were hunted during the rut, at the end of the autumn and during early winter (Le Gall 1984:175). Only one individual is claimed to have been killed during the summer, again supported by the herbivore material analysed by Delpech (Delpech & Le Gall 1983). The suggestion is made that some fish and parts of the large game (ibex) were smoked being transported to the main settlement. Meanwhile, the Grotte de la Gare de Conduche, like Pont d'Ambon, was occupied all year round. The bones derive primarily from Salmonidae and carp. There is little dace, however, as the water was too cold and turbulent for its liking. He goes on to concentrate attention on a single pike caudal vertebra, prepared as part of a necklace. This species was not available in the Cele, and we should therefore conclude that the single bone was brought from elsewhere.

Research into Magdalenian seasonality in France conducted by Gordon (1982a, 1982b, 1982c; 1986a, 1986b; 1988) was initially undertaken on the understanding that 'most hunting peoples of the world are and have been nomadic in order to intercept the migrating animals which provide primary sustenance'(Gordon 1982a:1). Thus by selecting reindeer material he implies that the species was in fact the resource providing 'primary sustenance'. He aims, through establishing the seasonality for a series of sites, to begin to map prehistoric herd movements, and perhaps predict the location of sites which would, by their geographic location, benefit the hunter.

Approximately 1000 caribou teeth of known age were collected from sites of known seasonality on the Kaminuriak and Beverly ranges in Canada. These incisors were sectioned, stained and examined under polarizing microscope at X100 or X40, depending upon state of preservation, for annuli related to seasonal growth periods. Given that the time of death was known, patterns of growth (dark, narrow bands in winter; lighter, wider bands in summer and autumn) could be related to the season of kill. They may however be related to dietary and metabolic change. Subsequent research has shown, however, that the situation is a relatively complex one. Calves (born in the first half of June) develop a 'winter restline' or increment only after their first winter (at the age of 7 to 10 months). Different increment thickness is often observed in male and female incisors as male teeth grow mainly between July and September , while females grow between July and December. Gordon (1986a) prefers to use M_1 / M_2 and PM_2 / PM_3 as 'concurrently erupting' lower teeth, with the added advantage of smaller size and simpler root systems.

Over 1000 Magdalenian reindeer teeth were studied from French contexts using the methodology devised for modern and historic Canadian material. In 1982 Gordon concluded that herd and associated human group movements were more limited in Magdalenian France than they are in the Canadian Barrenlands. He presents evidence for small-scale vertical migrations in the Pyrenees and Massif Central and discusses assemblages from cave sites which yield evidence for winter and early spring occupation eg: Lascaux (similar to Spiess (1979) at the Abri Pataud). Meanwhile, sites in close proximity to rivers, especially fords and other water crossings, are seen to show late spring and autumn occupation, times at which Gordon suggests intensive reindeer culling.

In 1986, and subsequently in 1988, Gordon (1986a, 1986b, 1988) produced a more detailed report of his findings, upon which his major 1988 monograph is based. The results of his incremental analysis of seasonality are summarised, for selected sites, in table 5.7.

Table 5.7 Seasonality of Selected Sites in S.W.France, based on data taken from Gordon 1988.

Site	Period	Season
Badegoule	Protomagd.	Winter
Bruniquel	Upper Magd.	Spring/Autumn
Cap Blanc	Upper Magd. ?	Early spring
Cavart	Upper Perigordian	Spring
Chaire a Calvin	Upper/Final Magd.	(Winter)/Spring
Combe Cullier	Mid./Upper Magd.	Spring/Autumn
Courbet	Upper Magd.	Spring/Autumn/Winter
Gare de Couze	Upper Magd.	Spring
Enlene	Upper Magd.	Winter/Spring
Fauroux	Final Magd.	Spring
Faustin	Upper Magd.	Spring/Summer
Flageolet II	Upper Magd.	Winter
Fongaban	Upper Magd.	Summer ?
Fontarnaud	Upper Magd.	Spring
Bois Ragot	Upper Magd.	Winter
Gourdan	Upper Magd.	Year round ?
Longueroche	Upper Magd.	Winter
La Madeleine	Upper Magd.	Winter/(Spring)
Roc de Marcamps	Upper Magd.	Winter
Morin	Upper Magd.	Late winter/spring
Placard	Early Magd.	Spring
Reignac	Upper Magd.	Late winter/(spring)
Fritsch	Early Magd.	Winter
St.Germain-en-Laye	Upper Magd.	Late winter
Vidon	Upper Magd.	Summer ?

Based upon such results as are presented above, Gordon (1988) identifies three reindeer herds in S.W.France: the western, central and eastern ranges (1984:96). The western herd, Gordon maintains, migrated during the late spring from the Gironde and Charente over the Landes into the western part of the present Lot-et-Garonne, the Pyrenees-Atlantiques and Hautes Pyrenees.

After wintering in the Vezere and Dordogne valleys the central herd (and associated hunter-gatherers) moved to the Lot, the easter Lot-et-Garonne, the Tarn-et-Garonne, the Haute-Garonne and the Ariege. Gordon suggests that the Lot canyons forced hunters south in spring and north during the autumn (Gordon 1988:96). Meanwhile, the eastern herd moved between the Aude and the Massif Central or eastern Pyrenees, wintering in the latter two regions.

Thus, winter ranges are to be found primarily in the Charente, and Dordogne (where late winter/early spring increments prevail), while evidence for summer calving can be found in the

Pyrenees. The distance between these areas he now considers to be of the same order as that covered by Canadian Barrenland Chipewyan herds (Gordon 1988:96). The Magdalenian migrations also resemble the Canadian system in that they are altitudinally orientated although a distance of 200 km is postulated between summer and winter ranges. Calving occurs at higher altitudes, with winter refuge sought in lower, sheltered forests. Massif Central calving is not accepted (contrary to Bouchud (1966:243) and David (1973:287)) due, he claims, to the fact that herds are known to return to spring sites during the autumn, a season for which there is no evidence in the Charente and Gironde. The Pyrenees is preferred as the calving ground. There are no summer sites nearer to the Charente-Gironde than the Pyrenees (Gordon 1986:10), and the Dordogne is a 'demonstrated' winter range.

Finally, Straus (1986) has suggested that there is evidence of winter and non-winter occupation of the Pyrenees and Landes, based upon tooth sectioning and antler analysis. Duruthy, Dufaure, Gourdan, Lespugue, Espeche and possibly Isturitz display evidence of winter occupation. Non-winter evidence is seen at Mas d'Azil and Espelugues. At Terascon-sur-Ariege ibex and lagoped specialist hunters are active during the winter, whereas to the immediate south of the Pyrenees (on the Spanish side), only summer evidence is available. In coastal Spain we see both summer and winter occupation. One should not assume however that this represents year-round occupation.

At the moment the picture of Upper Palaeolithic seasonality depends largely upon the Magdalenian analysis carried out by Gordon (1986a, 1986b). Work such as this and the conclusions reached by Binford (1973) have led us to critically question the results of others such as Bouchud (1966). However this valuable research is of relevance only to those sites which have been studied. It is tempting to make sweeping generalisations regarding the seasonal occupation of the Perigord and adjacent areas as a whole during the course of the Upper Palaeolithic but, given the nature of the fluctuations in climate and environmental conditions between 35 000 and 10 000 B.P. and associated changes in the faunal resource base available, seasonality patterns may be expected to change throughout the period. The demonstration of winter occupation of rock shelters in the Dordogne during the Magdalenian (Gordon 1986a) does not necessarily imply similar occupation patterns during the Early Upper Palaeolithic. Seasonality patterns must be considered in the context of demonstrable environmental change. Furthermore, most of the material studied so far derives from rock shelters and cave sites. Faunal material is rare (more often non-existent) from open air contexts and therefore we lack a more than patchy record of the season of occupation of these sites. However, our understanding of the seasonality of occupation in South West France is ever increasing as further studies are completed, employing alternative forms of data.

Chapter Six

CHRONOLOGICAL PATTERNING.

Shipman (1981:123) emphasises the importance of sequential fluctuations in faunal representation, which, she points out, may result from changes in associated environmental conditions, evolutionary trends or shifts in human behaviour. In the present chapter an attempt will be made to identify some of these chronological changes and to determine the extent to which they may be explained in terms of environmental variables.

In addition to the relative (%N.I.S.P.) frequencies of species, regional mean values are examined in which the denominator represents the number of levels in the area which date to the period under consideration, whether or not the species occurs at the sites in question. The intention is that such a mean value should provide a descriptive measure of faunal representation in the region concerned.

For the sake of data handling and in order to address the question of the degree to which data may be considered representative of South West France as a whole, the region has been divided into four zones, which in the present discussion are based largely upon river catchment:

Zone	I	Central - Southern	Dordogne, Vezere, Lot
	II	Northern	Isle, Dronne, Charente
	III	Western	Gironde
	IV	Eastern	Massif Central

The river catchment has been selected as a unit based upon the propensity of hunter-gatherers, of which palaeolithic man is assumed to be an example, to divide their landscape in such a way. The catchment also forms a useful ecological division. Our regional units are however arbitrary.

The mean percentage frequency of each of the major herbivore species is plotted in figs. 6.1 to 6.3, raw data being shown graphically in Appendix III. Although attention will be concentrated on mean frequency values observed, it was felt that it would be of value to present the original chronological plots in order that some idea of the 'spread' of data may be obtained. The regional assemblages have been coded in order to aid differentiation. Following the examination of the available quantitative data the non-numerical material is divided geographically and finally used in chronostratigraphic order in regions yielding sufficient material.

To present as complete a picture as possible, the chapter concludes with a synthesis concerning the chronological faunal patterns, aiming to collate the qualitative and quantitative data. The hope is that such a discussion will enable us to gain further insight from non-numerical material given what we have learned from the more detailed and specific nature of their quantitative counterparts.

6.1 Quantitative Data Patterning.

6.1.1 Reindeer.
Rangifer tarandus L. (Fig. 6.1(a))

Reindeer totals rise dramatically during the Early Aurignacian from substantially lower Chatelperronian levels (mean = 49.57%). In Zone I frequencies rise to 100% (at Abri Pataud eb.13/14, where n=410), averaging 86.31% with the species dominating in every level for which we have quantitative data. The Northern Zone displays a wider range of frequencies, with an average of 57.39% and a maximum of 78.90% (La Chevre). The sample is however substantially smaller than that of the Central-Southern region.

The Later Aurignacian sees a decline in reindeer totals in both regions, although the range of values in Zone I remains considerable. Frequencies average 45.79% in region I and reach 31.70% at the single site of La Chevre in zone II, each showing a substantial reduction.

The Upper Perigordian (Perigordian IV,Va,Vb,Vc) and Protomagdalenian levels observed in the Dordogne have been amalgamated. Only three sites of the later period have yielded samples of sufficient size for consideration (Lachaud, Roc de la Belle and the Abri Pataud) and the resultant 4-level data set was not considered to be large enough to warrant consideration on its own. Reindeer averages 73.57%. If one excludes Roc de la Belle, at which no reindeer survives (Delpech 1983), the mean value changes to approximately 78%.

It is at this point that we begin to observe the predominance of the reindeer which traditionally characterises the Upper Palaeolithic. The complete absence of reindeer from the Upper Perigordian levels of Roc de la Belle is remarkable and it becomes clear, given the nature of other sites of similar age, that this site should be considered as an exception to the rule. Indeed, of the 28 Upper Perigordian levels (Perigordian IV-VI included), reindeer exceeds 90% at 13 and 80% at an additional 5. The lowest Protomagdalenian frequency is 83.97% (mean = 89.59%).

Solutrean levels in the Dordogne retain high reindeer frequencies, a fall being observed, with the mean value attaining 85.46%. Initial Magdalenian levels at Pegourie (Lot) contain approximately 45% and 89% reindeer. In the more northerly area, frequencies range from 22% to 88%. With the continuation of high frequencies the overall dominance of the species in the Dordogne region remains unquestionable until we reach the early post-glacial. Only when we encounter Azilian levels do we see a marked decrease in reindeer totals, which suddenly fall to a regional mean of 19.71%, with a median of 21.91.

Meanwhile in the northern, western and eastern Zones reindeer frequencies of the Magdalenian are neither as stable nor as high as in Zone I.

In the northern region, reindeer peaks at 91.35% in the Vienne, at Bois Ragot, level 6, during the Upper Magdalenian. Mean frequencies are markedly lower than those observed in the Dordogne,

Chronological Trends in Faunal Frequencies in
Zone I (●) and Zone II (•)

Fig. 6.1

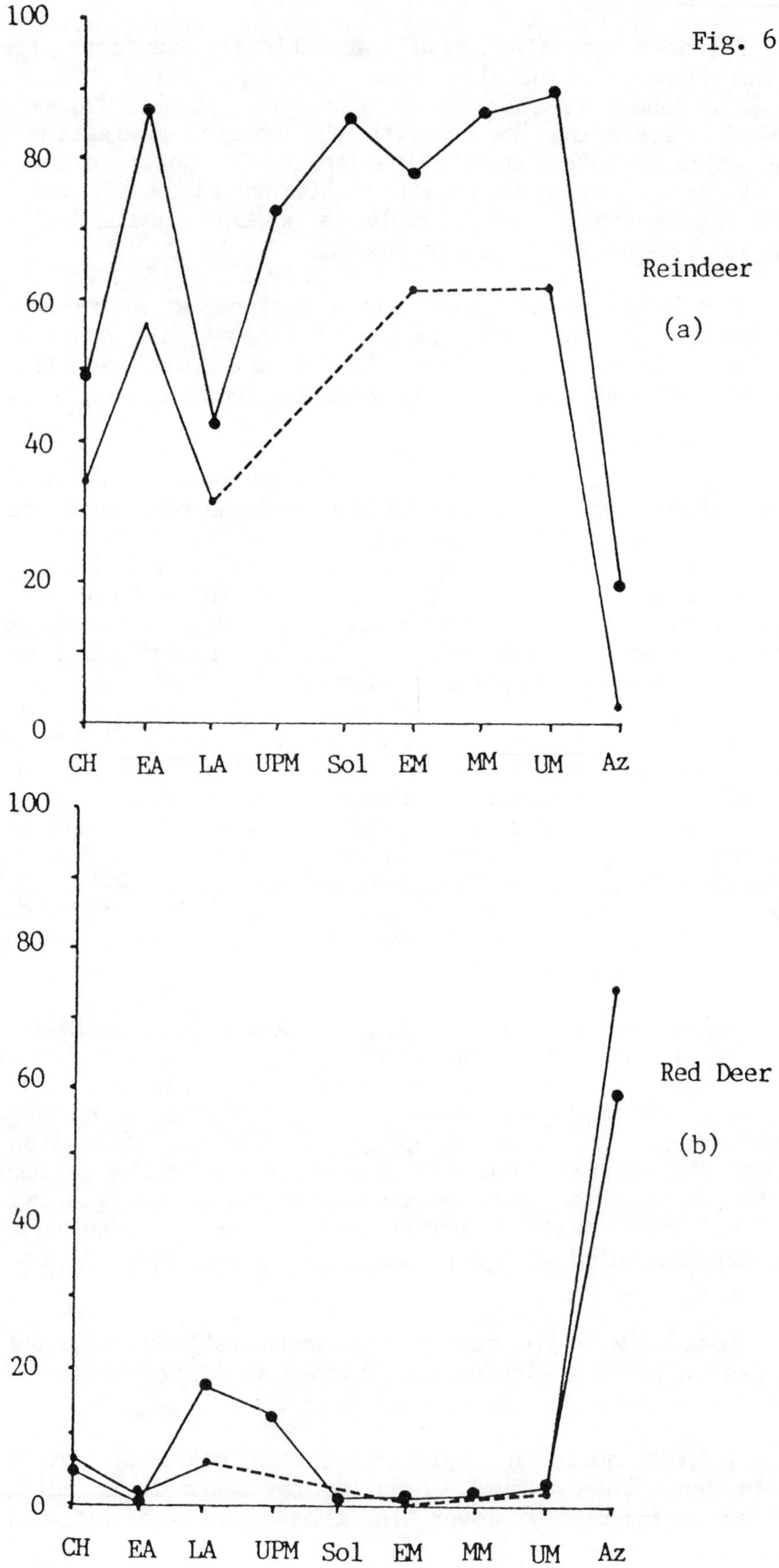

reaching 61.18% and 62.27% in the Lower and Upper Magdalenian respectively. Totals then fall dramatically to average 1.12% during the Azilian. It may, in fact, be argued that the reindeer occurs only in Terminal Magdalenian / Early Azilian levels in the northern region, and that by the 'Azilian-proper' the species had all but disappeared.

During the Upper Magdalenian in the Gironde, reindeer appears to take second place to alternative herbivores. The mean frequency of 29% contrasts significantly with that in the Dordogne, while in the east, during the Azilian, frequencies remain somewhat higher than elsewhere, with an average of 12.45%. This would suggest that conditions in the Massif Central remained such that the reindeer continued to provide an important resource for Palaeolithic hunters.

6.1.2 Red Deer.
Cervus elaphus (Fig. 6.1(b))

Red deer frequencies are low in the Chatelperronian in both Dordogne and eastern sites, attaining at Chatelperron (Allier) a maximum of 14.03% in level B4, but averaging only 3.97% during the Lower Perigordian as a whole. At the site of Roc de Combe (level 8) red deer totals 2.4%, and at La Ferrassie, 5 - 10%, the regional mean being 4.46%.

During the Early Aurignacian Cervus elaphus totals fall in Zone I, reaching a maximum of 4.17% at Battuts. In general, frequencies remain low, averaging .50%, while to the north, frequencies are lower still, with a mean of 0.24%. It should be borne in mind however that the northern sample comprises only four levels.

Later Aurignacian red deer figures in the Dordogne increase substantially. At Maldidier the species represents a maximum of 91% of the herbivore bone material recovered, although this sample is small (total n=12) and is not included here in calculations. At Le Flageolet I red deer equals 50.25% of the N.I.S.P. count recorded. However, mean values are not high (17.01%). Nevertheless, this does indicate a significant rise in species representation. To the north, only La Chevre yields any red deer (5.9%).

With the exception of Battuts, Le Flageolet I and La Ferrassie, where values of 47.25%, 50.25% and 80.10% are found respectively, Upper Perigordian-Protomagdalenian levels yield relatively low red deer frequencies. In general frequencies are low, and indeed continue to fall throughout the South West corner of France during the Solutrean and Lower Magdalenian periods.

As the Magdalenian progresses, Dordogne and Lot sites yield rising red deer frequencies and despite medians of 0% during both the Middle and Upper Magdalenian, mean values slowly rise throughout the region (see table 6.1).

By the advent of the post-glacial, red deer populations appear to be securely established in the region. Azilian frequencies rise considerably: in the Dordogne we have a mean of 59.04% and to the north, 74.34%, although eastern figures remain somewhat curtailed (c. 11%).

Table 6.1 Red Deer Mean Frequencies:

		Lower Magd.	Middle Magd.	Upper Magd.
Zone	I	0.540	0.929	2.430
	II	0.075		2.200
	III	3.970		
	IV	8.263		

6.1.3 Bovids

Bison priscus. Bos primigenius.
(Fig. 6.2(a))

Initial inspection of the chronological plots shows that bovids provide a relatively constant component in the fauna. There is no period during which such populations were not exploited by Palaeolithic man to some extent.

In the Lot and Dordogne (at Ròc de Combe and La Ferrassie respectively) bovids are relatively abundant in the first phase of Wurm III: the Chatelperronian yields frequencies in the region of 40% (mean=35.79%). In the Allier, to the north-east, frequencies do not attain this level but range between 7% and 32%, with a mean of 11%.

Bovid totals drop considerably during the early Aurignacian, although a relatively wide range of values is observed. At no point however, do figures exceed 22%. In Zone I the mean value has fallen to 4.84%. Figures fall within a similar (but more restricted) range to the north, with a mean value of 5.15%.

The Later Aurignacian at La Chevre yields substantially higher bovid totals (55.193%), suggesting, if this is 'typical' of the northern region, perhaps a change in humidity conditions as reflected by variations in forage availability. To the south however, the change from Early to Later Aurignacian is less extreme, mean values rising from 4.84% to 13.73%. In addition, the range and distribution of values remains similar: it is not a question of bias caused by one or two outstanding assemblages.

The Upper Perigordian, Protomagdalenian and Solutrean levels occurring in the 'Dordogne' yield low frequencies of bovids, with a mean value during the Upper Perigordian/Protomagdalenian of 2.55%, despite a few substantially higher frequencies (eg: Roc de la Belle, where sample size is small). Bovid levels remain relatively constant during both parts of this period; during the Upper Perigordian the mean reaches only 3.65% while the Protomagdalenian value falls slightly to 2.70%.

Solutrean figures are also low, although the Grotte de l'Eglise yields a percentage value of 11.81: overall, the southern region averages 1.48%. Initial Magdalenian (as defined by Hemingway 1980) levels at Pegourie (8,9), Lot, present a rise in bovid totals, averaging 29.34%. The frequencies at this Causses site contrast

Chronological Trends in Faunal Frequencies in Zones I (●) and II (•)

Fig. 6.2

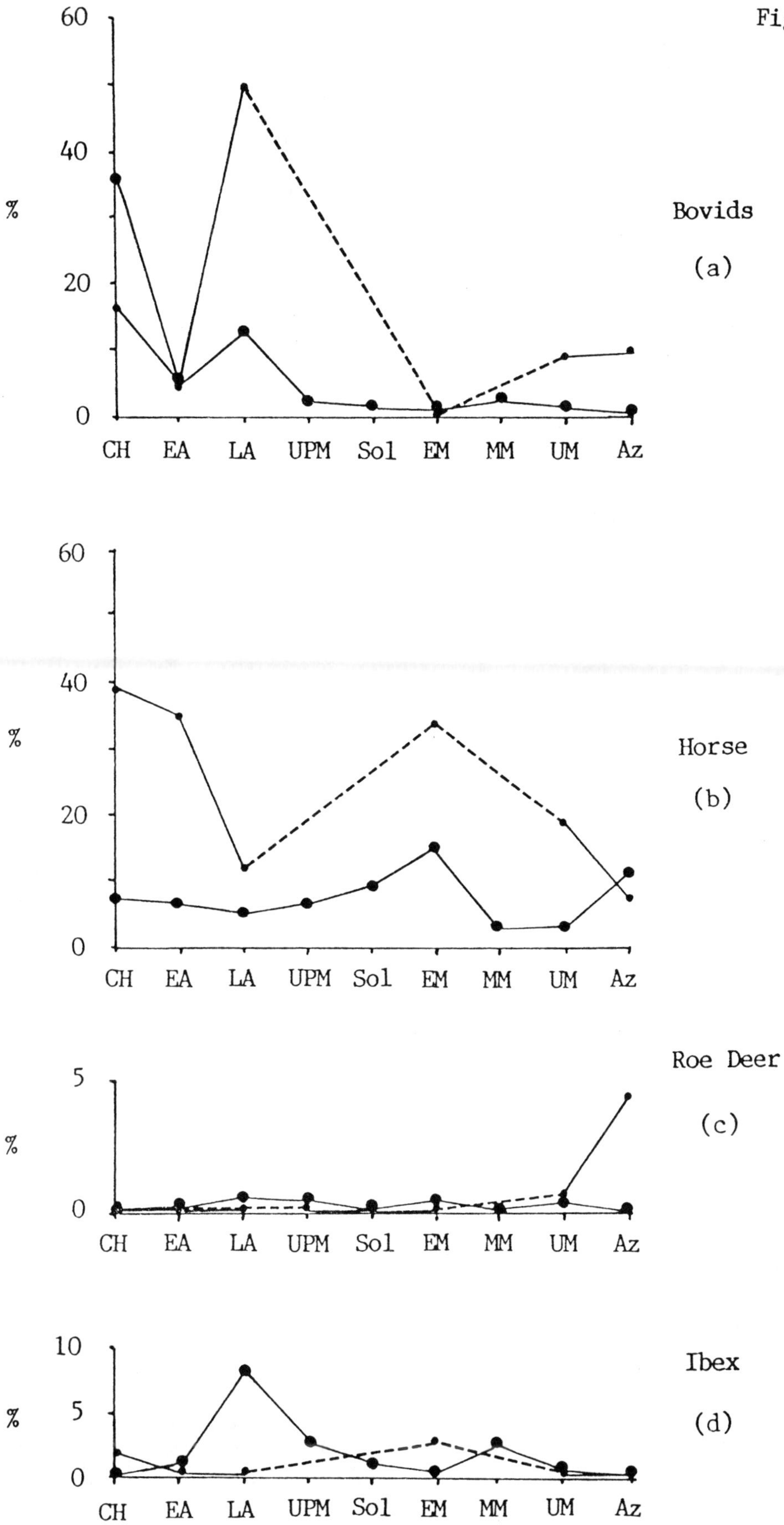

strongly with those in the northern zone, where the maximum at the Abri Fritsch (Indre) reaches only 3.85%.

Bovid frequencies in Zone II are consistently higher than or equal to those in Zone I, with the exception of the sample from Pegourie (Lot). Such a trend continues into the early post-glacial, so that during the Azilian we see a wider range of frequencies in the flatter north than in the valley and plateau landscape of the Dordogne and Massif Central.

It is to the western and northern Zones that we must turn in order to uncover relatively high bovid frequencies during the Magdalenian. Middle Magdalenian assemblages in the Dordogne average approximately 2.10% bovids, falling to 1.81% during the Upper Magdalenian. In the north, Upper Magdalenian deposits range from 2.80% to 26.70%, with a mean of 8.20%. To the east the species makes up about 15% of the large herbivore remains, while in the west (Gironde), it rises to 39% (with a median of 21.7%).

Table 6.2 Bovid Frequencies

Zones	I	II
Azilian (mean)	1.25	8.06
(median)	2.60	3.70

6.1.4 Horse
Equus caballus.
E.c.gallicus. (Fig. 6.2(b))

At the beginning of the Upper Palaeolithic, horse signifies as an important species in the eastern region. At Chatelperron the species comprises 50% of the large herbivore count. Meanwhile in the Dordogne and Vezere frequencies fall to 15.7% and 4-7% respectively, providing a regional mean of 7.49% and median of 6.36%. Meanwhile, to the north, the mean value for horse is 33.56%, perhaps reflecting conditions more similar in nature to those encountered in the east than in the south.

Early Aurignacian frequencies increase in the Dordogne, with a maximum of 35.86% (A.Pataud, c.11) and a mean of 6.72%. In the north frequencies are still higher, with a mean of 35.06%. Figures in both areas either fall or remain relatively constant as the Aurignacian continues, although frequencies in the northern Later Aurignacian, where La Chevre yields 11.80%, continue to exceed those from the south (mean = 5.81%). It is worth noting here the relative similarity in horse frequencies in the southern region from Early to Later Aurignacian levels. It is a species which, with the exception of the Badegoulian, represents only a small part of the total fauna prior to the Azilian (when the mean=11.70%).

The Upper Perigordian and Protomagdalenian see a decline in horse frequencies to a mean of 6.65% in the Dordogne, although the figures yielded by Roc de la Belle (39.13%) and Gavaudun (32.30%) would appear to be at variance with this figure.

Solutrean levels see a temporary rise in totals. At Laugerie Haute Est Upper Solutrean frequencies rise to 17.22% prior to a possibly Early Magdalenian total of 25.99%. However, at Laugerie Haute Ouest Final and Upper Solutrean levels yield only 0.38% horse. A regional mean of 9.37% is obtained, prior to a rise to an Initial Magdalenian mean of 14.36%.

In the Central-Southern region, throughout the Middle and Upper Magdalenian, horse frequencies average 2.74% and 2.94% respectively, no single occurrences reaching 20%. To the north, frequencies are higher, approaching 40% in the Upper Magdalenian, with a mean of 18.67% (median 16.67%), while in the east frequencies average 12.96%. In the Gironde however, frequencies reach 60% and fall in a larger range than elsewhere, with a mean of 17.85%.

Azilian frequencies of <u>E.caballus</u> are substantially reduced in the east, reaching only 5.8%, with similar figures in the south and north (11.7% and 6.10% respectively), as post-glacial conditions begin to make themselves felt.

6.1.5 Roe Deer
<u>Capreolus capreolus</u> (Fig. 6.2(c))

Roe deer is to be found only at Chatelperron during the early stages of Wurm III, where frequencies average 2.56%, reaching a maximum of 10.63%. It is not until the Solutrean that <u>Capreolus capreolus</u> appears in areas other than the Dordogne-Vezere valleys region. During the Earlier Aurignacian the roe deer appears in the latter area (at La Gravette, for which complete quantitative data were not recovered)and remains, albeit in very small numbers, throughout the course of the Upper Palaeolithic.

Table 6.3 <u>Roe Deer in the Central Zone</u>

(mean)	
0	Chatelperronian
*	Aurignacian I
0.500	Aurignacian II
0.510	Upper Perigordian/Protomagdalenian
	(0.75 - Upper Perigordian)
	(0.44 - Protomagdalenian)
0.041	Solutrean
0.313	Lower Magdalenian
0.080	Middle Magdalenian
0.298	Upper Magdalenian
0	Azilian

It is interesting to note that the roe deer does not occur in the Central Zone during the Azilian, despite the fact that the species is commonly associated with the temperate woodland conditions which are generally believed to have prevailed during the post-glacial. The implication may be that heavy snow fall became a more common occurrence once the climate became more oceanic.

Roe deer occurs in all four areas during the Upper Magdalenian, reaching 4.2% in the Gironde (Faustin), 3% in the North and 5.8% in the East. It is only in the Northern and Eastern Zones that the species occurs during the Azilian. At Pont d'Ambon frequencies reach 7.24%, while at Bois Ragot 5.88% of large herbivore remains belong to Capreolus capreolus, the mean in the northern zone being 4.25%. To the east, roe deer occurs at Campalou, but frequencies reach only 0.11%.

6.1.6 Ibex.
Capra ibex. (Figs. 6.2(d))

Somewhat surprisingly, given the presence of ibex in the Dordogne during the Chatelperronian, the species appears to be absent from the more 'mountainous' eastern region. In the former it is a constant component over the course of the Upper Palaeolithic with frequencies reaching a maximum of 29% during the Later Aurignacian (II) at Le Maldidier, explicable largely in terms of the local topography in the area.

Early Aurignacian figures average 1.35% in the Central region and during the Later Aurignacian, 8.06%, with frequencies declining steadily during the Upper Perigordian, Protomagdalenian and Solutrean, when regional totals remain relatively low (mean=2.93%). Figures further decrease towards the Initial Magdalenian in the Dordogne and Lot, during which it reaches only 0.24% (0.96% at Laugerie Haute Est). However, at this point the species makes a brief appearance in the North, with a mean of 2.65%. By the Middle Magdalenian it has disappeared, despite its presence in the Dordogne during both the Middle (mean = 2.23%) and Upper (mean = 0.48%) Magdalenian. In this area the species does not appear during the Azilian.

In the Massif Central frequencies rise at the end of the Upper Palaeolithic. Late Magdalenian frequencies average 19.75% in the east, rising substantially to 55.19% during the Azilian. At this point frequencies range from 11% to 79% of the large herbivore remains, with a median value of 60.19%. The species is second in importance only to the red deer, perhaps explained by the local topographical conditions in an area somewhat slower to respond to post-glacial warming.

The ibex does not occur in the west of our region during the Upper Magdalenian, topographic conditions there being unsuitable. Furthermore the species is absent during the Azilian throughout most of the region.

6.1.7 Boar
Sus scrofa. (Figs. 6.3(a))

Sus scrofa is relatively rare during the course of the Upper Palaeolithic in S.W.France, failing to reach major proportions during the Wurm. Only in isolated cases, eg: to the immediate east of our region at Rond du Barry, do frequencies increase substantially. During the Azilian however, frequencies do show increases which may

Chronological Trends in Faunal Frequencies in Zones I (●) and II (•)

Fig. 6.3

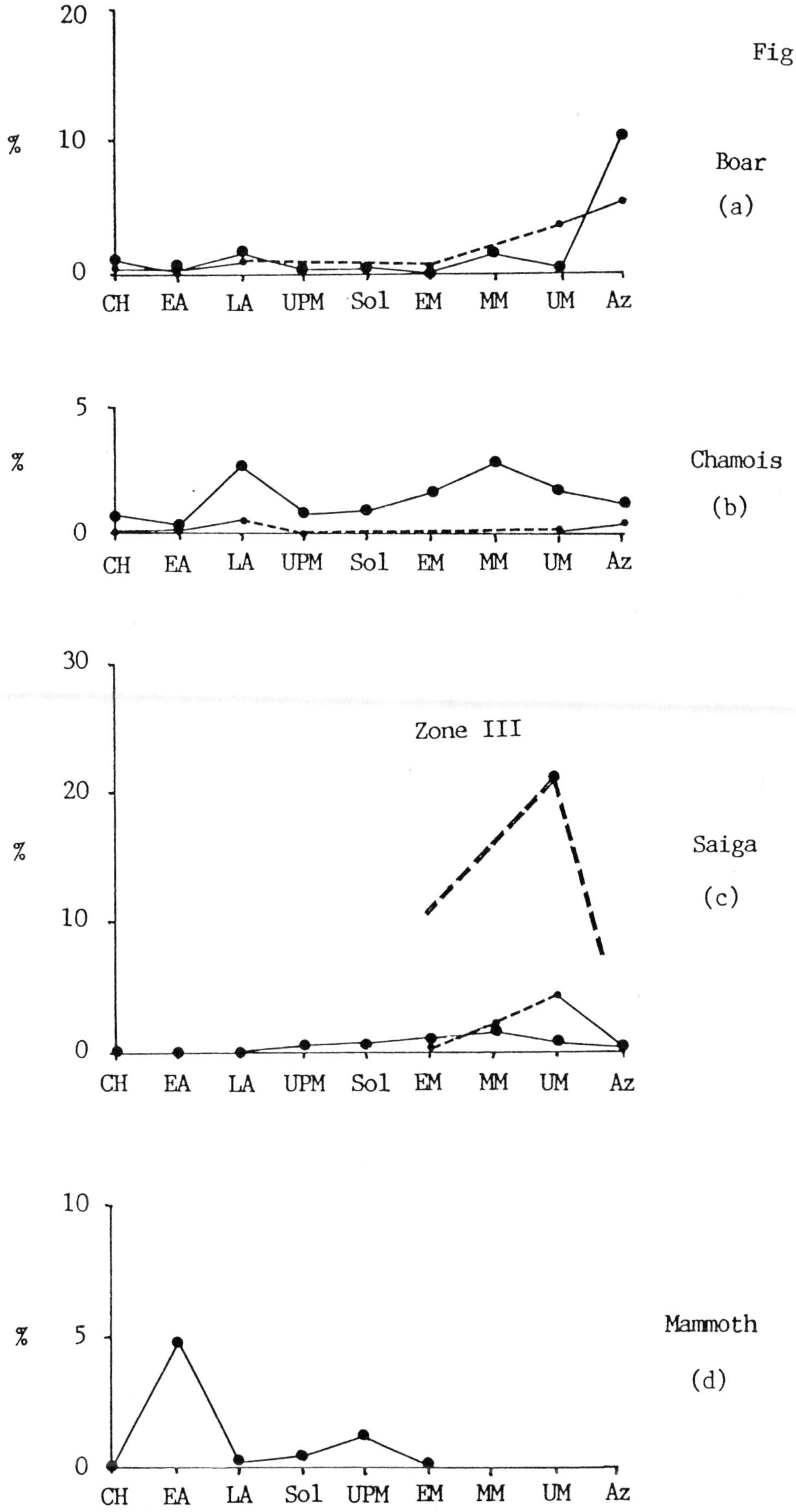

perhaps be explained in terms of changes in climate and increased woodland cover.

In the Dordogne-Vezere, boar is absent only during the Protomagdalenian and Solutrean, although at times, frequencies are pitifully low. Three periods can be identified during which the species comprises more than 1.00% of the herbivores recovered from sites: Aurignacian II (1.68%), Middle Magdalenian (1.18%) and Azilian (10.21%).

To the north, we can add to these, the Upper Magdalenian, as in the west and east.

Table 6.4 Boar Frequencies

Zones	I	II	III	IV
U.M.	0.25	3.63	8.68	17.25
Azil.	10.21	5.78	-	2.86

Boar frequencies observed during the course of the Upper Palaeolithic are particularly remarkable for their geographical variation and hence, explanation should be sought in spatial terms in addition to the chronological.

6.1.8 Chamois

Rupicapra rupicapra. (Figs. 6.3(b))

Chamois first makes an appearance at Roc de Combe, reaching 0.61% in the Lower Perigordian. Only during the Upper Magdalenian do chamois frequencies attain relatively high levels, eg: 20.32% at Bruniquel (Lot), Roc d'Abeilles (Dordogne) and Sainte-Eulalie (Lot), although regional means peak during the Later Aurignacian and Middle Magdalenian. All these sites are to be found in areas of 'rugged' terrain. At Roc de Combe, the Early Aurignacian, sees 1.18% chamois, at La Ferrassie approximately 5%. At Le Flageolet I however, totals reach only 2.07% and 8.33% during the Later Aurignacian at Le Maldidier (Delpech 1983:204). Such peaks are commonly explained in terms of increased humidity or suitable topography. To the north (La Chevre) chamois forms a negligible component of the fauna (0.45%) during the Later Aurignacian, being absent during the Earlier Aurignacian.

Chamois figures decline during the Upper Perigordian and Protomagdalenian (mean = 0.977%), maintaining a mean value of 0.99% during the Solutrean and rising to 1.50% during the Initial Magdalenian. However the Middle Magdalenian sees the return of the species with frequencies reaching 6.66% at Combe Cullier. We have a regional mean of 3.63%.

During the Upper Magdalenian totals generally decline, giving a mean of 1.673% in Zone I, although the figures from Roc d'Abeilles and Ste. Eulalie would appear to contradict this, local relief and vegetation conditions perhaps accounting for the higher values. To the north and east, chamois appears to be absent, whilst in the South

West (Gironde), figures vary from 0% to 0.4%, the regional mean equalling 0.059%.

During the Azilian the species appears in the north and east, albeit in low frequencies, with concomitantly low means:

Table 6.5 <u>Chamois Frequencies during the Azilian</u>

Zones I	1.190%	large herbivores
II	0.352%	
III	/	
IV	1.970%	

It is interesting to note however, that the species occurs only at Roc d'Abeilles in Zone I, where, as indicated above, topography is ideal for the chamois.

6.1.9 Saiga.
<u>Saiga tatarica</u>. (Figs. 6.3(c))

The saiga has a limited geographical and chronological distribution.

The first record of this species in S.W.France is that recovered at Le Piage (Lot). The Protomagdalenian level (C-E) yields two fragments (M.N.I.=1), totalling 0.15% of the herbivore remains (Beckouche 1981). It is during the Magdalenian however, that the species gains in importance.

Initial Magdalenian figures in Zone I attain 1.25% at Laugerie Haute Est (Magdalenian 0). By the Middle Magdalenian the mean value has risen to 1.496% in the Dordogne while in the Gironde, Upper Magdalenian saiga populations seem to reach considerable proportions, with a mean of 20.11%, although the number of sites from which material is available is limited. However the qualitative data available from the area (see p.138) would suggest that the sites for which we have data are typical, the Saiga being the dominant or co-dominant species. Le Morin however provides the major exception to this, there being a total lack of <u>Saiga tatarica</u>. By omitting Le Morin from our calculations, the regional mean rises to 58.91%, perhaps reflecting, to a greater extent, the importance of this 'short-lived' species.

Towards the end of the Magdalenian, frequencies decline, although the species remains a major one in the western half of our region while mean values have fallen to 0.141% in the Dordogne. At this point saiga appears further north, at Montmorillon (Vienne), where the species forms 4.81% of the bone count and 19.15% at Gabillou. The regional mean in the north equals 3.99%.

Saiga is absent in the Massif Central throughout the Upper Palaeolithic (a fact which is probably to be explained as a result of highly unsuitable topographic features and associated forage conditions in the area), and disappears as rapidly as it arrives over the region as a whole. There is no record of the species in the Azilian levels examined.

6.1.10 Mammoth

Mammuthus primigenius. (Figs. 6.3(d))

Mean mammoth frequencies peak during the Early Aurignacian (ca. 5.0%), quantitative data being recorded only in the southern/central zone (I). After falling sharply, totals rise slowly reaching 19.27% during the Middle Solutrean at Laugerie Haute Ouest (level 9). The most recent occurrence of mammoth, in zone I, is recorded by Delpech (1983) at Gare de Couze. This may be as late as 12 000 B.P. To the west, at Roc de Marcamps, mammoth is last found some 2000 years earlier.

6.2 Non-Numerical Data.

Having considered the available quantitative data for the major herbivore species, it is worth briefly considering the qualitative material. This material forms a database of considerable size and as such should not be ignored although of course insights to be gained are much more restricted. After discussing these data both types will be amalgamated in an attempt to provide some form of regional synthesis considering all species.

In general, only a very limited amount of information can be gleaned from the chronological examination of qualitative data: this is particularly true where the type of data varies, as in the present case. Tables 6.6 - 6.8 show the presence/absence and where possible the predominance of each of the major large herbivore species as these parameters represent those indications most commonly provided. It should be noted however that the lack of a marked predominant taxon means only that the literature source consulted did not stipulate such a species. It does not mean equal abundance or rarity.In addition, the total number of large herbivore species (N) is provided, the unlisted species being the woolly rhinoceros, mammoth and ass which the quantitative data tell us, are rarely represented by more than one or a few bones. In each case their presence is indicated by 'very rare','present' or a very low ranking. Also added are those levels for which quantitative data are available, but which comprise fewer than 50 bones or were deemed unsuitable for quantitative analysis.

However, despite the problems inherent in employing qualitative data, there is a quantity of data such that it would seem irresponsible to ignore the available material altogether.

The data in tables 6.6 to 6.8 (once again divided geographically), have been placed in as concise a chronological sequence as is possible, given the state of the literature from which the data were obtained. The tables run, from top to bottom, from the Chatelperronian to the Magdalenian/Azilian transition. Only by doing this can we hope to gain an insight into any observable chronological patterning in the available data.

6.2.1 The Central Zone (I). (Table 6.6)

In general the data indicate that four species comprise the major resources exploited during the course of the Upper Palaeolithic. These are reindeer (dominant in 60% of samples), horse (dominant in 14.29%), bovids (dominating in 5 levels, 7.14%) and red deer (dominant in only one assemblage). The additional species would appear to maintain a secondary role, with roe deer occurring only twice prior to the Upper Perigordian, saiga a total of three times and the chamois, ten.

Quantitative data indicate that the Chatelperronian is associated with low frequencies of reindeer and indications of mild, relatively wooded conditions. This is a phenomenon reflected to some extent by the qualitative data available. During the Early Upper Palaeolithic and especially the course of the Aurignacian, which in early documents was often undivided, reindeer is commonly the predominant species with horse in second position. The varied nature of the Dordogne landscape is emphasised by the occurrence of roe deer and boar (temperate, woodland species) in Lower Perigordian and Early Aurignacian layers, associated with a climate commonly perceived of as relatively cold and dry. During the Later Aurignacian the reindeer is still predominant, although to a lesser extent than in earlier levels. Meanwhile the more temperate red deer (woodland) element grows stronger, emphasised by a temporary increase in boar.

There is a marked lack of these temperate species during the Upper Perigordian and Solutrean. At only five of the sixteen occurrences do we see red deer, with boar at only one and the roe deer absent altogether. Only at Cavart does the red deer play an important role. Elsewhere, at this time, levels which are not dominated by reindeer are composed primarily of horse, reinforcing the notion of cold, dry, open conditions, while bovids occur primarily during the Upper Perigordian as opposed to the Solutrean.

Once we encounter Magdalenian levels the reindeer is almost unassailable, failing to occur only once but occasionally co-dominating assemblages with horse. In order to detect change in faunal assemblage structure we have to look for changes in the occurrence of alternative species, for only at Roche de Castelmoron and Longueroche is the reindeer supplanted prior to the occurrence of only horse and bovids at Jardel II in Final Magdalenian levels.

The Limeuil and Crabillat levels immediately appear as a feature of the sequence, each totalling 7 species. These show a broad spectrum of mixed character, despite the fact that reindeer continues to predominate. In both cases the occurrences are found in close association with other sites yielding high numbers of species, indicating perhaps a temporal change in the number of taxa easily available for exploitation (i.e. greater diversity) or a need to exploit additional taxa in order either to meet requirements or to vary the diet. It is noticeable that those sites at which higher numbers of species are observed (including, in addition, Longueroche) are some of those which occur in the Vezere valley and its immediate catchment.

Woodland species occur primarily during three episodes of the Magdalenian. Red deer is present, although rare, during the earliest

Table 6.6. Chronological Change in Available Qualitative Data from Zone I (Central)

	Rt.	Ce.	Bov	Ec.	Cc.	Ci.	Rr.	Ss.	St.	Nsp.
Chatelperronian - Aurignacian.										
Roc de Combe	*		**	*						3
Roc de Combe	**		*							2
Castanet	**		*	*		*		*		5
Cellier	**	*	*	*	*	*	*			7
Le Flageolet I 9	**			*		*				3
Roc de Combe	**	*	*	*						4
Caminade Est	**	*	*	*			*			5
Caminade Est	**	*	*							3
Castanet	**	*	*	*		*		*		6
Castanet	**	*	*	*						4
Pataud	**		*							2
Blanchard	*	*		*						3
Cellier	**	*	*	*		*	*			6
Lartet	**	*	*	*				*		5
Poisson	*		*	*						4
Pasquet	*	*	*	**				*		5
Caminade Est E			*	**						3
Caminade Est D	*	*	**	*			*	*		7
Pataud 8	*	*	**	**				*		5
Pataud eb.7/8	**	*	*							3
Pataud eb.6/7	*	*	*	**						4
Maldidier		**					*			2
Le Flageolet I	*	**	*	*	*		*	*		8
La Ferrassie		*	**	*				*		5
Upper Perigordian - Solutrean.										
Cellier	*		*	**						3
La Ferrassie B5-B3	*									1
Roque-St.-Chr.	**		*	*						3
Grotte des Morts	**		*	*						3
La Gravette	**	*	*	*		*		*		7
Cavart		**	*	*						4
Cavart	**	*								3
Cavart		*		**						3
Lacave	**			*		*				3
L.H.E. 34	**			*		*				3
L.H.E. 33	**			*		*				3
Liveyre		*		*						3
L.H.O. 11	**			*						2
Lestruque	**		*	*						3
L.H.E. 21	**			*						2
L.H.O. 1	*									1

	Rt.	Ce.	Bov	Ec.	Cc.	Ci.	Rr.	Ss.	St.	Nsp.
Magdalenian - Azilian.										
Cassegros 10	*		*	*		*				4
Cassegros 9	**	*	*	*						4
Cassegros 8	*									1
Jolivet	**			*						2
Cassegros 7	**	*	*	*						4
Crabillat	**	*	*	*		*	*	*		7
Montastruc	**	*	*	**		*				5
Montastruc	*									1
Combe Cullier 13b	**	*		*		*	*			5
Combe Cullier 10	**	*	*							3
Combe Cullier 7	**			*						2
Combe Cullier 4"	**		*							2
Combe Cullier 4	**						*			2
Roche de C'moron	*		**	**					*	4
Longueroche	*	*	*	**		*				5
La Madeleine 17	**			*						2
Longueroche	*	*	*	**		*			*	6
Longueroche	**	*	*	*						4
La Madeleine 13/14	**		*	*						3
La Madeleine 10/12	*									1
La Madeleine 9/11	*									1
La Madeleine 8/9	*									1
Longueroche	*	*	*	*						4
La Madeleine 3	**			*						2
Le Flageolet II	**			*						2
Le Flageolet II v	*									1
Le Flageolet II iv	*									1
Le Flageolet II ii	*									1
Limeuil	**	*	*	*		*	*	*		7
Villepin	**	*	*	*						4
Fontales	*	*	*	*		*	*			6
Gare de Couze A	**	*						*		3
Jardel II	**	*	*							3
Villepin	*	*		*						3
Jardel II		*	*							2
Le Martinet	**			*					*	3
Roc Allan	**							*		2
Roc Allan	*			*						2

Key: Rt.= Reindeer, Ce.= Red deer, Bov.= Bovids, Ec.= Horse
Cc.= Roe deer, Ci.= Ibex, Rr.= Chamois, Ss.= Boar, St.= Saiga.

Nsp. = number of large herbivores, including ass, mammoth and woolly rhinoceros.

* = present ** = dominant/co-dominant or abundant.

stages (Badegoulian) and occurs occasionally during the Middle Magdalenian. Prior to the Magdalenian VI at Limeuil we see red deer only at Longueroche - the only site at which species other than reindeer, bovids and horse are seen at this time.

Roe deer fails to register at all during the Magdalenian, a fact which is not surprising given the exceptionally low frequencies observed in our numerical data.

The boar primarily occurs in later levels, perhaps indicative of increasing humidity and forest vegetation cover which may be associated with early post-last-glacial conditions. This is supportedby apparent increases in red and roe deer.

Bovids and horse regularly occur in association, although the number of horse occurrences does outnumber that of the bovids. Nevertheless the taxa co-dominate at Roche de Castelmoron (Agenais), the horse being of major importance during the early Upper Magdalenian (Magdalenian IV). It is at this point that the saiga is established in the region, albeit in low frequencies, a phenomenon remarked from numerical data considered. The taxon appears twice at this point and once later in the sequence. Its earlier occurrences are associated with predominant horse and important (or co-dominant) bovids, indicating perhaps a brief episode of dry, steppic conditions.

Ibex and chamois (a montane component) are found mainly in the earlier portion of the Magdalenian, its chronological (and spatial) distribution reflecting the occurrence of suitable topography as much as climatic conditions. Both species however, prefer low temperatures, the chamois looking for higher humidity levels than does the ibex.

Finally, the predominance of reindeer begins to decline as we approach the early post-glacial, although cold, steppe elements are maintained in the Agenais, in the south of the region.

6.2.2 The Northern Zone (II). (Table 6.7)

The Early Upper Palaeolithic levels in Zone II are characterised by abundant horse and bovids, species supplanted by the reindeer during the course of the Aurignacian and Upper Perigordian for, as in the Central region, reindeer is the major species in the northern zone (35 occurrences) followed closely by Equus caballus (33 occurrences). The earlier stages are characterised by a mixture of horse, bovids and reindeer, with even red deer dominating at Fontechevade. By the Upper Perigordian reindeer is unassailable, until we see an incursion of steppic elements during the Magdalenian (horse and saiga) - the earliest Upper Palaeolithic occurrence of saiga being recorded at Chancelade during the Proto-magdalenian.

Unfortunately predominance is not indicated for the majority of Magdalenian deposits, although the material available gives us no reason to believe that reindeer was not the dominant species. However, during this period saiga and horse form a major part of the faunal

Table 6.7 Chronological Change in Zone II.

	Rt.	Ce.	Bov	Ec.	Cc.	Ci.	Rr.	Ss.	St.	Nsp.
Brouillaud			*	*				*		3
Chaise a Vouthon	*	*	*	**		*		*		6
La Quina	**		*	*						3
Festons	*	*	*	*				*		6
Vachons	**		*	*						3
Brouillaud	*		*	*						3
Fontechevade	*	**		*				*		7
Rois	*	*	*	*			*	*		10
Bernous	**	*	*	*		*				5
Les Plumettes		*	**	**				*		6
Vachons	**			**						2
Bonhomme	*		*	*				*		4
Fourneau du Diable	*	*		*						3
Trou de Cluzeau	*		*							4
Vachons	*		*	**						3
Vachons	*			**						2
Orgedeuil	*	?						*		5
Chancelade	*			*						2
Chancelade	*			*					*	3
Fourneau du Diable	**	*		*						3
Combe-a-Rolland	*		*	*						3
Le Roc, Sers	*								*	2
St. Queroy		*		*					**	3
Chateau Bernard									*	1?
Dalignac	*									1
Gte des Moradies	*							*		2
Montgaudier	*		*	*					*	4
Gte de Rochaudry	*					**	*			3
Jumeau	**			**					**	3
Chaire a Calvin vi	*		*	*					*	4
Chaire a Calvin v	*	*	*	*					*	5
Chaire a Calvin iv	*		*	*					*	4
Chaire a Calvin iii	*	*	*	*					*	6
Chaire a Calvin ii	*	*	*	*					*	5
Chaire a Calvin i	*	*	*	*					*	6
Le Placard	*	*	*	*		*		*	*	8
St. Queroy				*						1
St. Queroy	*			*						2
St. Queroy	*			*						3
Pont d'Ambon 6	*									1
Rochereil	**		*	*						3
Rochereil	**	*	*	*	*	*	*	*		8
En Face de Fieux			*	*						2

Key: as in table 6.6

spectrum, dry conditions being emphasised by the sudden importance of ibex.

With the exception of Magdalenian levels at Chaire a Calvin and Le Placard, red deer is more prevalent during the Later Aurignacian (a reflection of the situation observed in quantitative data). Similarly boar is more commonly seen prior to the Upper Perigordian, while roe deer is seen only at Rochereil (Final Magdalenian/Azilian), in association with red deer and boar, heralding perhaps early post-glacial conditions such as those reflected at the Abri en Face de Fieux where reindeer is lacking and the site assemblage comprises bovids and horse.

6.2.3 The Western Zone (III). (Table 6.8)

With as limited a chronological sample as we have in the Gironde, very little can be said regarding the development of faunas based upon qualitative data. Prior to the Magdalenian a total of only six sites are recorded, spanning more than 20 000 years:

Camiac (Ly - 1104) 35 100 +2 000 / -1 500 B.P. (Lenoir 1983)
St. Germain (Gif - 5478) 15 300 +/- 410 B.P. (Delibras et. al. 1987)
Vidon (Ly - 2701) 14 000 +/- 350 B.P. (Lenoir 1983)

Throughout the Magdalenian, assemblages are characterised by the presence and occasional predominance of saiga, a species which is associated with dry, steppic conditions. During the early stages the available, but limited, data suggest that the conditions may, in addition, have been cold, indicated by the presence of reindeer. Topographic features may be invoked to explain the presence of ibex at only 7 sites, its occurrence in the absence of chamois, re-iterating the dry nature of the environment, for the region is one of flat terrain broken by isolated areas of cave formation (eg: St. Germain de la Riviere). Roe deer is absent throughout the period, the boar occurring only once, in association with red deer.

6.3 Discussion.

The chronological significance of changes in faunal species representation has long been interpreted in terms of climatic and palaeoenvironmental conditions, for such a consideration allows one to attempt to trace fluctuations in the nature of animal communities in relation to their associated environment (McCown 1961). To claim that climatic conditions can be reconstructed from large herbivore remains is, however, a grossly simplified overstatement, for the fact remains that finding more of one species than another at a site means merely that more bones of that taxon were initially deposited than were those of any other, or that more were preserved (Levine 1977). When bone fragments are included in N.I.S.P. counts, the degree of bone fragmentation - intentional or otherwise - is also an important factor. Thus it should not be assumed that the archaeozoological data available simply reflect the environmental conditions prevalent at the time of assemblage formation. Nevertheless, given our understanding of these conditions from alternative data (eg: pollen, sediments) faunal species representation may be discussed within a chronoclimatic framework.

Table 6.8 Chronological Change in Zone III.

	Rt.	Ce.	Bov	Ec.	Cc.	Ci.	Rr.	Ss.	St.	Nsp.
Haurets	*	*	*	**		*		*		10
Camiac	?	?	*	*						5 @
Jolias	*		*	**				*		6
Chez Leix	*			*						3
Lespaux	*		*	*						3
Grand Moulin	*		*	**						4
Ste. Florance	*		*	*					*	4?
St. Germain	**					*			*	3
St. Germain	*	*	*	*		*			**	6
Piganeau	*	*	**	*				*	*	6
Garnet I	*								*	2
Baring									*	1
Guimberteau		*	*	*					*	4
Canere	*								*	2
Vidon	**	*	*	*		?		*	*	7
Jaurias	*	*	*	*					**	5
Bisqueytan	*								**	?
Moulin Neuf	*		*	**					**	4
La Lustre	*		*	*					*	4
Houleau C	**		**	*					*	4
Houleau D	*		*	**					**	4
Fongaban			**						*	2
Fauroux	*	*	*						*	4

Key: as in table 6.6

@= Unspecified cervids

Hence, we are able to inquire as to the extent to which the temporal patterning observed in faunal representation may be explained in terms of natural environmental factors (eg: temperature and relative humidity), but straightforward reconstruction of conditions from fauna should be avoided.

The limited distribution of our quantitative data in South West France imposes upon us a constraint as to the availability of data for a chronological study. Thus we shall concentrate upon zones I and II in the following summary. In considerations of this type it is essential to have a complete or near-complete sequence, a feature which is not yet found in zone IV and is only available in zone III, to a severely limited extent, based upon non-numerical data.

Given the nature of the environmental reconstructions provided in chapter 3, based upon non-faunal evidence, it remains for us to parallel (as far as is feasible) changes in faunal and other data. Fig. 6.4 summarises broad environmental conditions as described above and reconstructed by Laville et. al. (1983:227-230).

The Wurm II/III (Chatelperronian) transition is marked by Hengelo and Cottes mild but unstable (relatively humid and warm) climatic conditions, which are associated with a mixed faunal spectrum of reindeer frequencies in the range of 30% to 50%, elevated horse totals and some bovids with the addition of occasional red deer. To the south of our immediate area associated A.P. levels reach more than 60% (Laville et. al. 1986), a feature which is reflected by the relatively high mean value for Cervus elaphus in the Central and Southern Zone. During the Earlier Aurignacian conditions become more severe, falling temperatures and relative humidity being reflected by an increase in reindeer frequencies (to a mean of 86.31% in Zone I) and the occurrence of ibex in both Zones I and II, and chamois in Zone I. The presence of chamois and more temperate species (red deer, roe deer and boar) in Zone I may reflect local variations in temperature and higher moisture levels than to the north, where higher horse frequencies indicate that drier, more steppe-like conditions may well have prevailed.

During the Later Aurignacian reindeer figures fall, reflecting the mild conditions commonly associated with the Wurm III, Perigord III episode or the 'Arcy' Interstadial. As fig. 6.4 indicates relative humidity rises more significantly than do mean temperatures, accounting perhaps for the continued prevalence of reindeer in much of the area. However, the decline in reindeer which **is** seen throughout much of the region is mirrored by a significant rise in red deer in Zone I and bovids in Zone II. The major exception is Roc de Combe where reindeer levels remain high. In both regions (I & II) horse frequencies fall, with a concomitant rise in boar while roe deer rises only in Zone I. We see here relatively temperate conditions, associated with a wide resource base across both regions. In the Central zone only the rhinoceros appears to be absent and frequencies of such species as the mammoth and ass are very low - means 0.04% and 0.48% respectively.

Reindeer frequencies rise again during the Upper Perigordian and remain high during the Protomagdalenian, Solutrean and Magdalenian (especially at the point of the glacial maximum), although Early

Upper Palaeolithic Temperature and Humidity Curves

Fig. 6.4

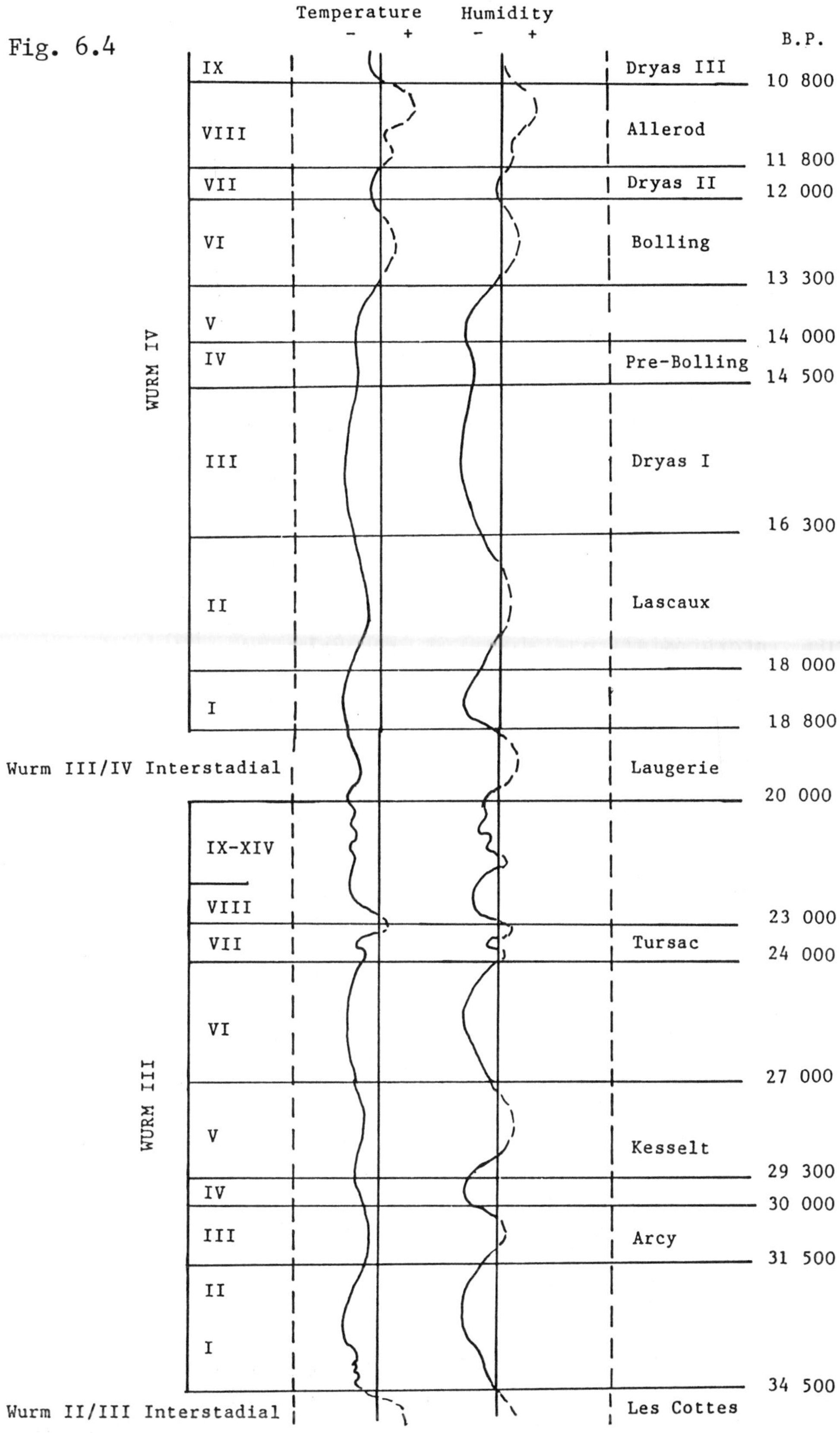

(Source: after Laville et. al. 1983:227-230)

Magdalenian figures show a slight reduction in zone I while conditions during Wurm IV become temporarily more humid. With the exception of a short warm episode at approximately 23 000 B.P., temperatures, although fluctuating, do not rise significantly until the Bolling (13 300 B.P.), perhaps accounting for the high reindeer totals recorded. Humidity however does fluctuate and may be seen to cause, at least in part, changing regional faunal assemblages.

Although sedimentological and palynological data do indicate the existence of brief, milder episodes during this relatively long time span, at a regional scale faunal species representation scarcely reflects these changes. This may be partly due to the short duration of the phases in question, during which, despite climatic and vegetational changes, the structure of animal communities inhabiting the region had insufficient time to register change.

The Laugerie Interstadial, marking the Wurm III/IV transition sees a rise in horse frequencies in Zones I and II despite the maintenance of high reindeer frequencies. The mild, more temperate conditions which one is usually led to expect, are not to be observed in terms of faunal representation, a picture which supports Hannon et. al.'s (1987) suggestion that the pattern of climatic change between ca. 20 000 and 10 000 B.P. associated with Northern Europe is not necessarily observed in its Southern counterpart. The interstadial is marked by an increase in humidity rather than marked improvements in temperature (see fig. 6.4). As noted in chapter 3 the existence of the interstadial has in fact recently been called into question (Laville 1988) and it is certainly the case that the large herbivore fauna do not suggest an improvement in conditions on the scale of the Wurm II/III Interstadial. In summary, faunal changes which might be associated with a Wurm III/IV Interstadial are not seen.

The nature of regional assemblages throughout the Magdalenian changes little, except for a surge in saiga (Dryas I), a species with a severely limited chronological range of which the first Upper Palaeolithic appearance may be observed during the Protomagdalenian. It is a species however, which reinforces the picture of cold, dry conditions. As the Magdalenian progresses, so conditions appear to become more extreme. After moving north at about 13 000 B.P. to yield milder Bolling conditions (associated with limited forest herbivores), the Polar Front moves south at approximately 11 000 B.P., to a position close to that of the glacial maximum. At this point reindeer is predominant, accompanied by a range of both open and woodland species - all of which attain only low frequencies. This return to cold conditions with A.P. less than 20% (Laville et. al. 1986) is emphasised by the importance of reindeer, ibex, chamois and saiga in Zone I and the predominance of reindeer (also associated with horse and saiga) in Zone II. (It should be noted here that such conditions are also reflected in the west and east at this time - especially in the Gironde where horse, saiga and bison frequencies are known, on occasion, to exceed those of reindeer).

As Wurm IV gives way to the early post-glacial, so develops the Azilian. It is at this point that some of the most significant climatic and faunal changes occur. An increase in precipitation - including snow - and the continuation of cold winters (due to the presence of anticyclones) in association with milder summer conditions, allows the development of a more temperate fauna. Reindeer frequencies fall during the Allerod perhaps due to an increase in snow cover, while red deer rises, in association with boar and, in Zone I, bovids (Bos primigenius). The presence of such species implies a marked increase in woodland cover, where, for example, winter protection may be sought. However, the apparent lack of roe deer remains an enigma in much of the region.

The faunal data available in Zones I and II reflect, to a limited extent, those environmental fluctuations identified from sediment, pollen and other analyses. In confirming palynological (and other) results, faunal assemblages provide some information regarding general temperature and humidity levels, and an indication of local snow cover, based upon present-day ecological characteristics of the species or their relations (Delpech 1973). Two points must be made however.

Firstly, selected valleys in South West France may have acted as refugia providing conditions which remain more hospitable for species with restrictive requirements than for their arctic counterparts. Thus, for example, while conditions are cold during the Upper Perigordian (Wurm III, Perigord IV-V) throughout much of the Dordogne and Vezere, the site of La Ferrassie (and to a lesser extent Le Flageolet I) displays a faunal assemblage which one might expect to encounter during the milder Later Aurignacian. Delpech (1983) attributes this to seasonal (summer) exploitation of faunal resources, invoking human selection as an explanation of an exception to the 'climatic rule'.

Secondly, as a result of the existence of possible refugia, faunal assemblages may include the remains of species which display severely limited geographical distributions at the time of procurement, emphasising the fact that we should not assume that assemblages reflect only the prevalent environmental conditions, for assemblages do indeed provide information regarding aspects of human behaviour in addition to the environmental background to cultural activity, a subject to which we shall return later.

Chapter Seven

CORRELATION ANALYSIS AND FAUNAL ASSOCIATIONS.

Biogeographic studies (of which zoogeography is an example) are, to some extent, exercises in pattern recognition. This chapter therefore aims to employ a variety of statistical techniques in an attempt to search for any significant patterns of association and covariation between different large herbivore species within South West French Upper Palaeolithic faunas. The starting point of analytical biogeography is the documentation of the distribution patterns of organisms (Myers & Giller 1988:xii) and thus the consideration of associated statistical and spatial patterns should prove to be of value in the present study. Two approaches are adopted in this chapter:

1. Simple bivariate correlation analysis using Pearson's correlation coefficient (r).

2. Multivariate correlation analysis seeking to identify groups of species which covary.

The principal aim of the approach was to identify groups of species which are linked for ecological reasons, ie: which occur together for the very fact that they were in natural ecological association. It is recognised at the outset that this is not the only reason for which two or more species could be associated in the archaeological record. Assuming that human agencies were responsible for the accumulation of faunal assemblages, human behavioural factors may be invoked to explain some association observed, such as differences in hunting and butchery strategies, or in some cases, long distance transportation of faunal remains from other ecological zones. However, it may be supposed that the repeated existence of similar herbivore associations at various points during the Upper Palaeolithic cultural sequence will reflect primarily ecological as opposed to other human or cultural factors.

The present chapter will therefore first describe the database employed and the relevant statistical techniques, and second seek recurrent patterning within the data from the Upper Palaeolithic. Finally, the extent to which the patterns appear to be ecological or otherwise is considered, and their spatial distribution is described.

The data with which we are here concerned are primarily numerical in nature, the percentage frequency of the major large herbivore species being considered in detail.

The calculation of Pearson's correlation coefficient is undertaken for each of the major cultural periods for which we have sufficient data (ie. a minimum of 4 sites and at least 10 occurrences): Chatelperronian, Early Aurignacian, Later Aurignacian, Upper Perigordian, Solutrean, Middle Magdalenian, Upper Magdalenian and Azilian. In order to ensure comparative size databases when more complex relationships are sought (ie: employing Principal Components Analysis) Pre-Upper Magdalenian and Upper Magdalenian data are considered. Thus we have at least 50 cases in each example, the earlier of which comprises assemblages dating from approximately 34 000 to 15 000 B.P. and the later from 15 000 to 10 000 B.P.

7.1. Bivariate Correlation Analysis.

7.1.1. Pearson's Correlation Coefficient.

Simple bivariate analysis is a useful descriptive statistical technique which allows us to identify levels of association between variables. Several statistical techniques exist which allow us to measure and determine the strength of association between two variables, either a dependent and an independent variable or two dependent variables, and so, in order to determine the relationships between the relative frequencies of the major large herbivore species during each of the cultural periods of the Upper Palaeolithic, Pearson's Correlation Coefficient (r) was calculated using the quantitative data available (see tables 7.1 to 7.13).

$$r = \frac{n\Sigma x_i y_i - (\Sigma x_i)(\Sigma y_i)}{\sqrt{\{[n\Sigma x_i^2 - (\Sigma x_i)^2][n\Sigma y_i^2 - (\Sigma y_i)^2]\}}}$$

where X and Y = frequency of selected species,
N = number of cases.

The Coefficients of Determination (r^2), measuring the proportion of total variance explained (Shaw and Wheeler 1985:154) or the proportion of variation in Y 'explained' by X (Silk 1979:223), are also presented where significant.

7.1.2. Results.

The Chatelperronian, as stated in chapter 3 is commonly placed within the first climatic phase of Wurm III in S.W.France, although elsewhere the period is associated with earlier (eg: Arcy-sur-Cure) and later (eg: Cueva Morin) deposits. The extent to which it is considered to cross the Middle/Upper Palaeolithic divide was shown in fig. 3.3 (see p.52) and needs to be borne in mind in the present consideration, in which the statistical patterns observed are far from typical of the Upper Palaeolithic as a whole.

Three significant correlations are observed during the Chatelperronian, between horse and reindeer (r=-0.826), horse and bovids (r=-0.711) and horse and chamois (r=-0.541). The highest positive value of r is to be found between horse and mammoth (r=0.459). All these correlations concern the horse, the predominant species at Chatelperron itself and of major importance (if not prime importance) in the Perigord region. The species occupies a position which will appear to be taken subsequently by the reindeer, as steppic conditions are replaced by colder, more moist (oceanic) tundra.

During the course of the Early Aurignacian, only the chamois is positively correlated with the reindeer, with a value of r which is highly insignificant (r=0.024). The predominance of the reindeer is thus established, replacing the steppic element of the Chatelperronian.

Table 7.1 Species Correlation during the Chatelperronian.

	R.t.	C.e.	Bov.	E.c.	C.c.	C.i.	S.s.	R.r.	C.a.	M.p.
R.t.	1.000									
C.e.	.124	1.000								
Bov.	.301	.038	1.000							
E.c.	-.826 .682	-.341	-.711 .506	1.000						
C.c.	.194	.058	-.432	.050	1.000					
C.i.	-.133	.266	-.029	.049	-.126	1.000				
S.s.	-.007	-.217	.310	-.138	-.142	-.136	1.000			
R.r.	.449	.258	.408	-.541 .293	-.167	-.057	-.198	1.000		
C.a.	-.439	-.059	-.126	.258	-.244	.047	.061	-.300	1.000	
M.p.	-.404	.133	-.467	.459	-.137	-.118	-.150	-.155	.019	1.000

Table 7.2 Species Correlation during the Early Aurignacian.

	R.t.	C.e.	Bov.	E.c.	C.i.	S.s.	R.r.	C.a.	M.p.
R.t.	1.000								
C.e.	-.331	1.000							
Bov.	-.558 .311	.369	1.000						
E.c.	-.891 .794	-.039	.232	1.000					
C.i.	-.108	.886 .785	.072	-.183	1.000				
S.s.	-.510 .260	.291	.485	.361	-.118	1.000			
R.r.	.024	.788 .621	-.045	-.241	.910 .828	-.153	1.000		
C.a.	-.480	.249	.417	.298	-.077	.722 .521	-.094	1.000	
M.p.	-.178	-.078	-.163	.315	-.072	-.043	-.081	-.058	1.000

The negative correlation of reindeer with bovids and horse may be seen to mirror the increased 'oceanic' nature of the tundra, these species providing the highest negative correlations (r=-0.558 and r=-0.891 respectively). Similarly the boar/reindeer relationship is a significantly negative one, indicating that the environment was too cold and open in areas of prolific reindeer for the boar populations to thrive.

Significant positive correlations between the red deer and both the chamois and ibex are observed and may imply that we are witnessing the results of valley as opposed to plateau, plain or interfluve exploitation by hunter-gatherers. The co-existence of red deer with these taxa, themselves very significantly correlated (r=0.910, r^2=0.828) and forming a montane or skeletal component, may be explained by invoking the simultaneous exploitation of relatively wooded valley bottom and lower-/mid-slope communities and higher valley slope or rockface faunas, perhaps seeking refuge from winter snows.

Finally, the boar/woolly rhinoceros correlation (r=0.722) may, in part, be explained by the fact that both species are to be found in flat terrain. Both are negatively correlated with ibex and chamois. Furthermore, according to Stuart (1982), the woolly rhinoceros may have sought woodland or even forest conditions - as does the boar today, while others have suggested affiliation with extensive grasslands and associated deciduous trees (Kurten 1968) - the latter particularly suitable for the boar.

During the Later Aurignacian (see table 7.3) the most significant correlations are the positive relationship observed between roe deer and red deer, where r=0.769, and that between chamois and ass (r=0.720). The former suggests the existence of a distinct woodland component in the fauna, although the lack of a significant correlation between forest-adapted boar and these species might be seen to suggest that the canopy was relatively broken. However, boar is significantly positively correlated with bovids. Bos sp., which prefers milder conditions than do many of the other species with which the boar is not correlated, may have occurred in wooded areas of the region, while up to 50% of the bison diet may comprise tree cover and shrub vegetation (Spiess 1979).

Also of significance during the Later Aurignacian is the positive correlation between red deer and chamois (r=0.538, r^2=0.289). As described in chapter 4 the chamois is commonly considered a montane species, but is in fact subject to regular, small scale seasonal altitudinal migrations (Hamr 1985; von Elsner-Schack 1985), frequenting open forest biotopes during part of the year - when heavy snowfall restricts forage availability. Perhaps, during the traditionally milder Later Aurignacian, precipitation, in the form of snowfall, increased in the Perigord, forcing the chamois to seek refuge and forage in open woodland areas close to rock and cliff faces. Meanwhile, the red deer may have been forced into deeper woodland cover in order to secure winter forage.

Reindeer is negatively correlated with all species except the ibex and mammoth, only the latter yielding a significant relationship (r=0.541, r^2=0.293). Significant negative species are yielded however, with three taxa, namely red deer (r=-0.613), bovids (r=-0.653) and

Table 7.3 Species Correlation during the Later Aurignacian.

	R.t.	C.e.	Bov.	E.c.	C.c.	C.i.	S.s.	R.r.	E.h.	M.p.
R.t.	1.000									
C.e.	-.613 .376	1.000								
Bov.	-.653 .426	-.094	1.000							
E.c.	-.250	-.282	.400	1.000						
C.c.	-.451	.769 .591	-.011	-.251	1.000					
C.i.	.033	-.002	-.320	-.206	-.322	1.000				
S.s.	-.390	.042	.592 .350	-.141	.026	-.301	1.000			
R.r.	-.648 .420	.538 .289	.217	.086	.529 .280	-.172	.059	1.000		
E.h.	-.480	.014	.442	.414	-.089	-.250	.303	.720 .518	1.000	
M.p.	.541 .292	-.444	-.231	.079	-.257	-.290	-.157	-.374	-.139	1.000

Table 7.4 Species Correlation during the Upper Perigordian.

	R.t.	C.e.	Bov.	E.c.	C.c.	C.i.	S.s.	R.r.	E.h.	M.p.
R.t.	1.000									
C.e.	-.782 .612	1.000								
Bov.	-.583 .339	.212	1.000							
E.c.	-.608 .369	.062	.637 .406	1.000						
C.c.	-.469	.759 .576	.027	-.089	1.000					
C.i.	-.532 .283	.152	.066	.438	-.112	1.000				
S.s.	-.522 .272	.678 .459	.127	-.010	.814 .663	.145	1.000			
R.r.	-.447	.107	.600 .360	.365	-.073	.212	-.022	1.000		
E.h.	-.626 .392	.626 .392	.576 .332	.350	.561 .315	-.148	.243	.296	1.000	
M.p.	.283	-.210	-.178	-.189	-.123	-.163	-.128	-.111	-.136	1.000

chamois ($r=-0.648$). The existence of tundra conditions might thus be suggested, given (1) the preference of red deer for mild, wooded, more temperate environments, (2) the suggestion that much of the 'bovids' is comprised of aurochs, and (3) the generally accepted association of the mammoth with the extreme cold of tundra-type conditions. The close association of chamois and both red and roe deer means that a negative correlation with reindeer is not surprising, especially given that the chamois is well adapted to the rockface and valley side conditions which reindeer avoid when possible.

Finally, bovids and horse are positively, although not significantly, correlated ($r=0.400$). If ranked data are considered however, the value of r rises significantly to 0.807 ($r^2=0.6512$), emphasising the importance of this possibly steppic element in the fauna, a group of species which will later be shown to differ significantly from the other taxa in terms of its spatial distribution (see chapter 8).

A total of 14 values of r exceed 0.500 during the Upper Perigordian, of which 6 are negative correlations with reindeer. Only the mammoth, which occurs solely at the Abri Pataud and Roc de Combe, is positively correlated with this species, both traditionally being considered typical of moist tundra communities to which bison, horse and ibex are not adapted. The existence of a woodland group is demonstrated by the series of positive significant correlations between red deer, boar and roe deer. These are in fact the highest positive correlations seen during the Upper Perigordian. The significant positive bovid/boar relationship suggests the possible existence of woodland dwelling bovids such as Bos sp. or the woodland bison, although the positive value of r between bovids and horse would also imply the existence of a distinct steppic group. The significant correlations between the ass and red deer/roe deer would also seem to imply that the species was encountered in similar wooded areas, perhaps in localities in which the ass sought refuge from heavy snow or strong winds.

In the course of the Solutrean we once again encounter a large number of significant correlations, even to the extent of a value of $r=1.000$ between boar and roe deer, artificial given the fact that both species occur only at one and the same site and are equally important.

The reindeer is the predominant species and is negatively correlated with all other taxa except the mammoth and ibex, both of which are associated with cold conditions, be they tundra or alpine. A steppe biotope is identified based on the positive correlation between bovids and horse, ($r=0.707$) and the important negative relationships with reindeer although the other major ecological group discerned is once again the temperate woodland community, based on correlations between red deer and bovids - not to mention the roe deer and boar.

Reindeer is positively correlated only with the ibex ($r=0.121$) and boar ($r=0.154$) during the Middle Magdalenian, yielding values sufficiently low to be considered insignificant in statistical terms (see table 7.8). The strongest reindeer correlation is that with the horse, the value of r being -0.580, with saiga following closely

Table 7.5 Species Correlation during the Solutrean.

	R.t.	C.e.	Bov.	E.c.	C.c.	C.i.	S.s.	R.r.	M.p.
R.t.	1.000								
C.e.	-.816 .666	1.000							
Bov.	-.764 .584	.809 .654	1.000						
E.c.	-.961 .924	.807 .651	.707 .499	1.000					
C.c.	-.685 .469	.745 .555	.986 .972	.635 .403	1.000				
C.i.	.067	-.327	-.173	-.169	-.230	1.000			
S.s.	-.685 .469	.745 .555	.986 .972	.635 .403	1.000	-.230	1.000		
R.r.	-.498	.262	-.011	.434	-.162	.411	-.162	1.000	
M.p.	.120	-.252	-.191	-.276	-.138	-.224	-.134	-.191	1.000

Table 7.6 Species Correlation during the Middle Magdalenian.

	R.t.	C.e.	Bov.	E.c.	C.c.	C.i.	S.s.	R.r.	S.t.
R.t.	1.000								
C.e.	-.306	1.000							
Bov.	-.383	.516 .266	1.000						
E.c.	-.580 .336	-.269	-.284	1.000					
C.c.	-.461	.588 .346	.953 .908	-.178	1.000				
C.i.	.121	-.024	-.235	-.390	-.270	1.000			
S.s.	.154	-.221	-.145	-.178	-.071	.395	1.000		
R.r.	-.440	-.021	-.385	.466	-.359	-.053	-.359	1.000	
S.t.	-.570 .325	-.323	-.217	.825 .681	-.126	-.421	-.126	.426	1.000

behind at r=-0.570. The relationship between the two dry steppe species (horse and saiga) is the highest statistically significant positive relationship observed (r=0.825). The apparently higher value of r encountered between bovids and roe deer is a result of the fact that only one roe deer occurrence is observed - at a site with high bovid percentages. The negative correlations between reindeer and horse, reindeer and saiga and the high positive r-value of horse and saiga point to the possible existence of two identifiable ecological groups, the cold, moist tundra and dry steppe biotopes which are given more detailed consideration below.

All Upper Magdalenian reindeer correlations, with the exception of chamois, are negative (see table 7.7), emphasising the paramountcy of the species, the chamois/reindeer relationship being very weak (r=0.033). Once again relatively high positive correlations exist between the red deer, roe deer and boar. Although that of red deer and boar is low (r=0.398) in comparison with the others, the value is second only to the red deer/roe deer correlation and may still be significant in ecological terms. It merely reminds us that the boar requires a relatively dense forest cover - a biotope type which probably did not prevail over much of S.W.France during the Upper Magdalenian.

The high positive correlation between boar and ibex (r=0.780, r^2=0.608) may, in ecological terms, be what Silk (1979:207) terms 'spurious' where both variables reflect the influence of a third, as yet undetermined, controlling variable. Given the apparently significant positive correlation between ibex and roe deer (r=0.531), we may be looking at the influence of some other factor such as the propensity of the species to occur only when a larger number of taxa are observed from a variety of habitat types. Perhaps these species represent low-ranking targets on the hunters' list of prey, for the role of socio-economic variables should not be forgotten altogether.

During the Azilian (see table 7.8), the highest positive relationship is found to exist between ibex and chamois (r=0.866), indicating perhaps the existence of a strong montane component in the fauna. Meanwhile, both the ibex and chamois are significantly negatively correlated with red deer (r=-0.818 & -0.674 respectively) - the major species encountered during the Azilian throughout the region as a whole. It is of interest to note that those sites at which the montane element in the fauna is of major importance are to be found in the east of our region, namely at Blassac, Campalou and the Grotte du Tai.

Positive correlations exist during the Azilian, between red deer and roe deer (r=0.594) and roe deer and ass (r=0.626), reflecting the growth in the importance of open woodland conditions during the early post-last-glacial. The fact that woodland conditions remained open is suggested by the fact that the importance of boar is not yet significant, indicating perhaps that tree cover was not sufficiently dense nor extensive to accommodate this forest-adapted species. Topographic, moisture and temperature requirements necessary for the growth of temperate woodland herbivore populations were, it would appear, more commonly to be found to the west of the Massif Central, in the Perigord proper, and it is here that such species slowly began to thrive and replace the reindeer in its position of predominance. The

Table 7.7 <u>Species Correlation during the Upper Magdalenian.</u>

	R.t.	C.e.	Bov.	E.c.	C.c.	C.i.	S.s.	S.t.	R.r.	E.h.
R.t.	1.000									
C.e.	-.256	1.000								
Bov.	-.681 .464	.011	1.000							
E.c.	-.555 .308	.171	.110	1.000						
C.c.	-.221	.547 .299	-.056	.170	1.000					
C.i.	-.247	.298	-.031	.000	.531 .282	1.000				
S.s.	-.308	.398	.028	.052	.635 .403	.780 .608	1.000			
S.t.	-.562 .316	-.147	.114	.195	-.149	-.097	-.097	1.000		
R.r.	.033	.105	-.080	-.143	-.112	.048	-.094	-.091	1.000	
E.h.	-.388	-.069	.005	.141	.027	-.078	-.069	.760 .578	-.076	1.000

Table 7.8 <u>Species Correlation during the Azilian.</u>

	R.t.	C.e.	Bov.	E.c.	C.c.	C.i.	S.s.	R.r.	E.h.
R.t.	1.000								
C.e.	-.540 .292	1.000							
Bov.	-.315	-.128	1.000						
E.c.	-.114	-.078	.714 .510	1.000					
C.c.	-.478	.594 .353	-.026	-.236	1.000				
C.i.	.229	-.818 .669	-.209	-.309	-.428	1.000			
S.s.	.394	.215	-.159	-.258	.102	-.444	1.000		
R.r.	.182	-.674 .454	-.209	-.395	-.267	.866 .750	-.364	1.000	
E.h.	-.245	.330	-.172	-.212	.626 .392	-.201	.133	-.203	1.000

significant negative correlation which is observed between reindeer and red deer may be considered to further point to the decline of the reindeer, suggesting that conditions were both warmer and more humid (reflecting an increase in precipitation) than previously during much of the Upper Palaeolithic. Although not statistically significant, the value of r for reindeer and roe deer (-0.478) is also of interest.

Meanwhile, the relationship observed between bovid and horse frequencies during the Azilian is significant. The positive value (r=0.714) is significant in that it reflects the secondary importance of the two species concerned in the north of our region at Pont d'Ambon and Bois Ragot where steppe-like communities may still have been of some, but reduced, importance.

7.1.3. Pattern Recurrence.

Having highlighted those significant correlations which exist during selected episodes of the Upper Palaeolithic, it remains to consider the extent to which such associations recur, and may reflect ecological variables.

Reindeer is the predominant species throughout most of the Upper Palaeolithic, the exceptions being the Chatelperronian and Azilian. Frequencies often exceeding 85 or 90% result in low percentages of other taxa and thus we more often than not see negative correlations with the reindeer. Only four species regularly display positive relationships with reindeer, and it is noticeable that, after the Chatelperronian, these mainly occur during the Azilian when the reindeer is no longer so abundant. Thus we see the following positive correlations:

Table 7.9. Upper Palaeolithic Positive Reindeer Correlations.

	Ch.	E.A.	L.A.	U.P.	Sol.	M.M.	U.M.	Azil.
Ibex			*		*	*		*
Boar						*		*
Chamois	*	*					*	*
Mammoth			*	*	*			
Roe deer	*							
Ass								
Red deer	*							
Bovids	*							
Woolly Rhino'	*							

Saiga is commonly associated with (relatively open) flat, dry terrain, a characteristic which is clearly supported in the present discussion by the positive correlations with the horse (Middle Magdalenian r=0.850) and ass (Upper Magdalenian, r=0.760) during the species's brief stay in South West France. Horse and saiga are in general negatively correlated with those taxa commonly found in rocky or montane environments (the ibex to a greater extent than the chamois). Emphasising the former group's presence in flat or gently undulating landscapes we see a statistically perfectly random correlation between the horse and ibex during the Upper Magdalenian (r=0.000) - species which display contrasting ecological tendencies.

Throughout the post-Chatelperronian Upper Palaeolithic negative correlations between such open, tundra or steppe species as reindeer, and woodland species (red deer, roe deer, Bos sp. and boar) generally occur. Red deer generally displays negative correlations with species such as reindeer, horse, saiga and ibex, although significant positive relationships do emerge with chamois, a montane or rock-face species requiring moister conditions than does the ibex. Meanwhile, significant positive correlations between roe deer, chamois and boar reinforce the notion of a more temperate woodland group. These relationships are seen sporadically throughout the course of the Upper Palaeolithic and may be thought of as implying wooded, humid conditions. Roe deer and boar are positively associated (often significantly) during all but the Chatelperronian and Middle Magdalenian when there may have been insufficient woodland cover and suitable forage to support boar populations, a hypothesis supported by the abundance of open, flat terrain horse populations.

A total of three negative relationships are observed between ibex and chamois, the two species encountered which are most commonly associated with montane conditions. The values of r which are obtained (-0.061 during the Chatelperronian, -0.076 during the Later Aurignacian and -0.053 during the Middle Magdalenian) are very low and allow us to continue to consider this montane group as a distinct component.

The bivariate correlation analysis undertaken leads us to conclude that four broad ecological types may be identified. These are based upon the degree to which positive correlations recur throughout the Upper Palaeolithic and a knowledge of the relevant aspects of large herbivore ecology (see Chapter 4).

1. (Arctic) Tundra: Reindeer.

2. Open, steppe: Bovids, horse, saiga.

3. Montane: Ibex, chamois

4. Woodland (Temperate ?): Boar, roe deer, red deer, Bos sp., (& woolly rhinoceros?)

7.2. Multivariate Correlation Analysis.

The aim of multivariate analysis is to "express data structure as faithfully as possible, with minimal expression of noise" (Gauch 1982:8). In the following discussion Principal Components Analysis (an ordination technique, henceforth P.C.A.) is employed partly in an attempt to reduce a large number of variables to a smaller, simpler, more manageable and easily explained set of data and also to assess the nature of patterning amongst the variables (Dunn and Everitt 1982:47). This is done in the belief that many of the variables considered (ie: the presence and frequency of large herbivore taxa) are in fact correlated with each other, and hence are measuring the same thing, such as prevailing snow or tree cover, temperature, humidity or the need for animal-derived raw materials such as antler or skins.

As Doran and Hodson (1975) point out, P.C.A. is of value to archaeological research in that it summarises relationships observed between variables and those between "units" (in the present study, levels or sites), while pointing to the main trends in the raw data and the major variables involved therein. For a detailed discussion of the technique the reader is referred to Shennan's "Quantifying Archaeology" (1988:263-270). The program used in the present research was part of the S.P.S.S.-X. package, modified by Dr. P. Callow of the Cambridge University Computer Laboratory to produce three-dimensional graphical summaries of the results which could be viewed stereoscopically.

P.C.A. of two large data sets, Pre-Upper Magdalenian and Upper Magdalenian, was undertaken in an attempt to identify groups of inter-correlated variables (taxa). In so doing, the aim was to reduce both data-bases to smaller sets without losing any information. The earlier Upper Palaeolithic culture-specific data sets were amalgamated in the present analysis so as to secure a sufficiently large data-base. In order for the results obtained to be 'representative' it was decided that a minimum of 50 cases (occurrences) be adopted.

By identifying the trends behind the raw data, it becomes possible to determine which variables are primarily involved therein, while the discernment of patterns is made substantially easier by the provision of correlation matrices such as those examined above. However, the provision of a matrix of uncorrelated variables (weighted representations of the original set) further facilitates interpretation (Johnston 1978:136).

Having computed the secondary (species by species) matrix from the species by occurrence data tables, eigenanalysis is performed, producing eigenvalues (the variance accounted for by each factor), the highest eigenvalue being that of the first principal component. Studies in community ecology have shown that the first three eigenvalues commonly account for 40-80% of the total variance, although in some cases, when the data set is large and 'noisy', the first few factors may account for no more than 5% of the variance, and yet still be informative (Gauch 1982:141). Alternatively, a high proportion of the variance may be explained in terms of a single component, which nevertheless remains meaningless in the context of the problem under consideration. In the following discussion components with eigenvalues exceeding 1.0 are considered. A value greater than 1.0 suggests that the component concerned 'explains' more of the total variance than does a single variable (Shaw & Wheeler 1985:282).

7.2.1. Pre-Upper Magdalenian.

Inspection of pre-Upper Magdalenian eigenvalues (see table 7.10) reveals five principal components with values exceeding 1.0, explaining (statistically speaking) a total of 69.3% of the variation.

Boar, roe deer, red deer and ass are positively correlated with component (p.c.) 1, reindeer being negatively correlated. As

Table 7.10 <u>Pre-Upper Magdalenian Factor Matrix and Final Statistics.</u>

	FACTOR 1	FACTOR 2	FACTOR 3	FACTOR 4	FACTOR 5
Reindeer(1)	-.82121	.06978	-.35098	.19430	.04837
Boar(7)	.77026	.03275	-.22308	.22876	-.21398
Roe Deer(5)	.72675	-.01061	-.23693	.25707	-.19858
Red Deer(2)	.68198	.44273	.06041	-.25044	-.04328
Ass(10)	.67965	.19056	-.05394	-.02563	.30173
Chamois(9)	-.07022	.81810	.22617	-.06095	.17640
Horse(4)	.10528	-.63728	-.02680	-.31358	.25769
Bovids(3)	.32600	-.23645	.72201	-.06729	.14882
Woolly Rhinoceros(11)	-.00102	-.29438	.63296	.27233	.11466
Ibex(6)	-.28744	.36630	.18099	-.59087	-.15526
Saiga(8)	-.23848	.39580	.33117	.57164	.08501
Mammoth(12)	.10323	.09054	-.32718	.02106	.83127
Communality	.84254	.72935	.71015	.58267	.68993
Eigenvalue	2.99531	1.75540	1.45894	1.07662	1.03563
Pct of Var.	25.00000	14.60000	12.20000	9.00000	8.60000
Cum Pct.	25.00000	39.60000	51.70000	60.70000	69.30000

what one may tentatively describe as the temperate (woodland) elements increase in frequency so we would expect a decline in the importance of cold, open landscape-associated taxa such as the saiga and reindeer. We have here a component which, representing relatively closed woodland or forest cover, accounts for exactly one quarter of the total variation encountered in our database.

Chamois is positively, and horse negatively, correlated with p.c.2, while the correlations between component 2 and the other taxa are too low to be considered statistically significant - with the possible exception of the red deer. This component, accounting for 14.60% of the variation, may perhaps be related to the occurrence of montane species (associated with 'skeletal' landscapes) for it is noticeable that the ibex is also positively correlated, although with a much reduced coefficient value (r=0.366). The chamois prefers to eat grass and herbaceous plants at all times of the year, but when deep winter snow is persistent and ground vegetation is difficult to obtain, the species is known to browse on several tree species, including <u>Pinus sylvestris</u>, <u>Abies alba</u> and <u>Juniperus sp.</u> (Perle & Hamr 1985:80). The necessary inclusion of some arboreal vegetation in the diet may go some way towards explaining the positive, although relatively weak, correlation of red deer with p.c.2. Meanwhile, those species requiring open, flat terrain are negatively correlated with p.c.2 (horse r=-0.6373, rhinoceros r=-0.2944 and bovids r=-0.2365), although the saiga is an exception (r=0.3958).

Both bovids and woolly rhinoceros are positively correlated with p.c.3, explaining 12.2% of the variation. Both species are commonly associated with open, relatively dry landscapes, a supposition which is supported in the present case by the positive correlation of saiga (r=0.3312). With the exception of the reindeer (negatively correlated with both p.c.1 and p.c.3), the red deer (positive in both cases, but with a very low value in the case of p.c.3 r=0.0604) and bovids (positive in both cases but a determining species in p.c.3) the taxa which are positively associated with the first component (woodland (?)) are negatively so with the third. Based upon this argument it would seem likely that we are dealing here with an open environment also suitable for the saiga. We have perhaps open, herbaceous vegetation with low-grazing grasses, a biotope unsuitable for the reindeer, which displays the highest negative correlation (r=-0.3509).

Cold, dry and flat (open) conditions are also indicated by p.c.4, with which saiga is positively and ibex negatively correlated. Thus low ibex frequencies are associated with higher saiga, a species which avoids rocky, mountainous terrain. This component explains 9% of the variation. However, when interpreting p.c.4 one must take into account the severely limited chronological distribution of the saiga during the Pre-Magdalenian IV Upper Palaeolithic. It occurs primarily during the later half of the Upper Palaeolithic, namely the Magdalenian and will thus be of little 'ecological' significance during the bulk of the Upper Palaeolithic.

Finally. only the mammoth is significantly correlated with p.c.5 (r = 0.8313). Woodland species and the ibex are negatively correlated with p.c.5, all of them with low values. Although the other taxa (excluding the rhinoceros) also yield low correlation coefficients, their being positive may suggest that we have an open environment rarely frequented by the ibex. However, given the overall low occurrence and rarity of the mammoth, we may be observing either the exploitation of a very rare taxon or perhaps one, of which the occurrence is the result of human behaviour (perhaps scavenging or co-operative hunting of large animals) or simply the statistical result of the occurrence of a relatively rare species. Larger species, such as the mammoth, may have been exploited by scavenging carnivore kills and natural deaths, or we may be witnessing the purposeful collection of bone and ivory as sources of raw material, for it is in the form of ivory that most of the mammoth recovered exists (Delpech 1983:227) while the species itself was of a size which might have dissuaded hunters from undertaking a kill without the necessary personnel and organisation.

7.2.2. Upper Magdalenian.

Running a P.C.A. program on Upper Magdalenian faunal data reduced a 54 x 10 matrix to four principal components (see table 7.11 which shows the factor matrix and final statistics upon which the discussion is based).

Boar, red deer and roe deer are highly positively correlated with p.c.1, reindeer being highly negatively associated. This tells us that high reindeer totals may be associated with low frequencies of the other taxa mentioned. The horse is also positively correlated with p.c.1, replacing the ass (during the Pre-Upper Magdalenian) as a member of the first factor, which may be thought of as representing an axis of woodland cover (indicating the extent to which temperate (?) woodland cover occurs.) P.c.1, highly similar in nature to p.c.1 of the earlier phases, explains 29.40 % of variation in our data.

Table 7.11 Upper Magdalenian Factor Matrix and Final Statistics.

	FACTOR 1	FACTOR 2	FACTOR 3	FACTOR 4
Reindeer(1)	-.75148	.62857	-.09592	.00607
Boar(7)	.74344	.33530	.01673	-.31817
Red Deer(2)	.70257	.41353	.12564	.19160
Roe Deer(5)	.68463	.52624	-.22870	-.02275
Horse(4)	.52529	-.25130	-.13351	.46066
Saiga(8)	.23782	-.75679	-.03767	.10687
Bovids(3)	.47388	-.66319	.04609	-.24489
Chamois(9)	-.09376	.05477	.73174	.61344
Ass(10)	.25571	.20243	-.56814	.50256
Ibex(6)	.48224	.31499	.52862	-.20793
Communality	.96906	.71711	.72648	.56911
Eigenvalue	2.94353	2.17436	1.23660	1.09421
Pct of Var.	29.40000	21.70000	12.40000	10.90000
Cum Pct.	29.40000	51.20000	63.50000	74.50000

While reindeer is relatively highly positively correlated with p.c.2, (the only high reindeer positive correlation during the whole Upper Palaeolithic), both bovids and saiga are negatively correlated. In addition we see a negative coefficient for the horse (-0.2513). These three negatively correlated species are those for which we would expect to see a western bias (see chapter 8), the major reindeer range being found to the east, in the Vezere / Dordogne valleys. It may be suggested therefore, that p.c.2 relates to the faunal spectrum as observed in the east of our region where conditions may be described as 'patchy' or mosaic-like. Stretches of open landscape are present but not sufficiently dry or extensive to support continuous saiga and bovid (mainly bison) populations. Reindeer dominated p.c.2 explains 21.7 % of the total variance observed in our data, and hence the first two components account for more than half of the variation.

Chamois and ibex are positively correlated with p.c.3, indicating perhaps a montane or rock-face group. The ass, which represents the highest negative correlation with p.c.3, is a species which, in addition to flatter terrain, may inhabit areas of steep valley slopes, although at lower altitudes than the true montane species. The negative correlation indicates that we may be dealing with a sheer rockface community occupying areas similar to those occurring near to Roque Gageac and Roque St. Christophe in the Dordogne and Vezere valleys respectively. The elements of a montane group may have occurred in pockets throughout South West France, displaying a patchy distribution which reflects the existence of suitable topographic features, eg: cliff-faces and steep, treeless valley sides. P.c.3 appears to explain only 12.4% of the variance in our data, bringing the cumulative total to 63.5%.

Chamois and ass are positively correlated with p.c.4, covarying in the same direction and perhaps indicating that those species which commonly inhabit areas close to woodland edges (albeit on a seasonal basis) are represented, for the chamois prefers higher pastures close to pockets of woodland in which it can take refuge when disturbed (Delpech et. al. 1983). P.c.4 explains 10.9% of the variance and may represent conditions somewhat similar to p.c.2, for once again we have open grassland (meadow) which may have been exploited on a seasonal basis (eg: snow-free pastures are sought during the winter months by the chamois (von Elsner-Schak 1985)) for the negative correlation between p.c.4 and the ibex suggests that we do not have a second montane group <u>per se</u>. P.c.2 is here equated with species which are relatively sure-footed, well able to exploit (dry ?) valley slopes and some wooded steppe edges.

7.2.3. Pattern Recurrence and Interpretation.

P.C.A. reveals the existence of five factors with eigenvalues exceeding 1.0 during the Upper Palaeolithic prior to Magdalenian IV, and four factors during the Upper Magdalenian (IV-VI). Two components clearly occur in both data sets: a woodland type (p.c.1 in both sets) and a montane component (p.c.2 in the former, p.c.3 in the latter). The constituent species of each group vary to some extent but the general nature of the groups is sufficiently similar to allow us to accept a broadly ecological interpretation. We must, however, accept that a mixture of biotopes may be observed. For example, during the Upper Magdalenian the woodland group (p.c.1) is positively correlated with horse, bovids and saiga (the first significantly so). One possible explanation of this "mixed" group may lie in the exploitation of resources which are rarely found to exist together but, given seasonal shifts in range, may appear to be archaeologically contemporaneous.

The only significant reindeer correlation during the Pre-Upper Magdalenian is with p.c.1 where a negative value is given, emphasising the closed woodland nature of the environment represented. The correlations between reindeer and other factors are low, and may be largely a result of the very high percentage frequencies of the taxon. High percentages of one species necessarily result in low relative frequencies of another and it is noticeable that we have no component with which the species is significantly positively correlated.

We see the existence of three possibly open environmental types:

p.c.3 - Bovids, woolly rhinoceros, saiga. Cold steppe (?)
(Dry) open herbaceous parkland.

p.c.4 - Saiga. Dry steppe, flat.

p.c.5 - Mammoth. Open, cold, grassland (scavenged ?)

Reindeer is positively correlated with p.c.4 and p.c.5 although the value of the coefficient is not high enough to be considered statistically significant.

During the Upper Magdalenian four principal components are identified. In addition to the woodland (1) and montane (3) groups, we see two open components.

P.c.2 is unique amongst all the Upper Palaeolithic factors in having a significant positive correlation with reindeer (r= 0.6286). Conditions in this moist tundra group may be likened to those encountered in Scandinavia (despite day-light differences resulting from latitudinal position), termed, by Remmert (1980) 'Warm' Arctic and High Arctic Oceanic. In both environments, where summers are relatively warm and winters cold, reindeer represents the most abundant large herbivore species adapted, as it is, to 'wet moss tundra' (Remmert 1980:194), where population figures fluctuate on a cyclical basis in response to and as a cause of varying vegetational conditions.

P.c.4 may be thought of as representing a group comprising those species which are sure-footed on steep valley sides and likely to undertake some seasonal, vertical migration. Both chamois and ass fall into this category whereas the ibex remains in rocky cliff-face zones.

In general the later fauna is more clearly divided than is the Pre-Upper Magdalenian, although woodland and montane components emerge in both cases. Pockets of woodland and forest-cover existed throughout the Upper Palaeolithic, their spatial extent fluctuating to a limited degree in response to climatic conditions. However, the nature of open, tundra and/or steppe environments is more likely to have varied, for slight changes in the climatic regime will result in more noticeable alterations in the vegetation of open landscapes, where changes in the rates of evapotranspiration, run-off and surface erosion are more quickly effected. Dense vegetation cover protects the surface from weathering, provides cover, breaks the force of rain, prevents rapid snow melt, regularises the amount of rainfall and prevents the occurrence of sudden floods. It also provides forage for species which may otherwise be forced to leave the area (Holmes 1969:76).

7.3 Conclusions.

The bivariate and multivariate correlation analyses undertaken result in the division of the faunal data into various categories. Interpretation of such categories is largely speculative, and although in the present study ecological meaning is generally assigned, it is realised that other, non-ecological interpretations may be preferred by some. However, given the recurrence and persistence of the factors, ecological meaning is assigned where relevant.

In the following discussion, a third statistical technique will be added (although not discussed in detail and used only as a corroborative technique), namely that of Cluster Analysis. The technique was employed using primarily quantitative data, these being the percentage frequencies of species in assemblages. Thus the Euclidean Coefficient of Distance was adopted, some of the results being presented here. In addition Jaccard Coefficient of Similarity was employed, enabling us to use the non-numerical data available. This techniques disregards 'negative matches' and concentrates upon occurrences of species.

Results of Cluster Analysis of Late Magdalenian Faunal Data

Fig. 7.1

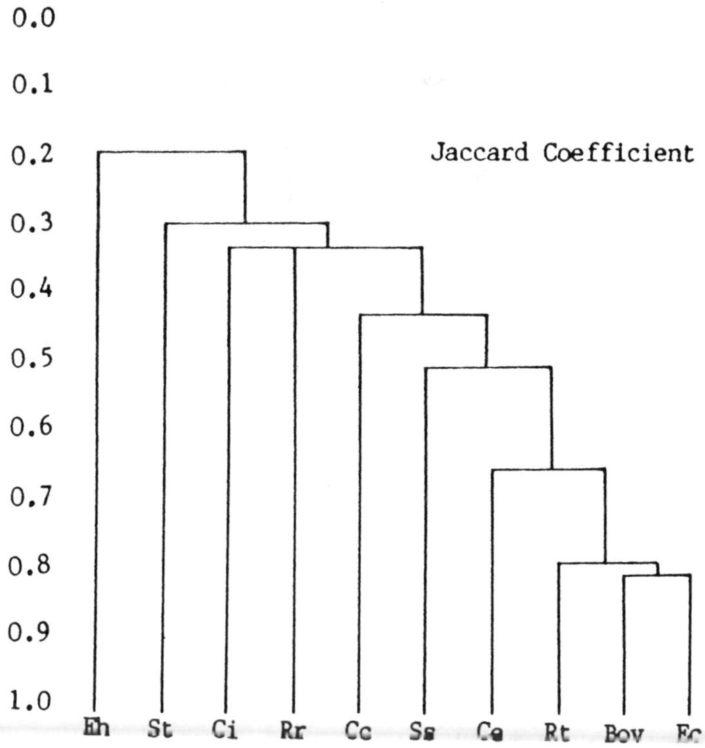

Jaccard Coefficient of Similarity

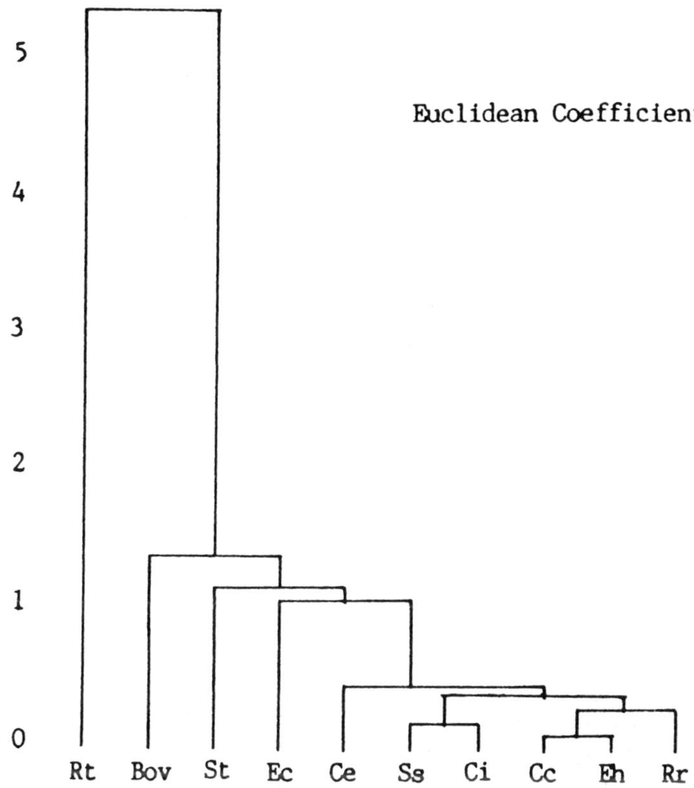

Euclidean Coefficient of Distance

Euclidean Coefficient of Distance.

$$\sqrt{\Delta_{jk}} = \Sigma(X_{ij} - X_{ik})^2$$

where X_{ij} = frequency of sp.i in assemblage j
X_{ik} = frequency of sp.i in assemblage k

Jaccard Coefficient of Similarity.

S = a/(a+b+c)

where a = the number of positive matches (ie: present in both (of 2) samples)
b = the number present in the first 'sample'
c = the number present in the second 'sample'

The success of Cluster Analysis depends upon the assumptions made, the algorithm used and the choice of coefficients of similarity or distance, different techniques producing different results - results which to some extent reflect the statistical procedures followed rather than the data employed. For example, in the present study, use of the Euclidean coefficient with Upper Magdalenian data isolates reindeer, as one would expect given its overwhelming abundance in much of the region (see fig. 7.1). Using the Jaccard Coefficient of Similarity (see Shennan 1988) with only presence/absence data, no account is taken of the frequency of species - only of their occurrence. In this instance equal importance is attributed to taxa whether representing 95% of the fauna or only 5%. Thus the reindeer fails to emerge as an important isolated element. Instead it is grouped with horse, bovids and red deer, losing its overall primary position.

We can see, from bivariate and multivariate correlation analysis, that three broad sets of environmental conditions are isolated during the Upper Palaeolithic prior to the Upper Magdalenian. Firstly, a mild, temperate woodland group emerges which comprises red deer, roe deer and boar, with the occasional addition of chamois, occupying forest edges and nearby pastures. This group may also be seen during the Upper Magdalenian, based upon the association of red deer, roe deer and boar. During the earlier period we see the addition of ass, during the later, horse (both being equids).

Cold open conditions are implied by the existence of significant correlations between reindeer, saiga, woolly rhinoceros, horse and bovids (prior to the Upper Magdalenian). In a recent study Delpech et. al. (1983) identified two open environmental types, namely the open, arctic and open, non-arctic. Two of the three species which she considers to reflect open, arctic conditions have, in the present study, been placed in a separate montane or cliff-face (skeletal) group, i.e.: chamois and ibex. These species emerge as members of component 3 during the Upper Magdalenian and component 2 prior to this.

During the Upper Magdalenian two open environmental types are identified in addition to the montane group. We have a moist (oceanic) tundra group which is dominated by the reindeer (Remmert 1980) and a cold, steppe (meadow) environment in which horse, saiga and bovids (primarily bison) thrive. As was noted above, the primary role

Fig. 7.2

Dendrogram of Upper Magdalenian Faunal Assemblages from South West France

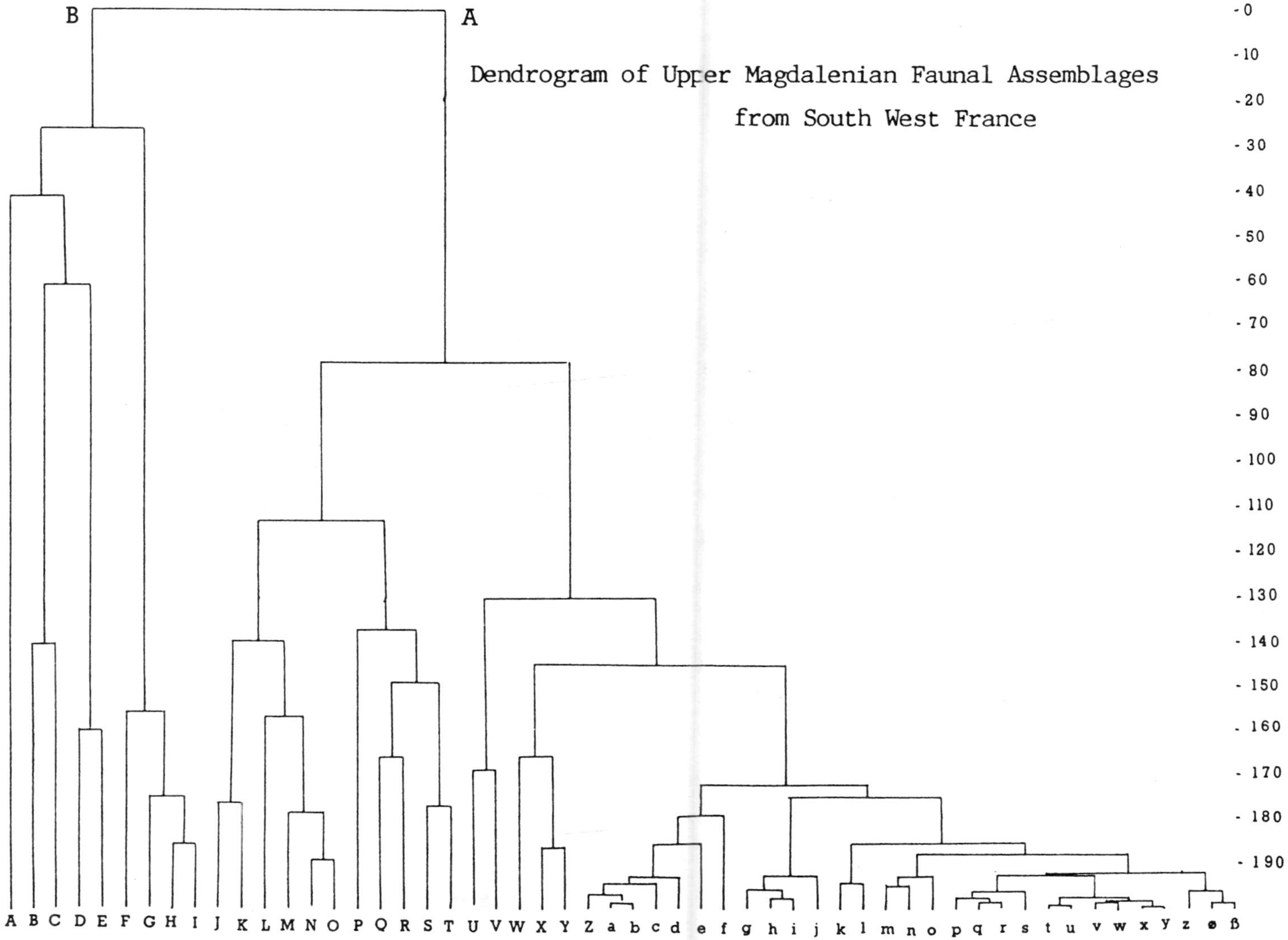

Key to Upper Magdalenian Sites in Figure 7.2

A Rond du Barry 53
B Grotte des Fees
C Roc de Marcamps
D Fontarnaud
E Faustin
F Fongabon 4-6
G Fongabon 3
H Fongabon 2
I Fongabon 1
J La Mairie
K Le Morin Bi
L Rond du Barry 54
M Le Morin IV
N Le Morin Aiii
O Le Morin Ai-ii
P Gabillou
Q Bois Ragot b
R Bruniquel
S Montmorillon
T La Madeleine 14
U Ste. Eulalie I
V Roc d'Abeilles
W La Mege
X Le Morin ?
Y Le Morin BII
Z Gare de Couze G_0

a Gare de Couze G
b Gare de Couze E
c Gare de Couze D
d Gare de Couze C
e Gare de Couze B
f Ste Eulalie 3
g Bois Ragot 6
h La Madeleine 17
i La Madeleine 16
j La Madeleine 15
k Cap Blanc
l La Madeleine 13
m Gare de Couze F
n Gare de Couze A
o Le Flageolet II (IX)
p Gare de Couze G_1
q Gare de Couze H
r La Madeleine 6
s La Madeleine 5
t La Madeleine 9
u La Madeleine 8
v La Madeleine 11
w La Madeleine 10
x La Madeleine 7
y La Madeleine 4
z La Madeleine 12
Ø Le Flageolet II (VIII)
ß La Madeleine 3

of the reindeer is illustrated in fig. 7.1 where, among Upper Magdalenian data, the species emerges as an isolated element, dominating the other taxa encountered, while the division of open communities can also be seen in fig. 7.2, in which the results of a site Cluster Analysis are presented. Group A (subdivided) may be seen to represent tundra-associated assemblages - perhaps of the Vezere and Dordogne interfluve and valley landscape, while group B denotes sites of a western distribution at which steppic elements predominate.

The relationship between the various components may be seen in figures 7.3 to 7.6 in which the most important results are shown. It may, for example, be seen that, in fig. 7.3(a) and 7.3(b) (Pre-Upper Magdalenian, p.c.1 'woodland' & p.c.5 cold, open grassland with some associated riverine or gallery (wooded) vegetation; Upper Magdalenian, p.c.1 'temperate' woodland & p.c.2 moist oceanic tundra), species 2, 5 and 7 (red deer, roe deer and boar) are located in close proximity, as one might expect from their relatively high correlations. Species close to the origin play little part in the 'two-dimensional summary' (Doran and Hodson 1975:195) whereas those occurring far from the centre are playing a more important role. In both cases therefore the woodland clusters may be significant, the association of Pre-Upper Magdalenian roe deer and boar (5 and 7) being emphasised by their close positions in fig. 7.3(c) as it is in fig. 7.4(a) in which Pre-Upper Magdalenian woodland and cold, herbaceous parkland components are presented.

In fig. 7.4(b) boar and roe deer are again closely associated, clustering with reindeer and mammoth, all four taxa preferring moister conditions than are represented by component 3 (cold steppe and dry, open herbaceous parkland). During the Upper Magdalenian only species 1 and 9 (reindeer and chamois) display negative woodland correlations, as can be seen in fig. 7.4(c), chamois and ibex (9 and 6 respectively) emerging as the primary species in the montane communities. Ass, roe deer, horse and saiga all display negative correlations with montane conditions, preferring, as they do, flat terrain.

In fig. 7.5(a) two groups of Pre-Upper Magdalenian species emerge. Those with high woodland values (red deer (2), ass (10), roe deer (5) and boar (7)) comprise one group, red deer and ass having the higher p.c.2 values. The other group comprises horse (4), bovids (3) and woolly rhinoceros (11), each having negative values of p.c.2. (montane). Given the positive (but low) woodland values, flat, dry, forest-steppe may be indicated. These three species (especially (3) and (11)) again cluster in fig. 7.5(b), representing cold, open grassy vegetation on flat terrain, while roe deer and boar, although found on flat terrain, prefer less open conditions.

In fig. 7.5(c) those taxa which are associated with montane or rockface conditions are of major importance, others being either 'insignificant' or negatively correlated. Equids (horse and ass) are negatively correlated with montane conditions although they display more significant (positive) correlations with p.c.4 (dry valley slopes).

In fig. 7.6(a) we see evidence of a tundra/steppe or cold-moist/cold-dry division of fauna which will become apparent in the next

Scattergrams showing Principal Component Relationships

Fig. 7.3

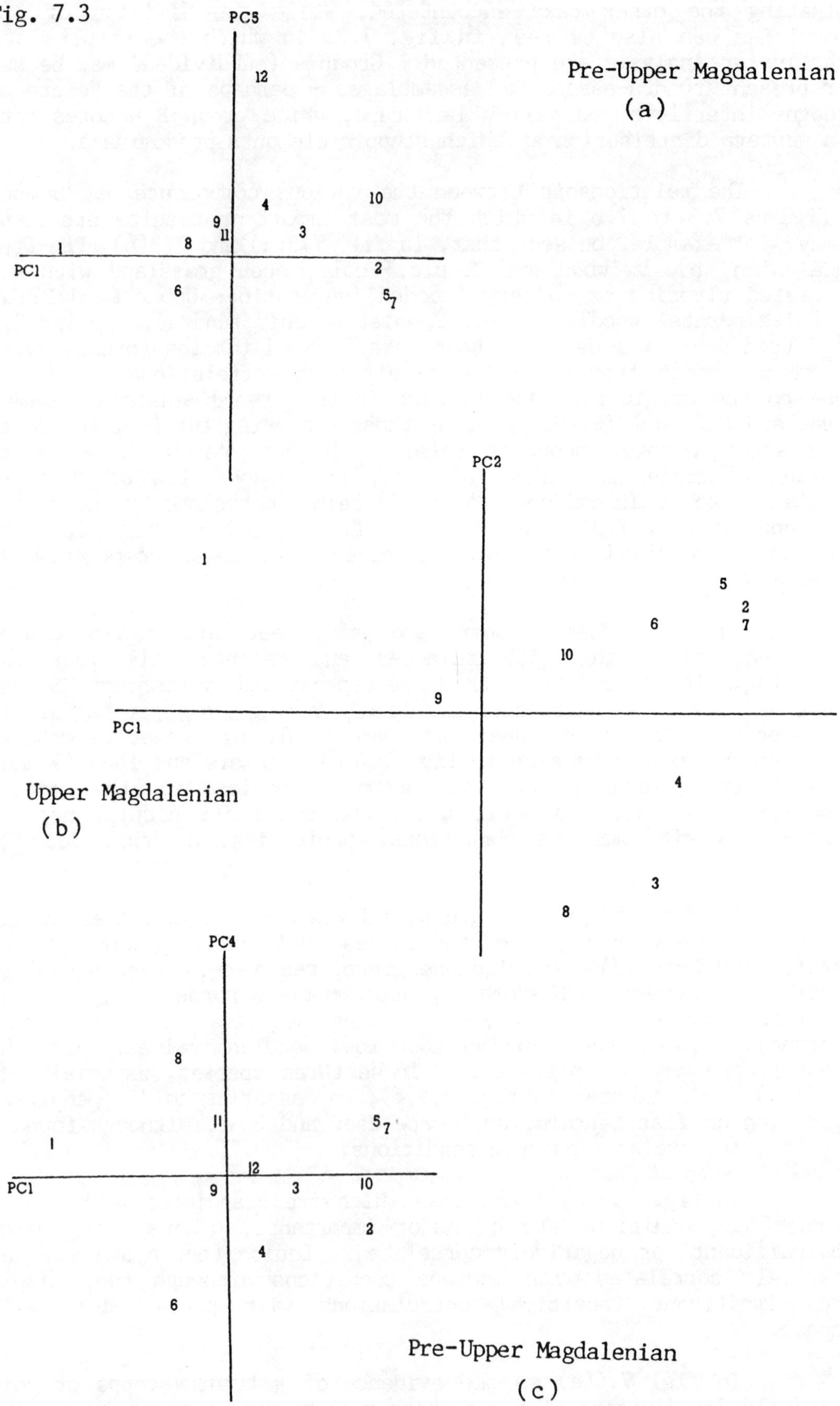

Pre-Upper Magdalenian (a)

Upper Magdalenian (b)

Pre-Upper Magdalenian (c)

Scattergrams showing Principal Component Relationships

Fig. 7.4

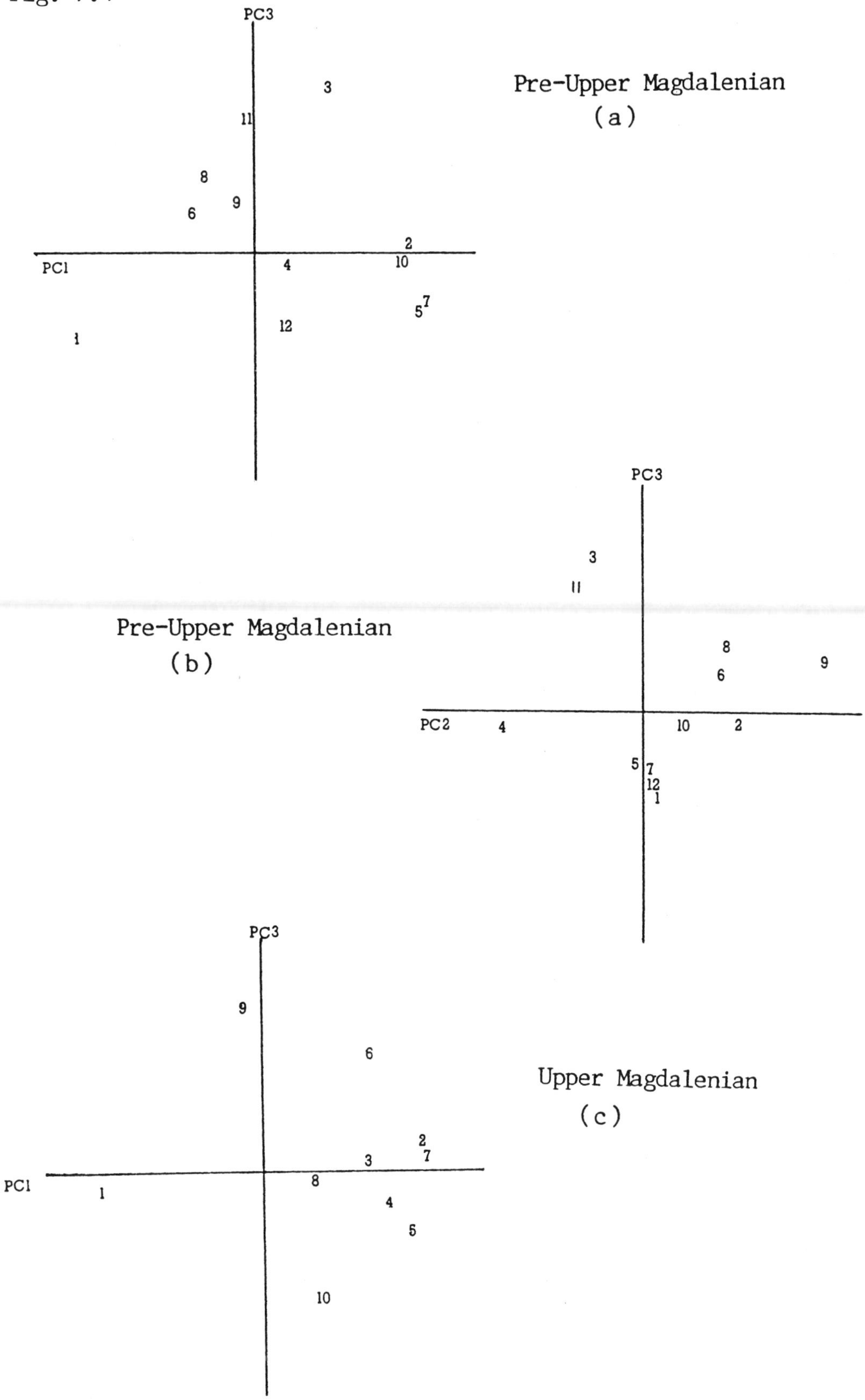

Pre-Upper Magdalenian (a)

Pre-Upper Magdalenian (b)

Upper Magdalenian (c)

Scattergrams showing Principal Component Relationships

Fig. 7.5

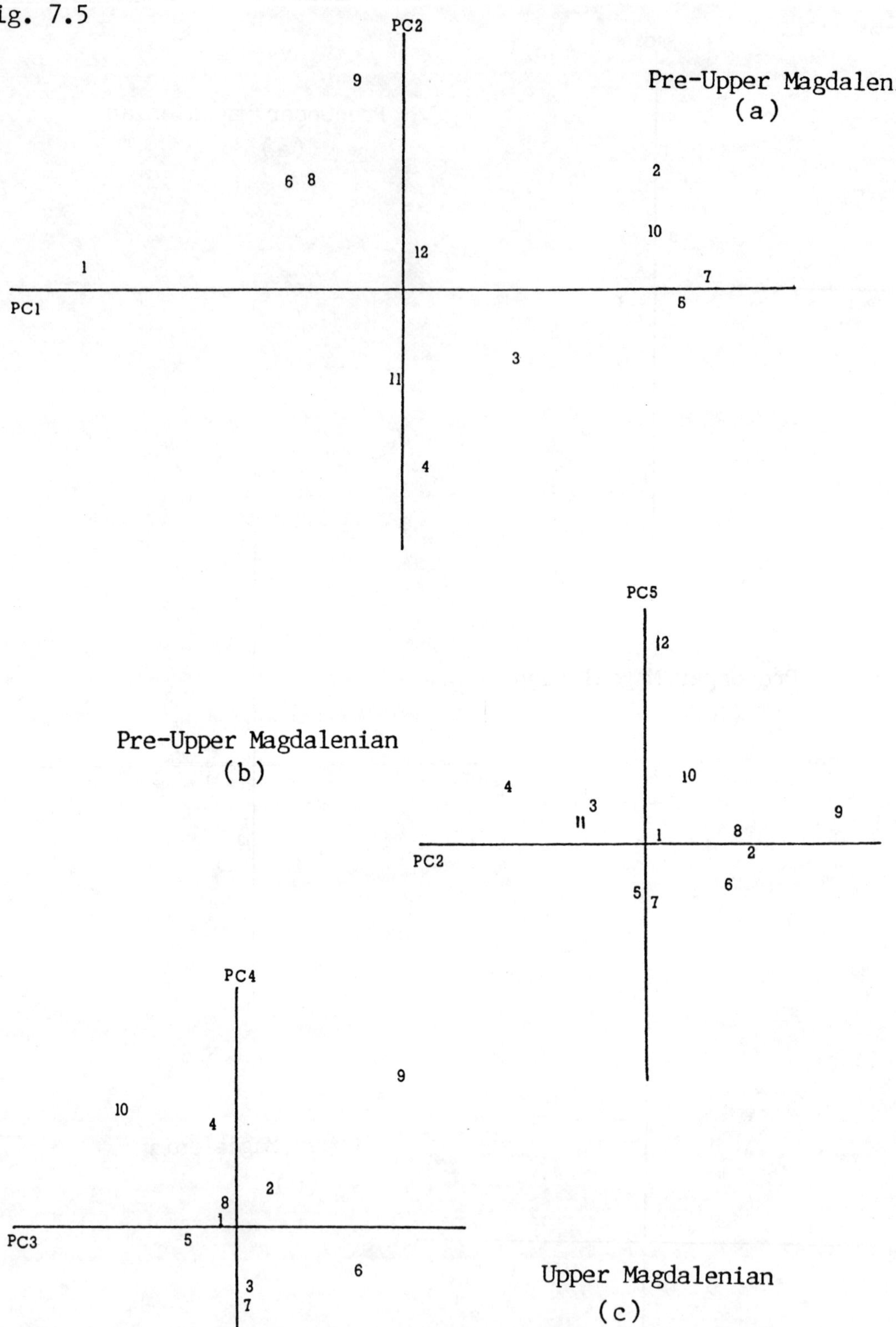

Pre-Upper Magdalenian (a)

Pre-Upper Magdalenian (b)

Upper Magdalenian (c)

Scattergrams showing Principal Component Relationships

Fig. 7.6

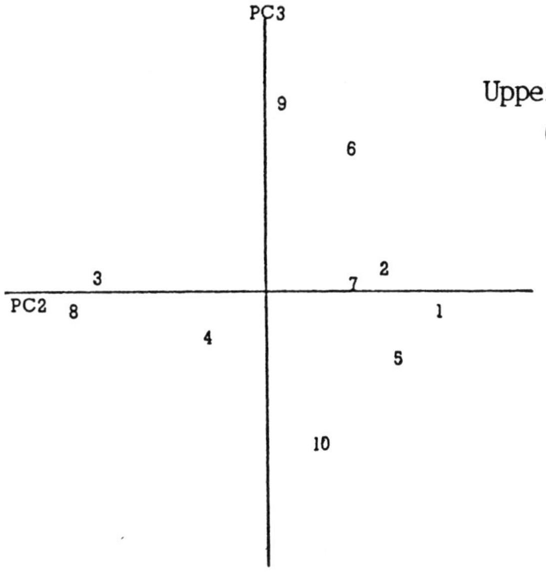

Upper Magdalenian
(a)

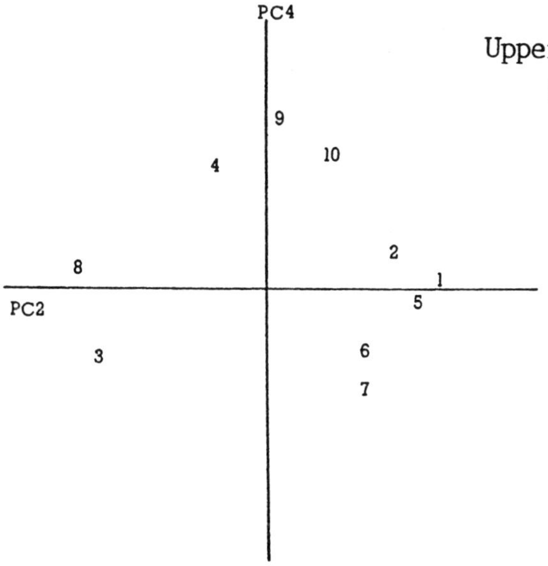

Upper Magdalenian
(b)

chapter, namely the occurrence of a cluster of broadly steppe species (saiga, bovids, horse) in opposition to the tundra reindeer and mosaic populations of red deer, roe deer and boar. This dichotomy of species is also seen in fig. 7.6(b). Meanwhile, fig. 7.6(a) shows the relative isolation of ibex and chamois as a montane or skeletal landscape component, only chamois emerging in fig. 7.6(b).

Having considered our data by plotting 'faunal scores', each occurrence of faunal data (ie: site or level) may be considered in relation to selected components. A combined score is obtained for each occurrence (relevant component weights and standardised scores being multiplied by each other and totalled) and may be employed as co-ordinates allowing the production of further summary plots such as those seen in figs. 7.7 to 7.9. It is possible, on these plots, to isolate groups of sites from which the faunal data may be seen as reflecting certain environmental conditions. For example, during the Pre-Upper Magdalenian those occurrences which are found in the upper portion of fig. 7.7(a) may be associated with relatively dense tree growth. Those sites occurring in the lower half of the diagram (68.75% of occurrences) reflect open conditions, those to the right being more characteristic of the mosaic of karst and valley vegetation in the Perigord.

In fig. 7.8(a) woodland communities are divided according to the degree of cover (or openness) rather than topography as in fig. 7.7(a) and 7.7(b), only seven occurrences being found in 'forest' as opposed to more open woodland conditions. Steppe and scrub conditions are identified, based on the lack of woodland, steppe occurring where conditions are more open and ground flora (grasses, forbs, Compositae etc) predominates. Most occurrences fall within the scrub zone - typified perhaps by a relatively consistent shrub component in addition to the abundant ground flora. During the Upper Magdalenian, when open conditions predominate (see fig. 7.8(b)), open woodland or taiga occurrences outnumber those in the forest zone.

In conclusion, throughout the Pre-Upper Magdalenian, faunal assemblages which may be deemed to represent open environments outnumber those reflecting forest community exploitation. Only when compared with the montane or skeletal component does woodland gain in significance.

During the Upper Magdalenian open conditions again predominate. Examination of fig. 7.9(b) shows sites which occur in areas where tundra conditions predominate. In fig. 7.9(a) we see evidence of the exploitation of 'lowland' (forest or grassland) species in preference to montane (woodland or meadow) communities. Only 33% of occurrences display positive values of p.c.3 (montane or skeletal), and much of this is the result of exploitation of small pockets of rockface communities in the south and centre of S.W.France. The major group of sites comprises open, flat conditions which are encountered away from the rockface landscape of the Lot and Vezere/Dordogne - namely cold, steppic grassland.

As noted above, in fig.7.9(b) we witness the importance of reindeer-dominated resource exploitation. Here flat, open conditions are indicated which may, to some extent, be equated with the tundra

Pre-Upper Magdalenian Biotopes, as defined by Combined Scores

Fig. 7.7

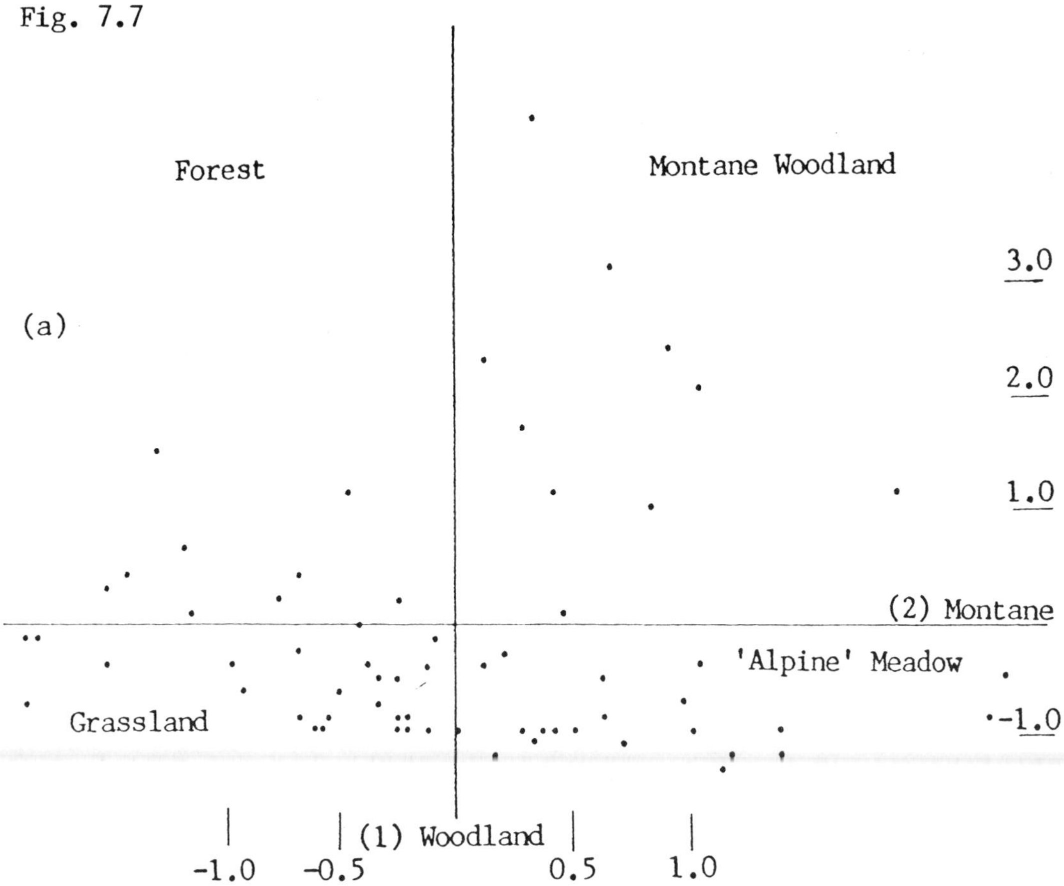

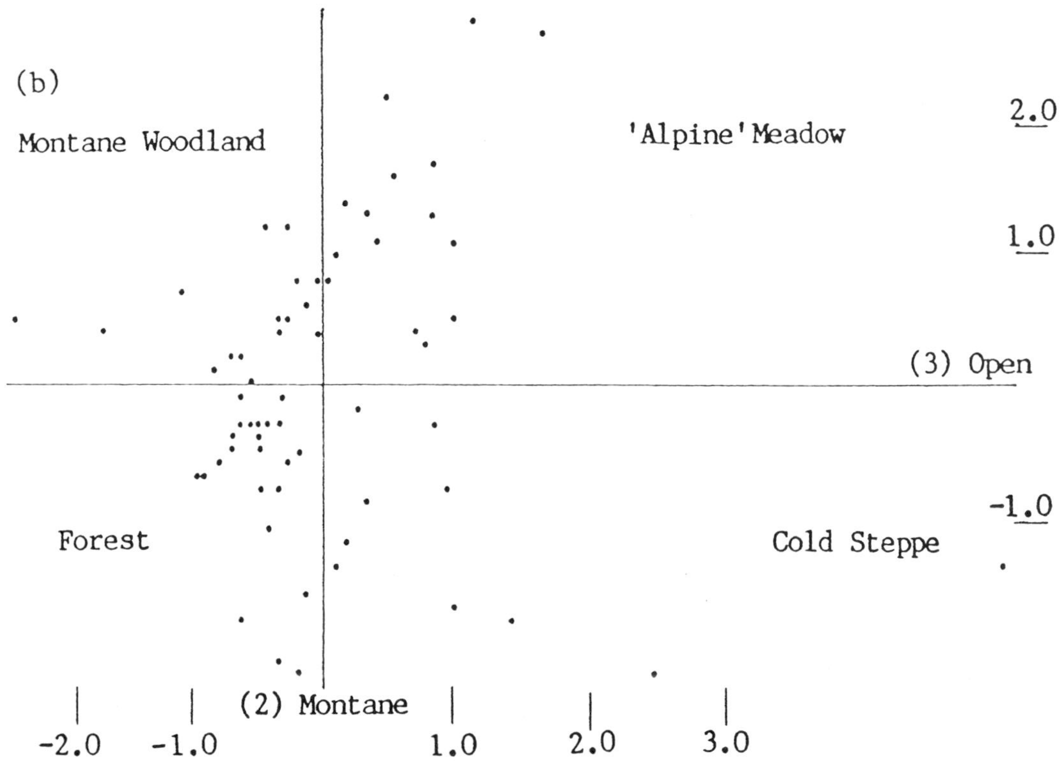

Pre-Upper and Upper Magdalenian Biotopes, as defined by Combined Scores

Fig. 7.8

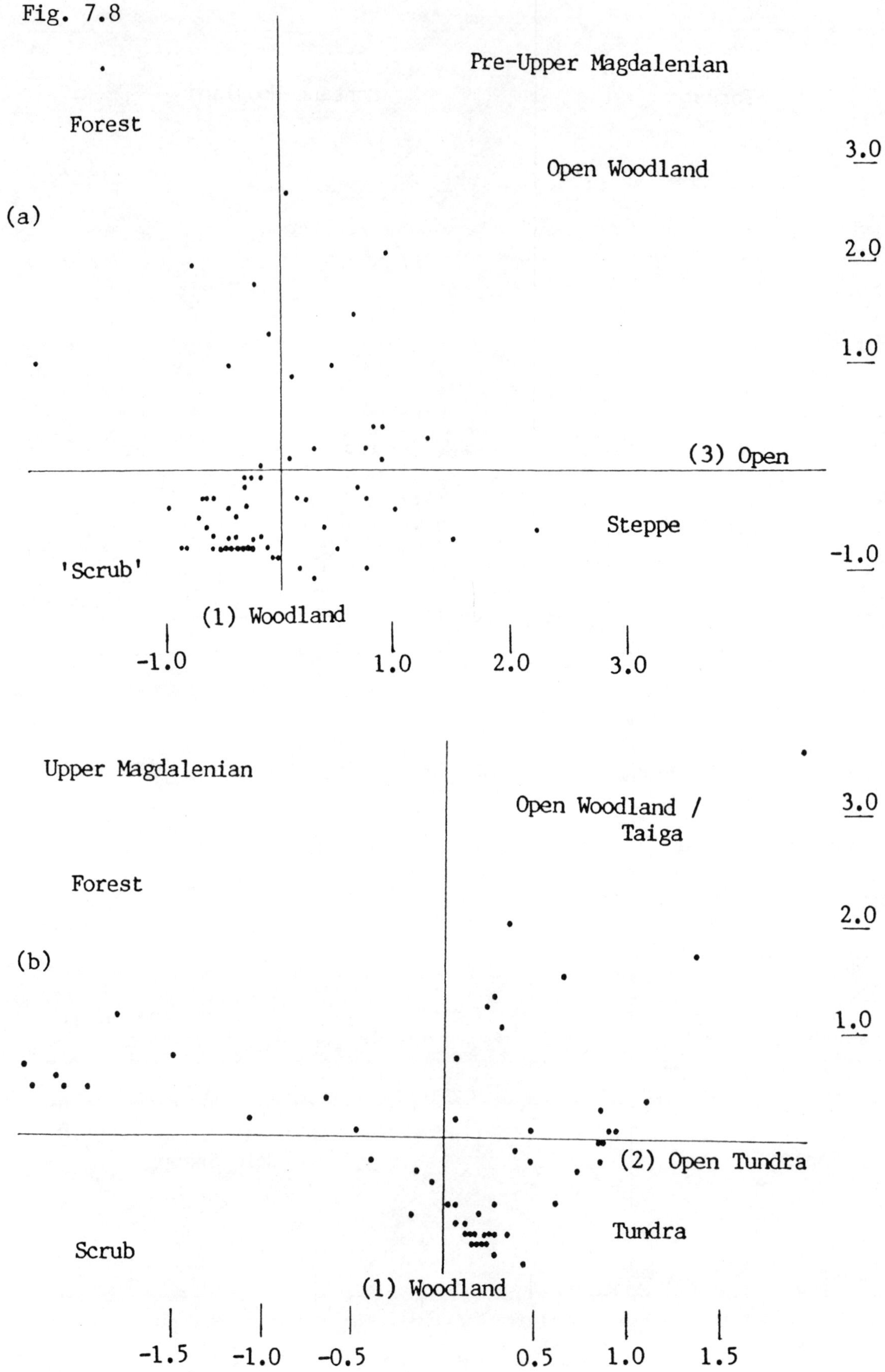

Upper Magdalenian Biotopes, as defined by Combined Scores

Fig. 7.9

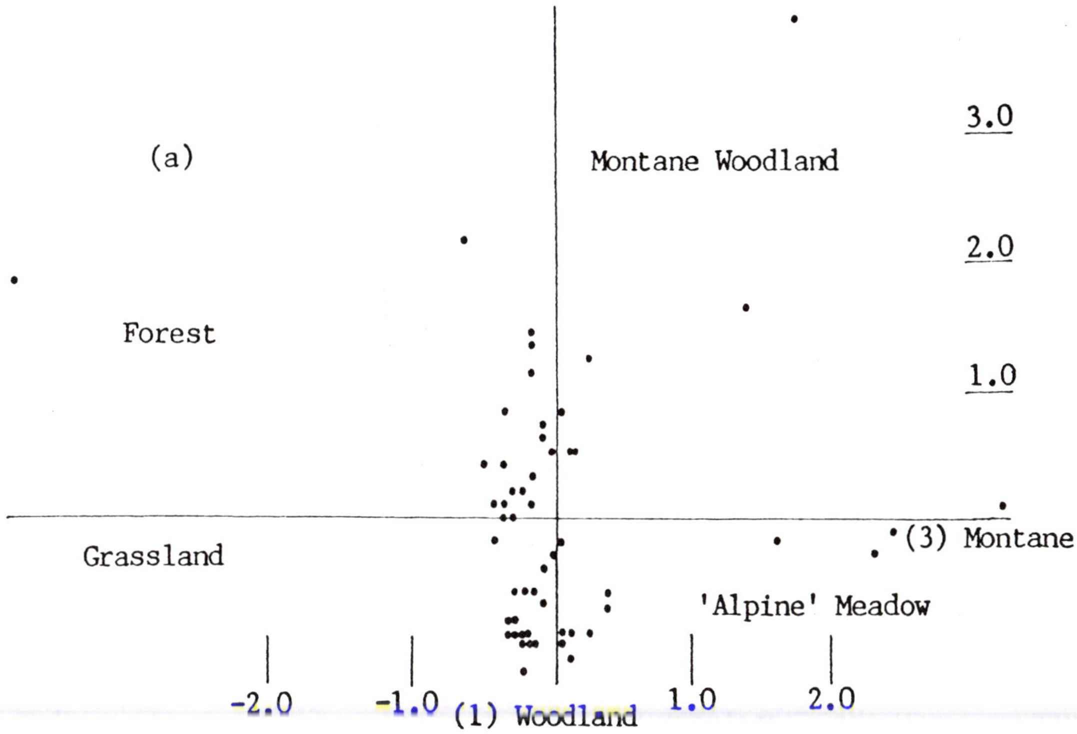

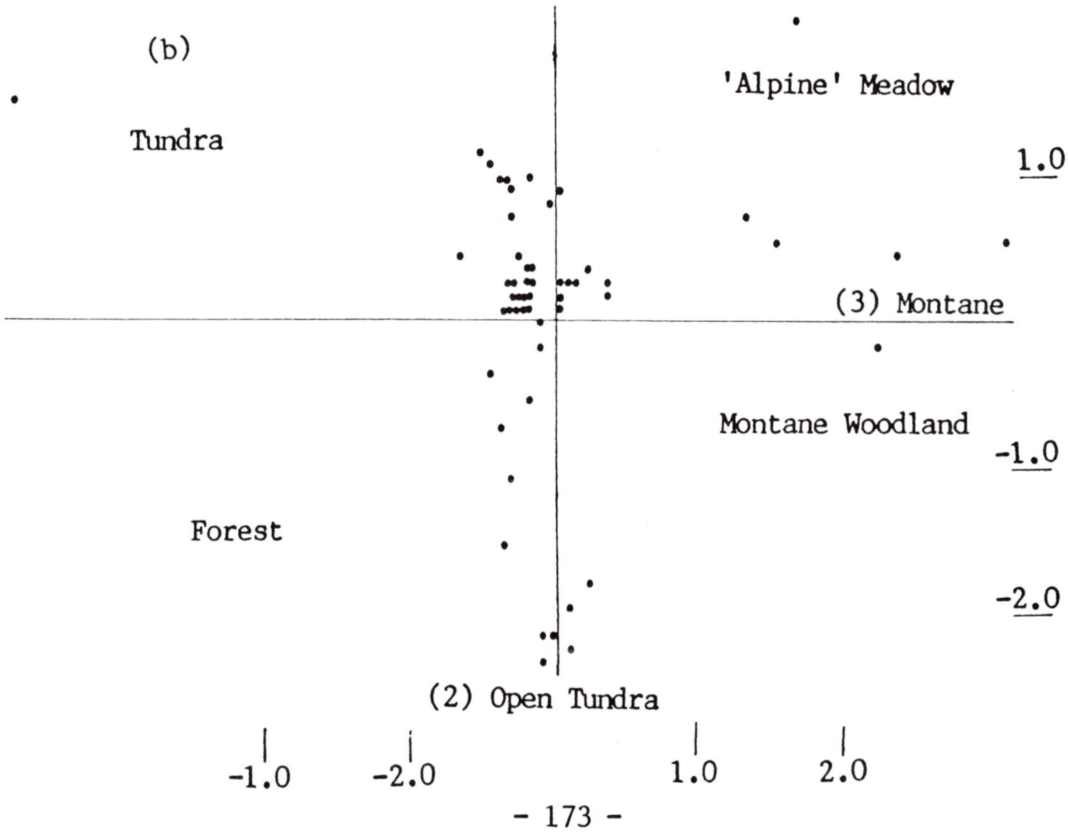

landscape traditionally associated with the Upper Palaeolithic of South West France.

In conclusion, some brief comments may be made concerning the chronological and geographical distribution of ecological components encountered during the Upper Palaeolithic.

Dividing the Central Zone (I) (as defined in Chapter 6) into two parts, Ia (Dordogne, Vezere etc) and Ib (Lot, Agenais ...) and considering them in conjunction with the Northern region (Zone II), marked similarities and differences in the chronological evolution of faunal assemblages emerge. Zones III and IV are not considered due either to their limited chronological range (Zone III) or to the restricted number of occurrences upon which patterning and associated conclusions would have to be based. Suffice it to say that a steppe community predominates in the Western region and a montane element occurs at every site - often predominating - in the East during the latter half of the Upper Palaeolithic. However, basing conclusions on only one eastern E.U.P. site (Chatelperron) is dangerous and generalisations will not be made. Let us simply remark that the consistent skeletal element is more likely to be a reflection of local topographic conditions than it is of climatic change.

In the Central and Northern Zones, tundra elements appear to predominate throughout the bulk of the Upper Palaeolithic prior to the development of woodland communities during the Azilian. The known exceptions are the Chatelperronian and Later Aurignacian in the north of the region, when mean values fail to reach 40% (see fig. 7.10) and steppe faunas predominate. At other times, the tundra component, which largely comprises reindeer and often mammoth, totals at least 50% (and often 75%) of the fauna - with the apparent exception of the Upper Perigordian in the southern half of the Central region. Even here, however, the tundra is the dominant group, despite low frequencies of reindeer at Battuts (as little as 0.60%). The high frequencies of reindeer commonly associated with the Upper Perigordian (IV-VI) and ProtoMagdalenian are, to at least some extent, replaced by a strong montane element adapted to the topography of the region.

By the Early Magdalenian the tundra and steppe elements comprise over 90% of the fauna, the importance of reindeer continuing to grow until its sudden decline during the Azilian. It is during the Early Upper Magdalenian that we see the probable disappearance of the mammoth, the last certain observation attributed to the Magdalenian III at Le Roc de Marcamps, although a single fragment of a molar has been recovered from Gare de Couze which Delpech (1983:227) cautiously dates to 12 000 B.P..

Reindeer-dominated, moist tundra environments, predominant until the final stages of the Palaeolithic, occurred, in the east of our region, on the interfluve surfaces of the Perigord (see fig. 7.11, showing the distribution of the predominant ecological groups during the Later Magdalenian). Cold dry steppe environments (characterised by horse, bison and saiga) also occur, on a smaller scale, within the same area, but more importantly to the west of this region where topography is more uniform and flatter. The rock-face component (ibex and chamois) displays a patchy distribution which reflects the nature of the local relief and altitude, while woodland pockets also display a

Fig. 7.10

Temporal Distribution of Ecological Groups during the Upper Palaeolithic of South West France.

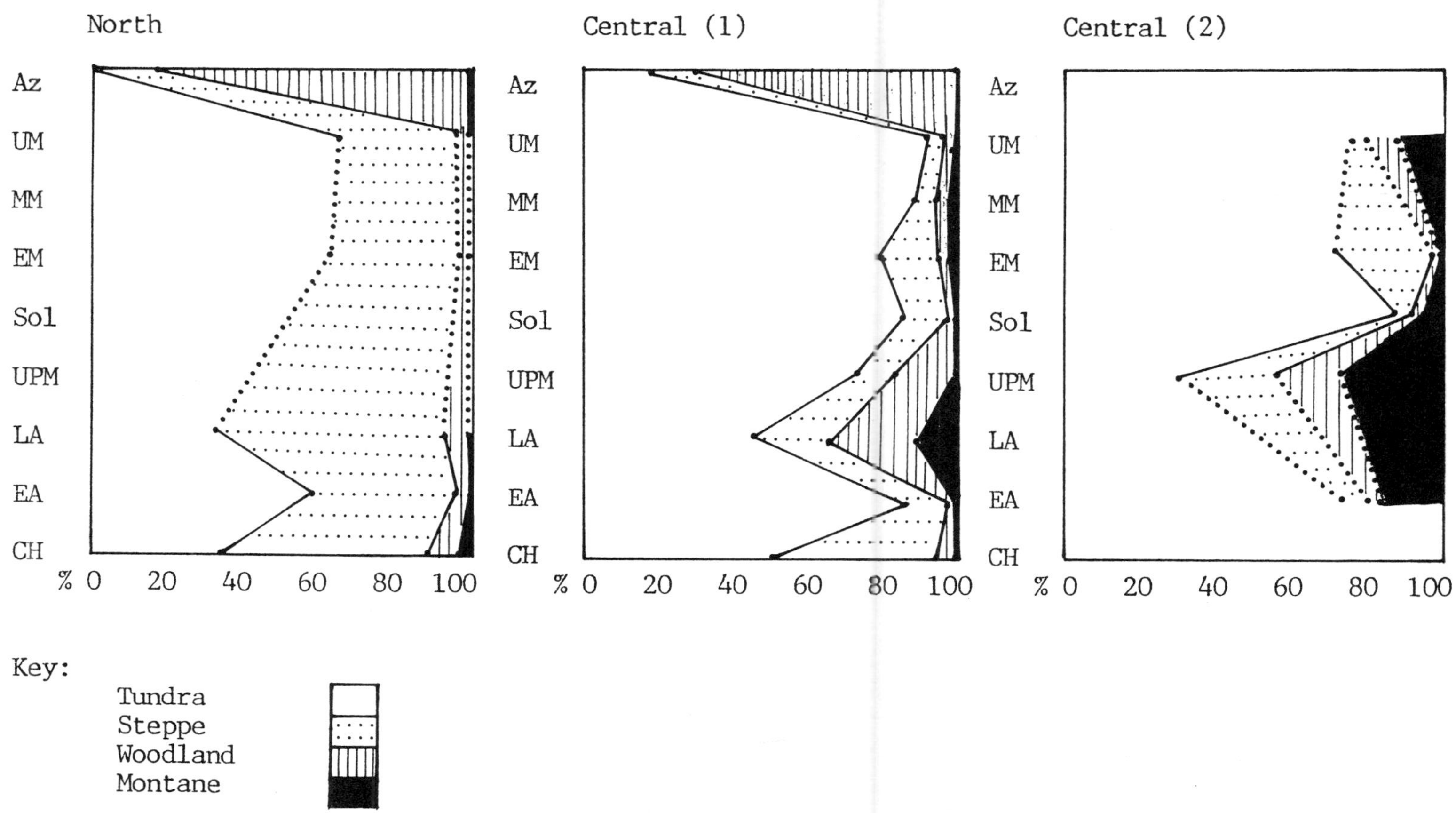

Fig. 7.11

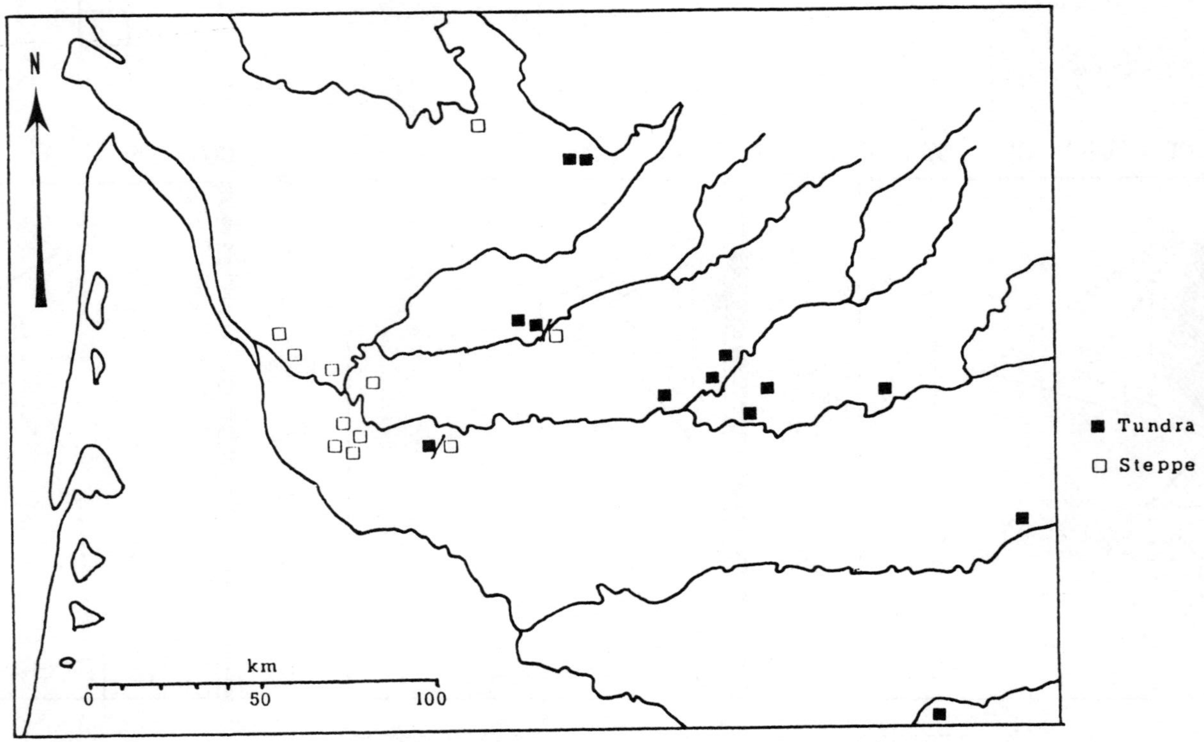

Map showing the Distribution of Dominant Ecological Groups during the Upper Magdalenian scattered distribution, implying the existence of small forested refugia. By the time we reach the Azilian the picture has changed considerably.

7.4 Non-Ecological Explanations.

Various explanations may be offered for those groups which do not fit a 'simple' ecological framework. For example, the mammoth may have been subjected to sporadic scavenging rather than systematic hunting, which would account for its occurring in low frequencies. The large carcase of a mammoth is of a size such that it must have been virtually impossible to transport to camp in one piece (ie. without a large, able labour force). Therefore only limited parts, deriving from high meat yielding segements of the body, may have been retrieved. If ivory were the major target, then only a small portion of the animal is likely to be recovered from any assemblage, especially if the ivory were obtained from a carcase some time after the animal had died. The ivory may represent a highly prized article which was sought after and retained, thus finding its way into a mixture of assemblage and environment types.

A model of multiple biotope exploitation may be offered in order to explain positive correlations and associations between such species as boar, ibex, ass etc.

Despite the fact that the ibex is commonly associated with mountainous and rocky areas it may be possible that the species descended to the Perigord and Aquitaine slopes during the winter, thus supplementing the resource base available during the winter months in our region. Alados (1985:138) considers two habitats which are occupied by the ibex, the second of which may be suitable for small boar populations:

1. **Open habitat**: low vegetation density, due to rocky surfaces with oak and other bushes in hollows, or due to grassland with few or no trees.

2. **Closed habitat**: high vegetation density, bushes and trees being less than 8.6 m apart.

Positive correlations between ibex, roe deer, boar and red deer during the Upper Magdalenian, indicate that two topographic zones may have been exploited simultaneously from one site. Given the location of a cave or abri (from which most faunal material is derived) on or close to a cliff-face along a river-bank, a minimum of two ecological groups may have been taken:

1.Valley bottom, rich vegetation (woodland, marsh) species, eg: roe deer, red deer, boar.

2. Montane, rock-face species, eg: ibex

In addition, those species found in the interfluve grasslands (reindeer, horse, saiga and bison) were also available for exploitation. Such a model should not be limited to the Upper Magdalenian, for sites such as Le Maldidier (or Roque Gageac) on the

Dordogne provide good examples of mixed assemblages in which a variety of biotope is represented during the Earlier Upper Palaeolithic.

Chapter Eight

SPATIAL VARIATION IN THE FAUNAL RECORD: UPPER PALAEOLITHIC ZOOGEOGRAPHY

"To do science is to search for repeated patterns, not simply to accumulate facts, and to do the science of geographical ecology is to search for patterns of plant and animal life that can be put on a map." (MacArthur 1972:1)

In this chapter an attempt is made to discern those spatial patterns which may be observed in the faunal data presented. Both qualitative and quantitative data are considered, derived from Aurignacian (Early and Later) and Later Magdalenian (III-VI) sources. These data were chosen for the present study due to their relative abundance, for, in order that we may observe patterning, a minimum size constraint was placed on the data set. It was decided that quantitative maps should be based upon a sample of at least 10 sites (locations). Only the above periods yielded sufficient material, at the time at which doctoral research was completed, from the area west of the Massif Central and north of the Pyrenean foothills. The addition of other data (both quantitative and qualitative, but especially the former) would of course enable similar pattern recognition study to be undertaken for other periods.

Patterns in species frequencies are sought in the belief that the distribution (spatial) of animal taxa is not haphazard, but, as has been demonstrated by biogeographers over the last century, ordered in some way. Alfred Russell Wallace (1823-1913) was amongst the first to draw attention to the apparent regularities in the distribution of animal, as opposed to plant, life (1860, 1876), his "The Geographical Distribution of Animals" (2 vols), appearing more than 20 years after the publication of de Candolle's plant geography text, "Géographie botanique raisonnée ou exposition des faits principaux et des lois concernant la distribution géographique des plantes de l'époque actuelle" (1855) and a full 70 years after von Humboldt and Bonpland's "Essai sur la Géographie des Plantes accompagné d'un Tableau Physique des Régions Equinoxiales" (1805).

Zoogeographic research is based on the premise that taxa occur in specific associations in particular contexts, the work described here being no exception. By recognising conditions in which a species consistently occurs and relating it to the distribution of the taxon in question, we should be able to shed some light upon the possible distribution of Upper Palaeolithic animal resources. If, as de Beaufort (1951) remarks, we plot species occurrence on a map, the area covered may be considered to represent the species' range - or area of distribution. When attempting to do this we are making the important assumption that species are adapted to a stipulated range of environmental conditions and will only occur where these requirements can be satisfied. Therefore their distribution is determined by such factors as temperature, humidity, topography, surface water and winds and is further limited by competition for resources and the threat of predation. However, human behaviour will distort the picture, hunter-gatherers perhaps transporting 'prey' (whether killed or scavenged) some distance from kill or find to camp. Nevertheless, in a region such as S.W.France where the resource base is believed to have been not only rich but also relatively predictable, long distance transport of prey, on a scale to seriously distort the picture, would seem to be unlikely.

In order to examine the spatial variations in faunal species abundance, large scale maps of the Aquitaine plain and Perigord region of South West France (1m x 1m covering the Dordogne, Charente, lower reaches of the Garonne, Lot and tributaries thereof) were constructed. All major and tributary rivers shown on the Geological Map of France (scale: 1/1 000 000) were recorded. On each of these the relative numerical (%) frequency of a stipulated species, during a selected chronological period, was recorded, the percentage value being noted - and circled where derived from M.N.I. counts. In situations where multiple layer sites are encountered all frequencies were recorded in stratigraphic order.

The quantitative maps produced were then summarised on smaller scale equivalents. Standardised coloured dots were used, each equating with a range of percentage values. The patterns emerging were finally summarised in the form shown in the present chapter, with the selected ranges of values showing patterning most clearly. The maps indicate only general tentative trends for the discovery of further sites and the analysis of additional data are quite likely to necessitate a reconsideration of the material.

In addition to the quantitative analysis, non-numerical maps were produced. These took three forms.

1. Presence/absence maps. For each period, the presence of a species at a site was recorded by a chosen symbol (●). These are here summarised at a reduced scale with the additional record of 'absences' (sites at which the species is not known to have occurred). The occurrence of an 'absence' tells us merely that the species is not recorded as present at the site in question. It neither tells us that the taxon was in fact not present at any time, nor that it was absent from the surrounding environment. For this reason the presences are more significant than are the absences.

2. Maps showing the occurrence of dominant animals, again colour-coded. These are summarised in figs. 8.15 and 8.26. When dealing with non-numerical data, usually deriving from pre-1945 publications, problems are encountered. So-called 'co-dominant' species are unlikely to have been present in exactly equal quantities but, as in the consideration of chronological patterning, lacking an alternative approach, their 'co-dominance' must be recorded.

3. Maps showing the distribution of numbers of species yielded by archaeological contexts. This value is employed as a gross indication of the diversity of assemblage structure, and is followed up by a more detailed discussion of diversity based upon numerical data (also mapped, see figs. 8.29 and 8.30)

Fig. 8.1

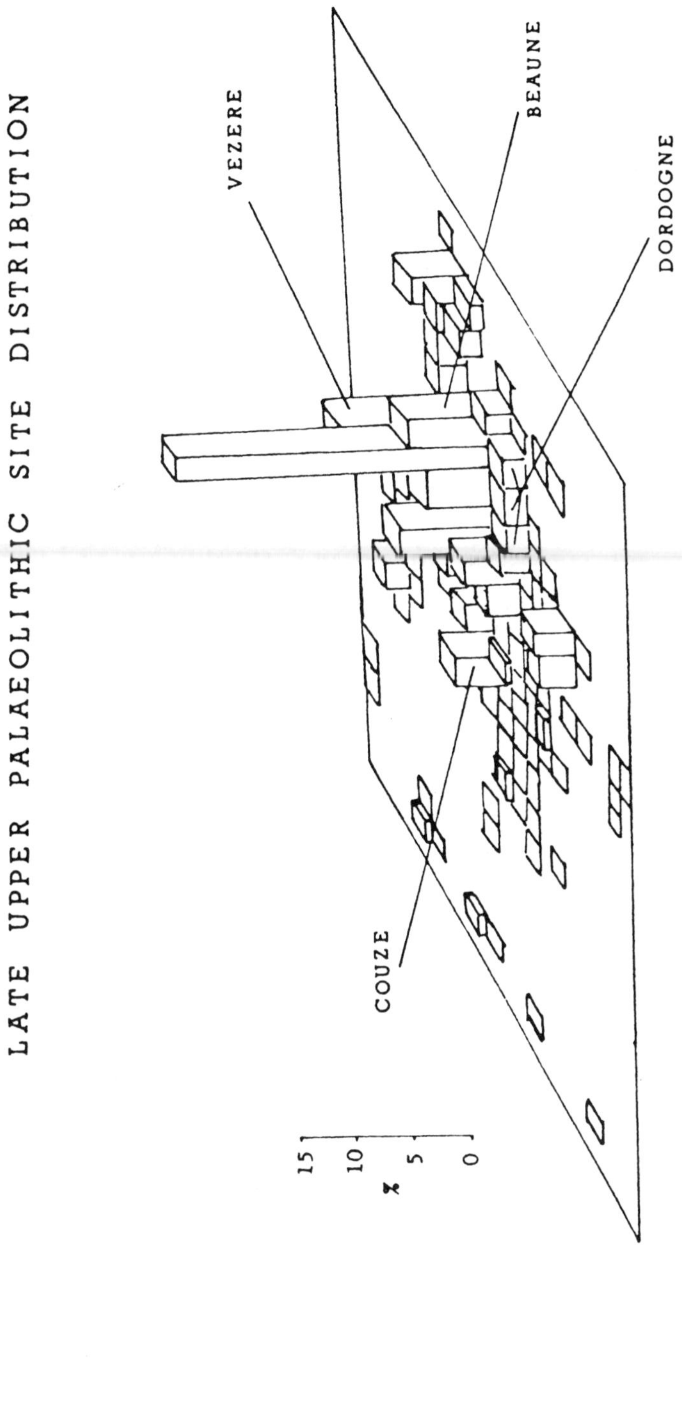
LATE UPPER PALAEOLITHIC SITE DISTRIBUTION
VEZERE
BEAUNE
DORDOGNE
COUZE
15
10
5
0
%

Qualitative maps were constructed for Upper Magdalenian and Aurignacian contexts. The latter period was often considered as a whole due to the fact that much of the qualitative data comes from early papers in which the period is not sub-divided, and where such division does occur the results are often dubious. Care must therefore be taken when using these data. However, complete exclusion of the material would be irresponsible - when there is no reason to doubt the taxonomic attribution reported by the author of the relevant publication.

Associated with the quantitative maps, a three-dimensional spatial plot of 'Later Upper Palaeolithic' sites known to the author at the time the analysis was undertaken. The plot was constructed using a grid (10km x 10km) overlying the base maps. The aim of the plot is to act as a reference point in the discussion (see fig. 8.1).

8.1 Aurignacian.

Due to the relative paucity of relevant, well dated material the decision was made to consider the Aurignacian as a whole, dividing the period into 'Early' (Basal and Aurignacian I) and 'Later' categories when suitable.

As fig. 8.2(a) shows, all Aurignacian occurrences from which faunal data have been obtained yield reindeer, indicating that the species was to be found over a wide geographical area, extending beyond the region covered in this study, for example into the Vienne where the species occurs at such sites as Cottes.

During the 'Early' Aurignacian (see fig. 8.2(b)) reindeer frequencies reach a peak in the Dordogne valley where the species makes up over 80% of the large herbivore remains. Throughout the region frequencies are high, for the reindeer is commonly the dominant species, gradually declining in number towards the north and west. The Early Aurignacian distribution displays a largely NW-SE orientation, broadly following the course of the main river valleys, ie: the Dordogne and Isle. In general, sites may be divided into two main categories, the vast majority of sites being found in the first of these:

1 - where reindeer exceeds 40% and almost invariably predominates
2 - where reindeer represents less than 40%, but may still be the major species occurring.

Sites in the first group are commonly found in the heart of the region (ie. the Dordogne and Vezere confluence and surrounding area), and are further subdivided according to species frequency: (a) greater than 90% and (b) above 40%, often exceeding 70% of the large herbivore total. In general therefore, reindeer seems to be most abundant in the centre and towards the east of the region, perhaps where a mixed, cold environment (tundra-like?) prevails.

During the Later Aurignacian reindeer frequencies peak in the south east of the region, although with reduced values, 83.8% at Roc de Combe, gradually declining in a north-westerly direction. The species remains a major component of faunal assemblages at Vezere and Dordogne

Maps showing the Distribution of Reindeer during the Aurignacian

Fig. 8.2

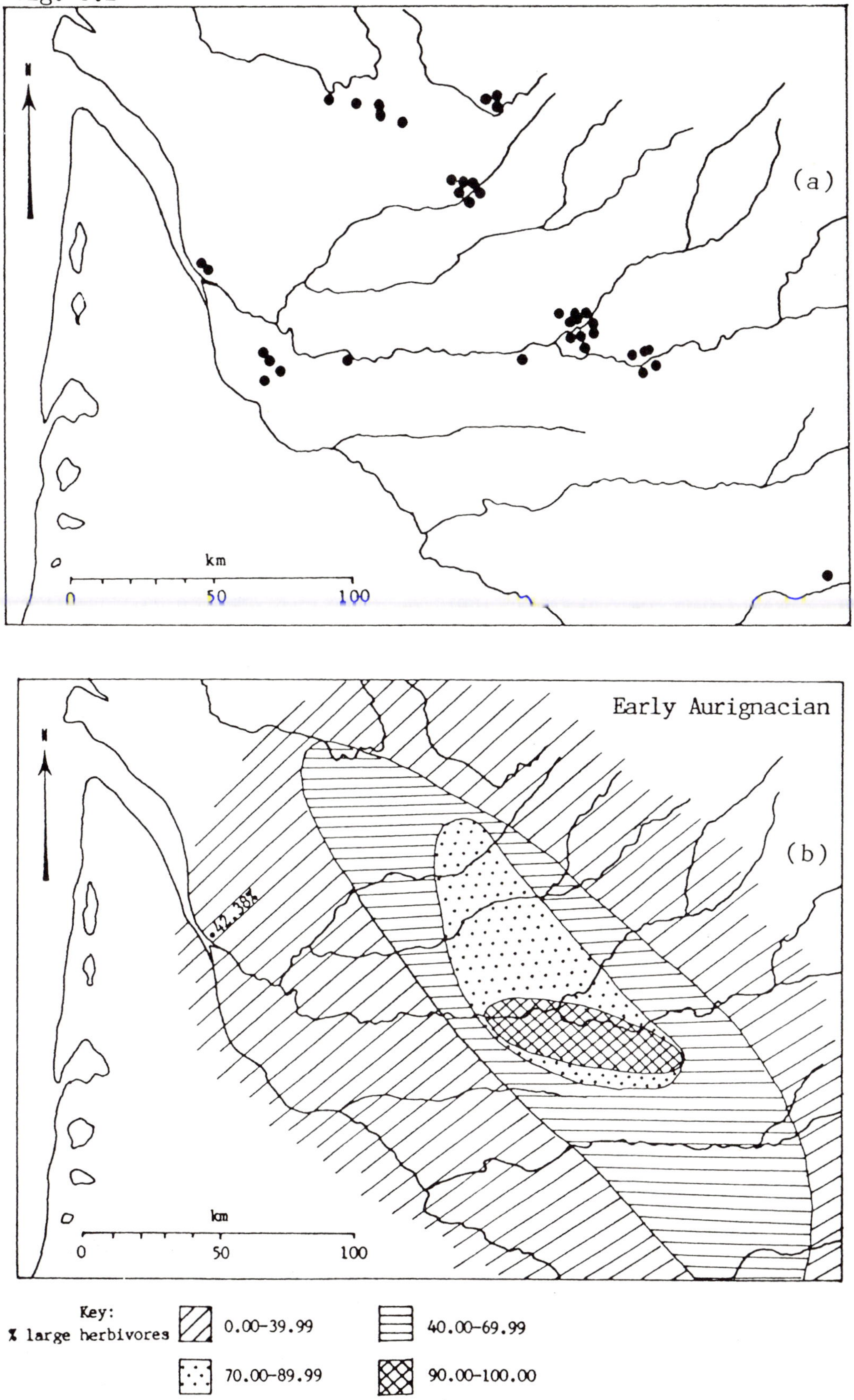

Map showing the Distribution of Reindeer during the Later Aurignacian

Fig. 8.3

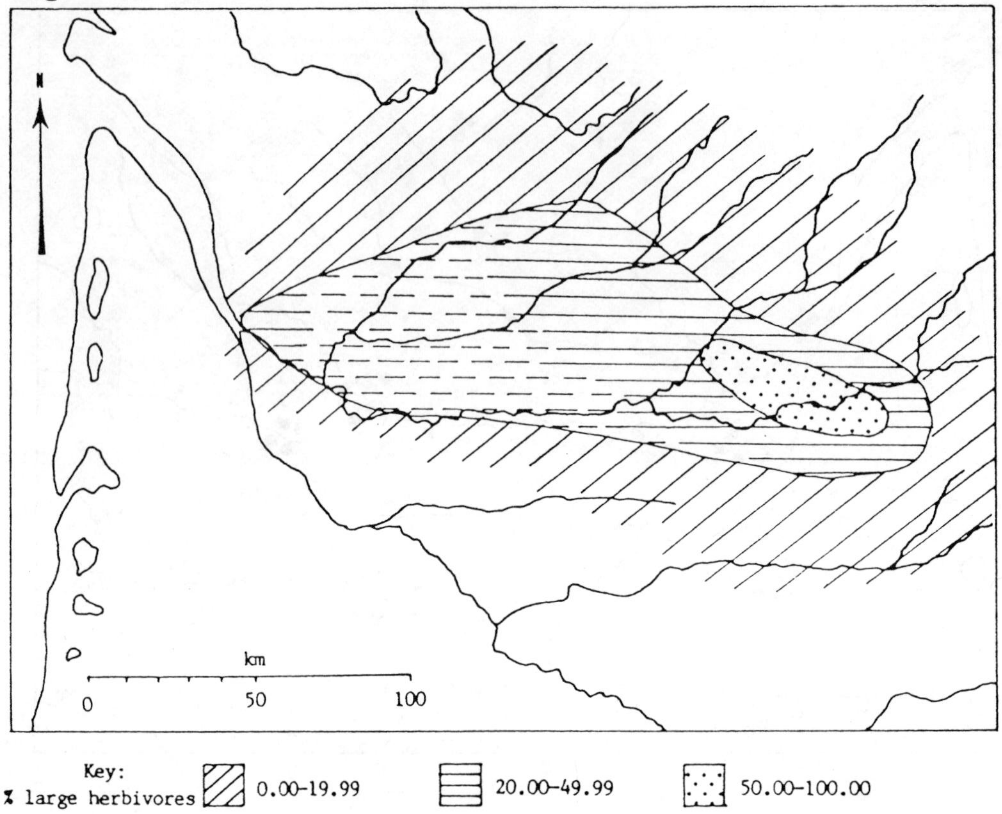

Map showing the Distribution of Bovids during the Aurignacian

Fig. 8.4

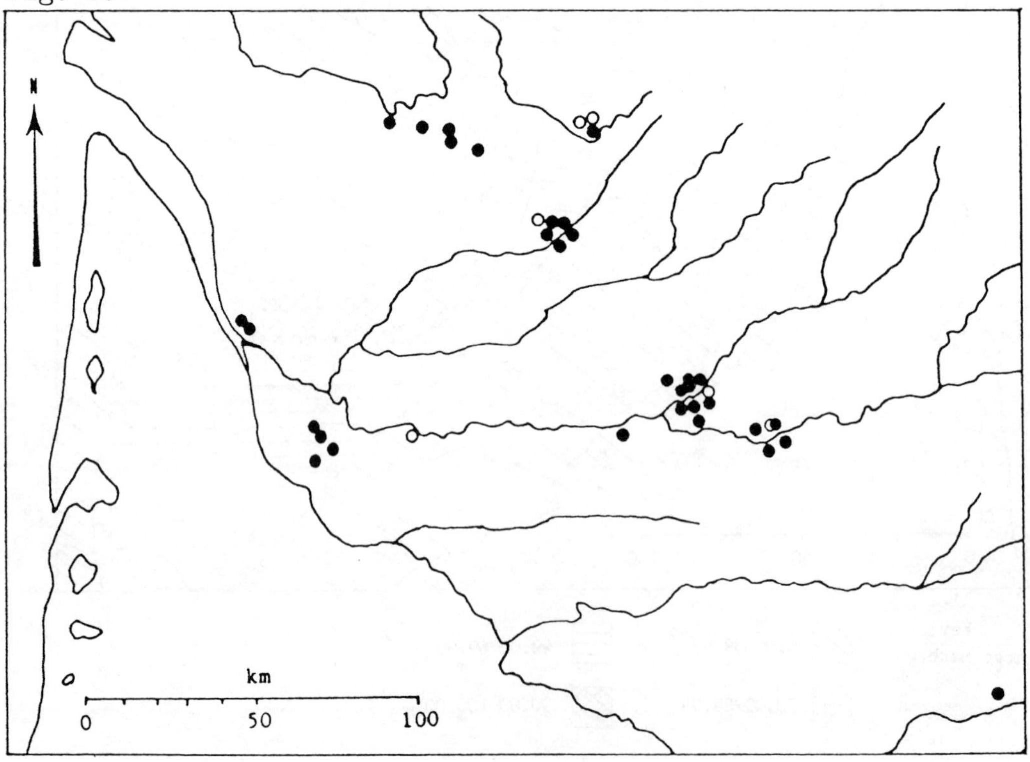

valley and interfluve sites, although it is noticeable that those sites at lower altitude (above river level) yield higher reindeer percentages. Towards the northeast and east percentage values appear to fall off more rapidly and, although the lack of quantitative data from Later Aurignacian sites to the south of the area shown in fig. 8.3 result in the distinct boundary shown, to the south percentages may fall more slowly or remain relatively high, especially in the Agenais.

Aurignacian bovids display what appears on initial inspection to be a distribution pattern broadly similar to that of reindeer (see fig. 8.4), although it does not equal the latter in terms of either frequency of occurrence or abundance, the species being absent at six sites, including at least one in each of the main clusters (north, west and centre). Despite the smaller totals of the species, the fact remains that bovids do form a significant part of the faunal record, for, as Brown (1988:59) remarks "rare species are usually as interesting to biogeographers as common ones."

Forming, as they do, only a relatively minor, but constant, component of the Early Aurignacian landscape (see fig. 8.5(a)), bovids are common and occasionally dominant at La Ferrassie and Caminade Est, the only two sites at which the species might be considered to play a major role. Both sites lie some distance from major rivers, La Ferrassie 3750m from the Vezere, Caminade Est 2750m from the Dordogne (White 1985). They are, however, both found in sheltered tributary valleys, with springs at a distance of 25m and 100m respectively, Caminade Est occurring in the wider of the two valleys (150m, as opposed to 100m in the case of La Ferrassie). It might be suggested that such valleys provided localised areas in which small, scattered groups of <u>Bison</u> sp./<u>Bos</u> sp. were available for exploitation as and when required. Solitary bison might, for example, seek shelter in such tributary valleys during the winter months, although for such speculation to be possible seasonality data, of the type presented for the reindeer by Gordon (1988), is needed for bovids and other species from Upper Palaeolithic contexts.

During the Later Aurignacian, average bovid frequencies peak in the Isle Valley at La Chevre (see fig. 8.5(b)). To the south of this area, frequencies fall to about 25% on the north side of the Vezere (again at La Ferrassie where the maximum value of 51% occurs), falling still further in the Dordogne valley and surrounding areas to values averaging approximately 3% if we omit Caminade Est from the discussion. Once again these two sites, Caminade and La Ferrassie, prove somewhat enigmatic, occurring away from the reindeer-dominated river valleys, while the occurrence of 27% bovids at Pair-non-Pair (Gironde) extends the range of the species westwards towards the Aurignacian coastal plain.

The horse occurs at all but one Aurignacian site in South West France (see fig. 8.6(a)), the absence being found in the north of the region. The occurrence of this species at every reindeer-yielding site in the Vezere and Dordogne area, at markedly reduced frequencies, suggests that it takes second place to the reindeer in a rank-order of species. It is also apparent that the taxon is of greater importance in the wider valley of the Dordogne than the Vezere and Lot, selecting, in addition, the more open, flatter regions of the Charente and Gironde.

Maps showing the Distribution of Bovids during the Aurignacian

Fig. 8.5

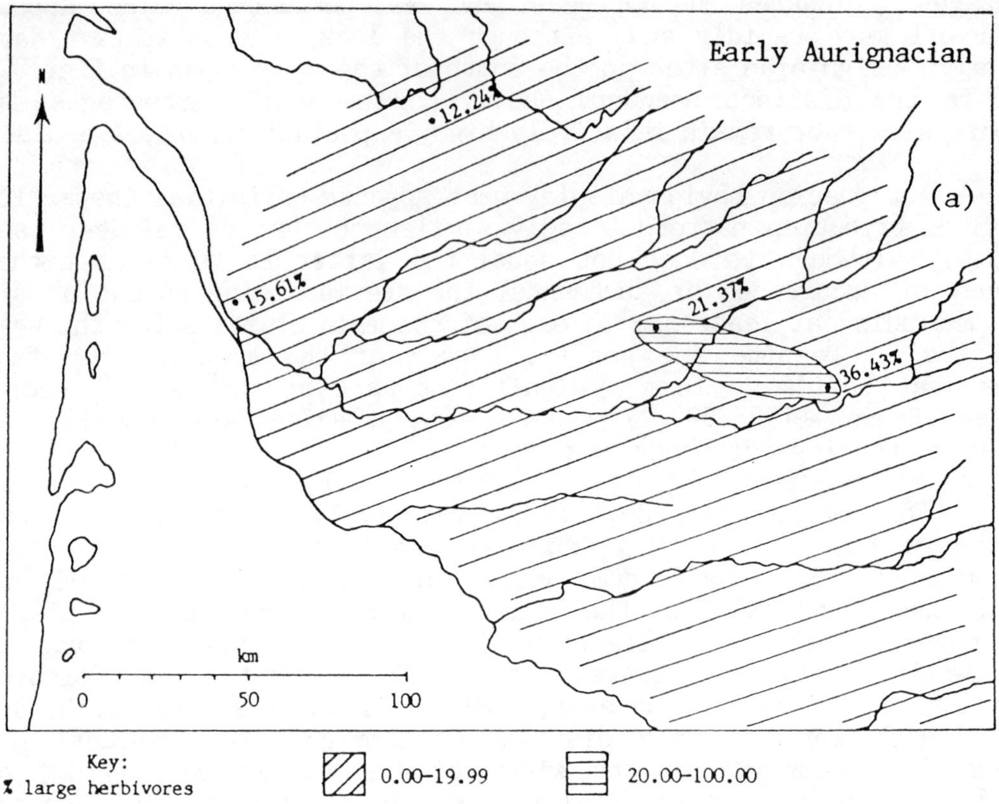

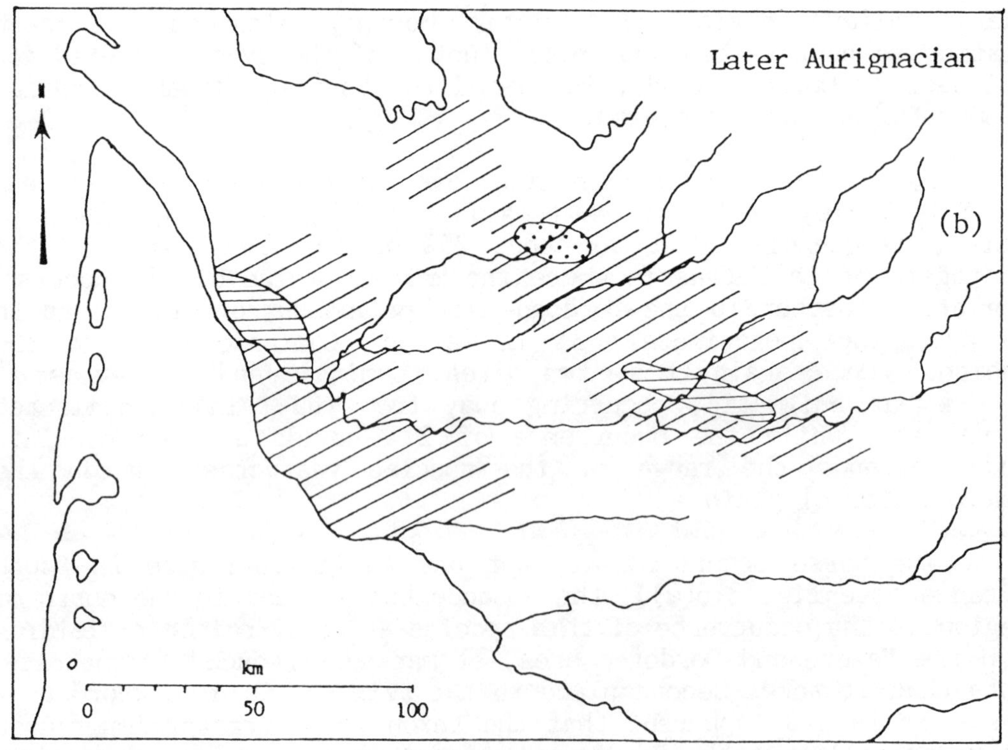

Maps showing the Distribution of Horse during the Aurignacian

Fig. 8.6

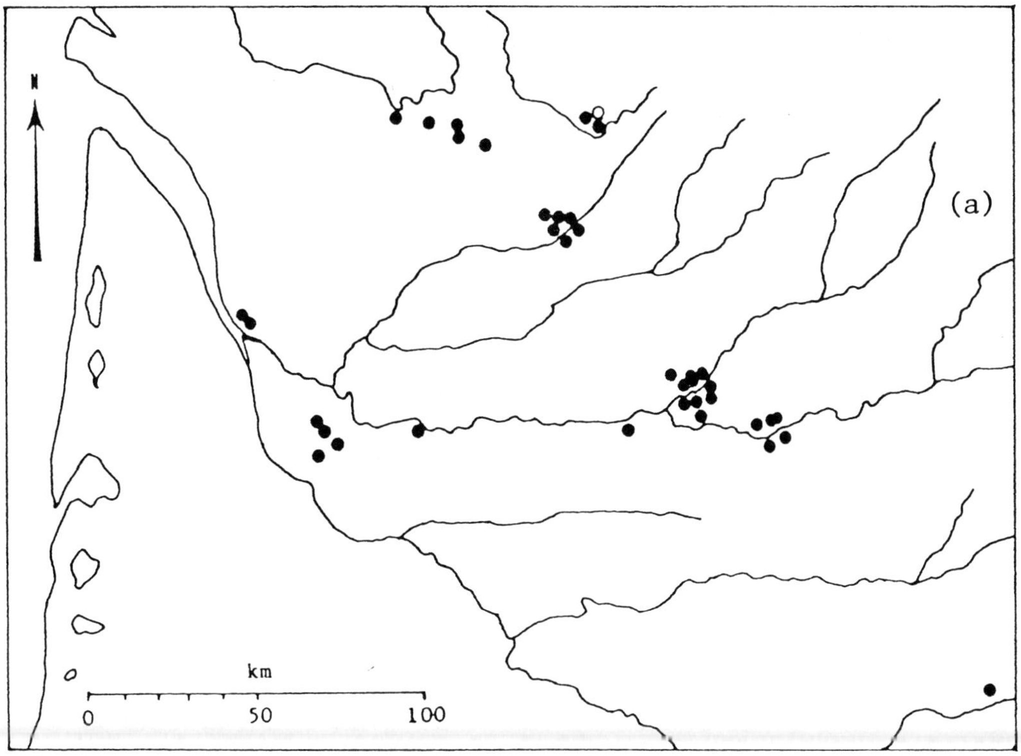

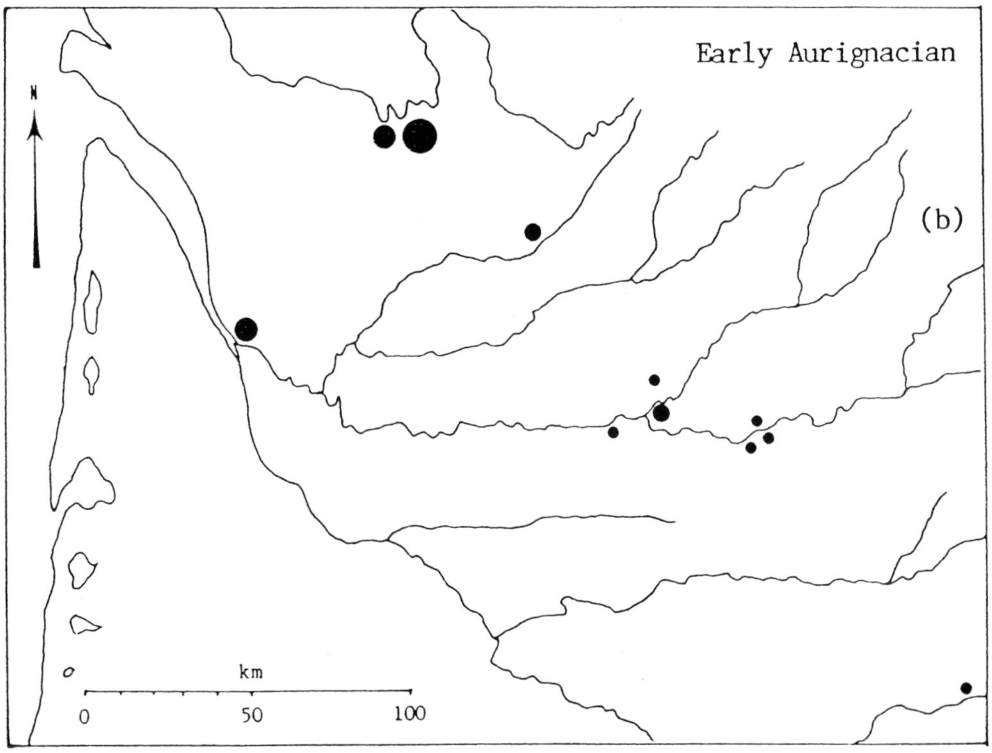

Key: • 0.00-9.99 ● 10.00-24.99 ● 25.00-49.99 ● 50.00-100.00

% large herbivores

A distribution pattern which is completely different from that of the reindeer and bovids is seen for the horse during the Earlier Aurignacian (see fig. 8.6(b)). Maximum values are found in the Charente region, frequencies exceeding 60% at Bourgeois-Delauny, and reaching 30% at Fontaury. In the west, at Pair-non-Pair, totals exceed 25% and reach 15% on the Dronne. Although mean frequency exceeds 10% at Pataud, in the Dordogne values fall dramatically to a regional mean of 5.80%, while at Battuts, just off the southern edge of the map, the horse total falls still further to 0.83% of the large herbivore count.

Given the apparent distribution of horse frequencies, it might be suggested that the Dordogne occurs on the edge of the range of the species. Frequencies are higher in the north and west, regions where the landscape may have been more suitable for the species, being relatively dry, open, flat and comparatively uniform. A species generally comprises a larger component of the fauna in that part of its range where conditions are most suitable than it does elsewhere, frequencies declining with distance, and associated environmental change from these 'best' conditions and the 'heart' of the range. Thus, perhaps we are looking at evidence of the exploitation of horse populations in both their principal range (to the north) and on the edge of their territory (to the south), for a species does not occupy the whole of its range evenly. Instead it is usually found to be common only where habitat requirements are easily met (Allee & Schmidt 1951:147) - in the present study, in the north and west. In the region centred on the Dordogne and Vezere, conditions are more suitable for the reindeer, other taxa being encountered in number only at isolated occurrences, perhaps where alternative ecological conditions prevail - in small local-scale pockets.

During the Later Aurignacian, the distribution pattern displayed by the available horse data bears some resemblance to that of the earlier period in that lowest values are again found in the south of the region. However, average frequencies have, in general, risen, the highest value now being seen in the west of the region, at Pair-non-Pair. It would appear that during the Later Aurignacian, conditions which may be deemed to have been suitable for large horse populations to have been exploited are encountered to a greater extent (ie: increased horse occurrence, or a decline in alternative species). An increase in horse population levels, although based on sites distributed over a smaller area (see fig. 8.7), is observed, for it is possible that contraction of range may have resulted in higher population density in this smaller area.

The patchy distribution of ibex and chamois as seen in figs. 8.8-8.10 largely reflects the occurrence of suitable topographic features, namely cliff faces, either continuous or broken. Both species are commonly regarded as members of a montane group, the ibex to a greater degree than the chamois, occurring primarily in the Dordogne, Vezere and Lot valleys. During the Early Aurignacian, south-facing Dordogne valley rock-faces yielded relatively constant, although not necessarily prolific, ibex populations. The species avoids flat, undulating terrain occurring only sporadically in the north of our region and at only one site in the west. Indeed, frequencies generally rise as one moves east, towards the Massif Central (Fig.

Map showing the Distribution of Horse during the Later Aurignacian

Fig. 8.7

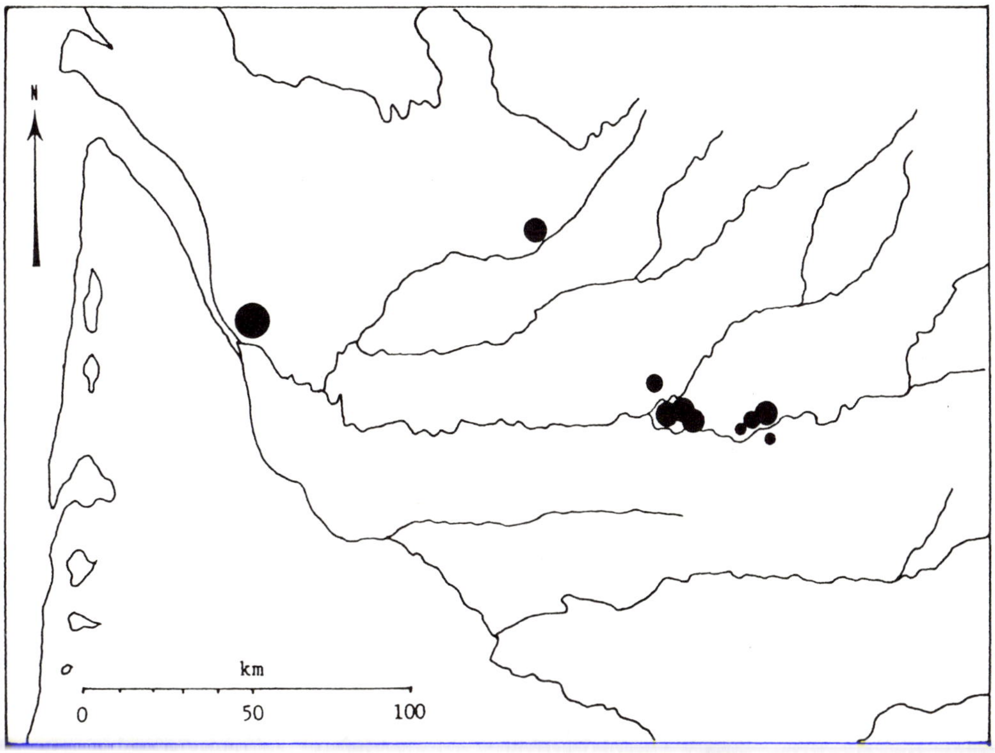

Key: ● 0.00-4.99 ● 5.00-9.99 ● 10.00-19.99 ● 20-100.00

% large herbivores

Map showing the Distribution of Ibex during the Aurignacian

Fig. 8.8

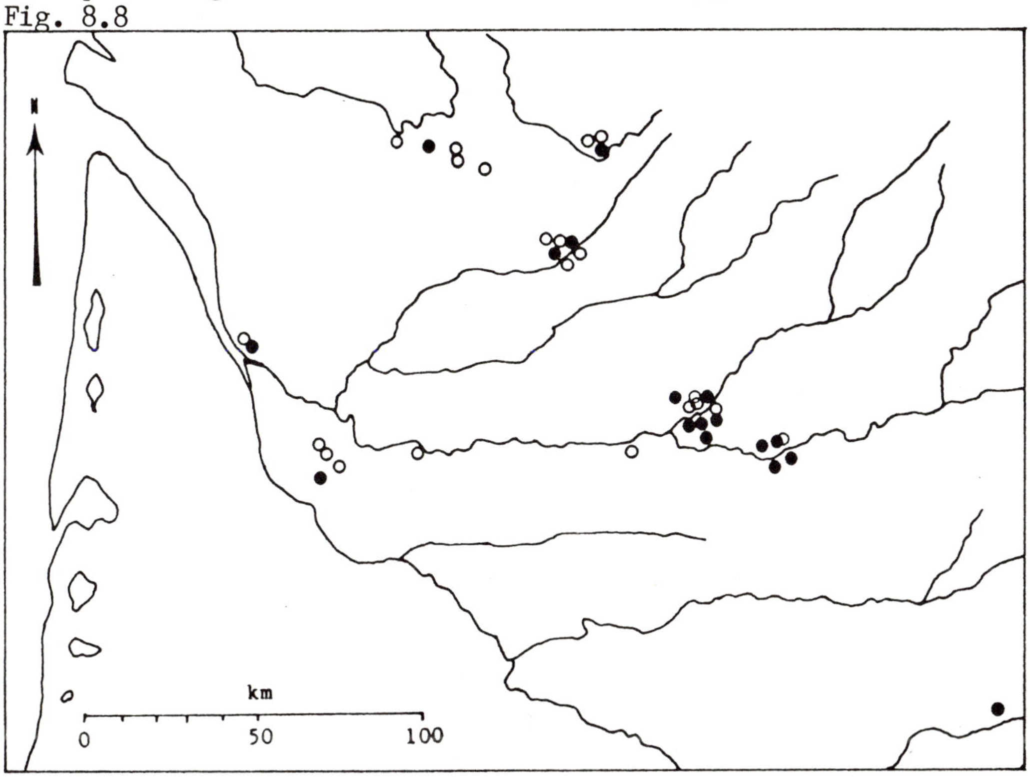

Map showing the Distribution of Ibex during the Early Aurignacian
Fig. 8.9

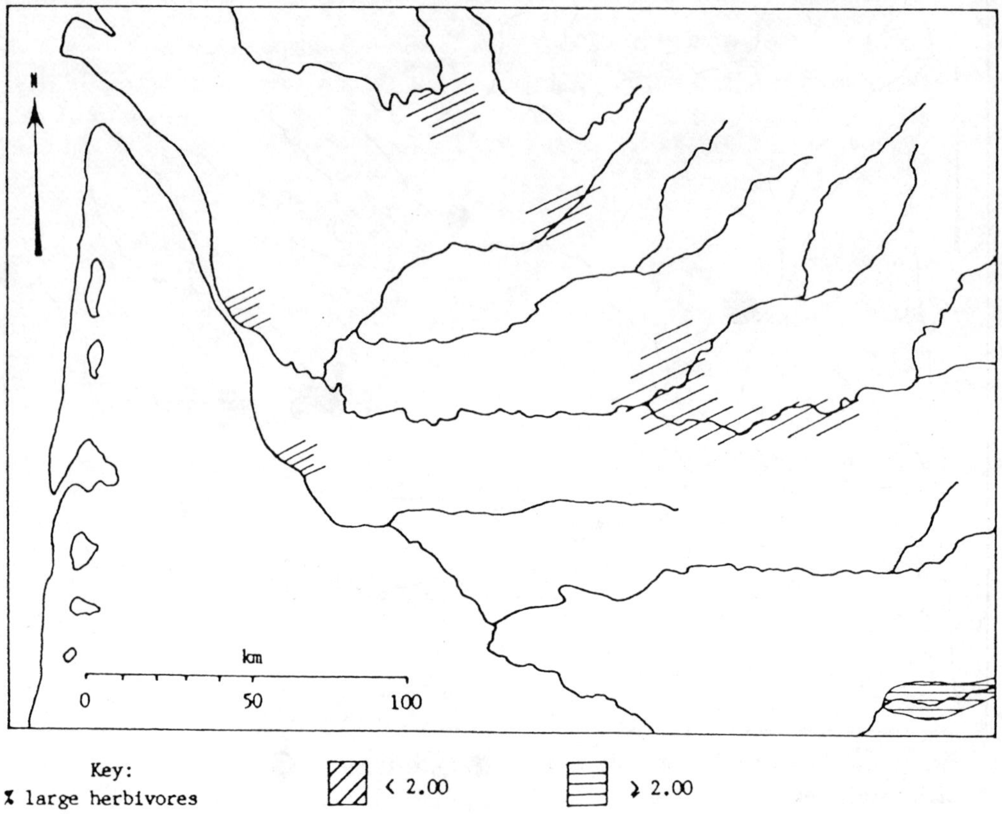

Map showing the Distribution of Chamois during the Aurignacian
Fig. 8.10

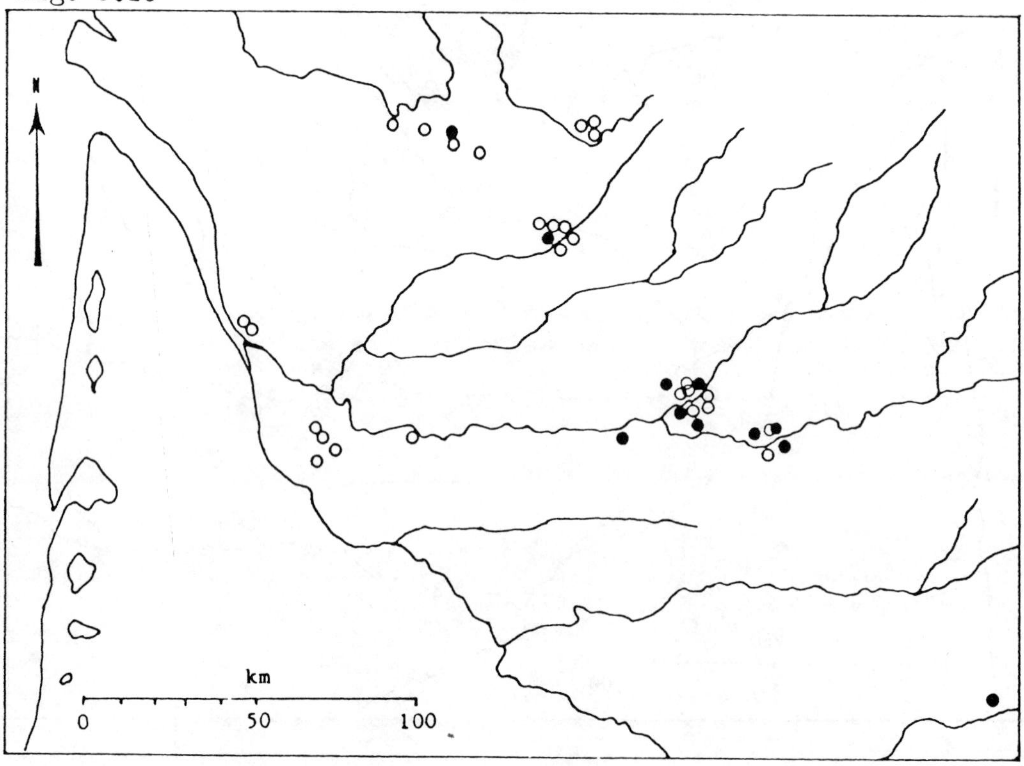

8.9). Meanwhile, the chamois (fig. 8.10) exhibits a more restricted geographical range than the ibex, and is present in lower frequencies.

The rhinoceros is observed only eight times during the Aurignacian. Several Early Aurignacian occurrences are to be found in the western half of the region, in association with Equus caballus, and bovids, and also in the Charente and Dronne areas. Meanwhile, the species occurs at only two of the seventeen sites in the Dordogne, Vezere and Lot (fig. 8.11(a)).

The mammoth occurs at nine sites during the Aurignacian (see fig. 8.11(b)). Of these, seven occur in the west and north of the region. In general, their distribution is similar to that of the rhinoceros. Indeed, of the nine occurrences, six coincide with woolly Rhinoceros observations, The co-occurrence of these two taxa suggests that the Aurignacian in the west and north of the region was characterised by extensive, relatively open vegetation - namely grassland with a low shrub cover, perhaps associated with widely scattered patches of woodland.

Hare and rabbit, treated as 'lagomorphs' in the present discussion occur in both the Vezere and Dordogne region: of its 11 occurrences 6 are to be found in this area, the others being widely scattered (see fig. 8.12) across the west and north. Lagomorphs are also to be found at the Abri Battuts (Lot) in the south of our region.

The traditionally more temperate component of our faunal assemblages yields a slightly southern orientated distribution. More than half of red deer occurrences are located in the Vezere/Dordogne region where it occurs at least once at every site and forms a central cluster (see fig. 8.13(a)). The area may have offered some protection to species which, like the bovids, were able to find suitable conditions in sheltered tributary valleys. During the Early Aurignacian the distribution of Cervus elaphus is restricted almost completely to the Vezere/Dordogne interfluve with the exception of a few individuals recovered at Pair-non-Pair in the west and Bourgeois-Delauny in the north. The omission of Caminade (on grounds of small sample size) renders the maximum value at Battuts, where totals reach 4%.

During the Later Aurignacian, despite a rise in frequencies, red deer totals remain relatively low, with the exception of the Vezere site of La Ferrassie (see fig. 8.13(b)). In the lower reaches of the Vezere valley, 'parallel' stretches of the Dordogne and the adjoining interfluve region, frequencies commonly exceed 10%, where shelter and forage conditions may have been suitable, while to the north and west the persistence of open, dry conditions is likely to have prevented the establishment of large, stable red deer populations.

The roe deer displays a severely restricted distribution (see fig. 8.14(a)) during the Aurignacian, occurring at only 6 sites, 3 in the Vezere valley and 3 in the Dordogne. Such a limited occurrence re-emphasises the potential of the area regarding the provision of sheltered ranges for species requiring relatively mild and wooded conditions. However, the boar, despite its traditional forest affiliation, has a much wider distribution than the roe deer, occurring at 23 sites (57.75% of Aurignacian occurrences), indicating that

Maps showing the Distribution of Woolly Rhinoceros and Mammoth during the Aurignacian

Fig. 8.11

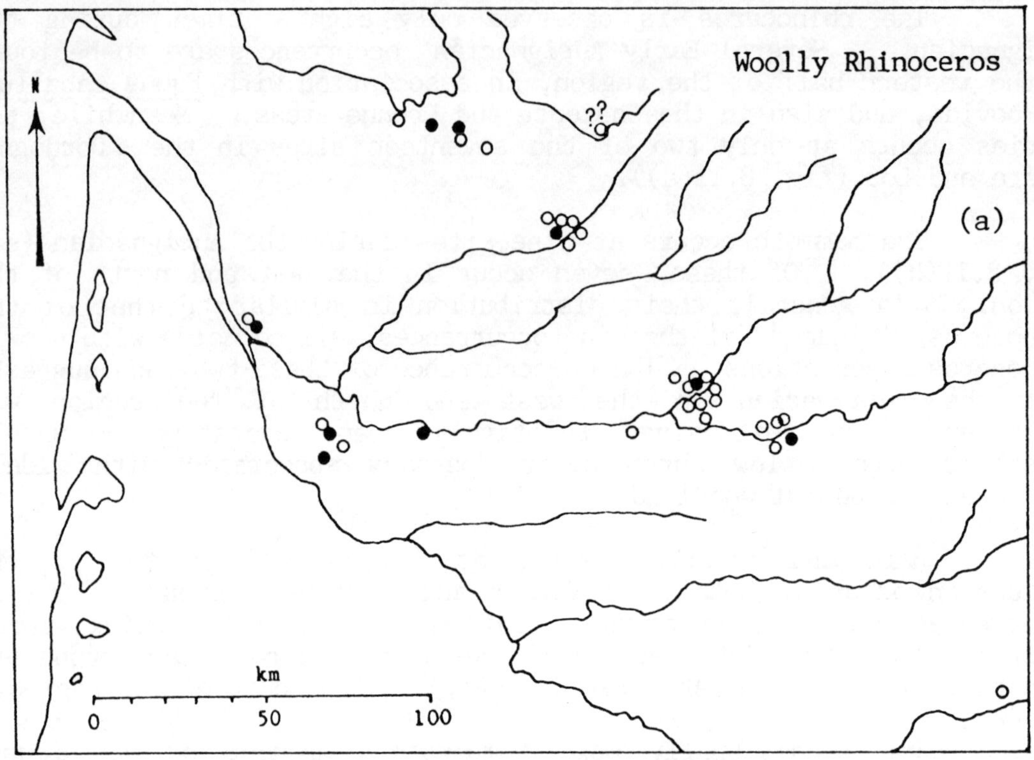

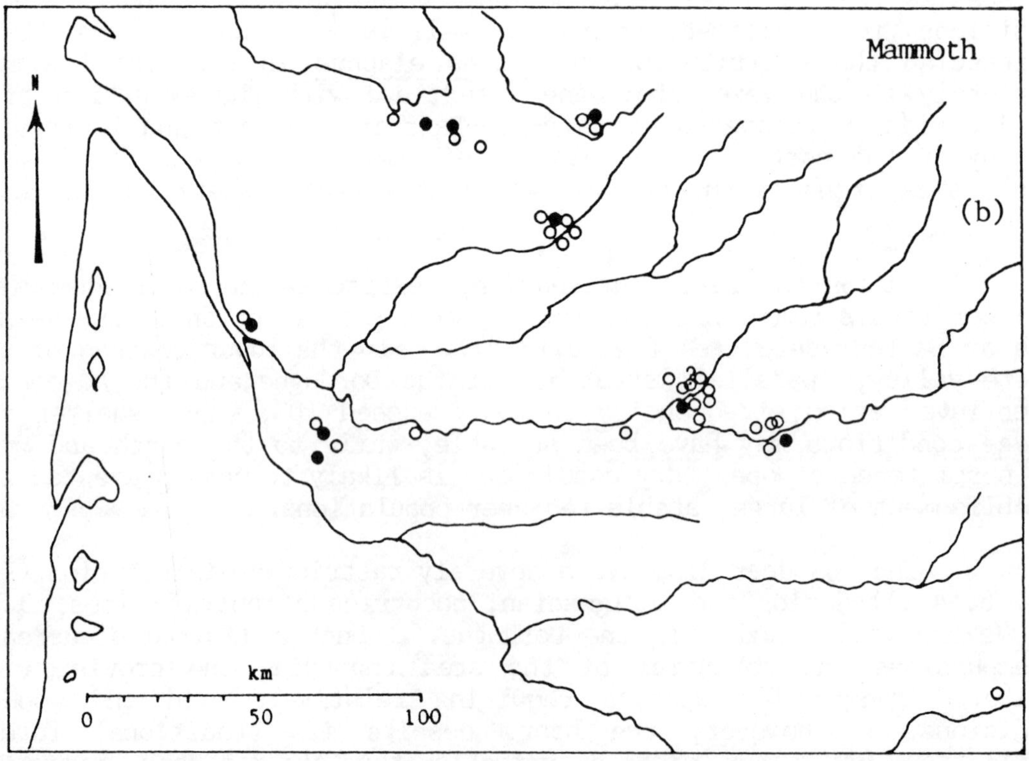

Map showing the Distribution of Lagomorphs during the Aurignacian

Fig. 8.12

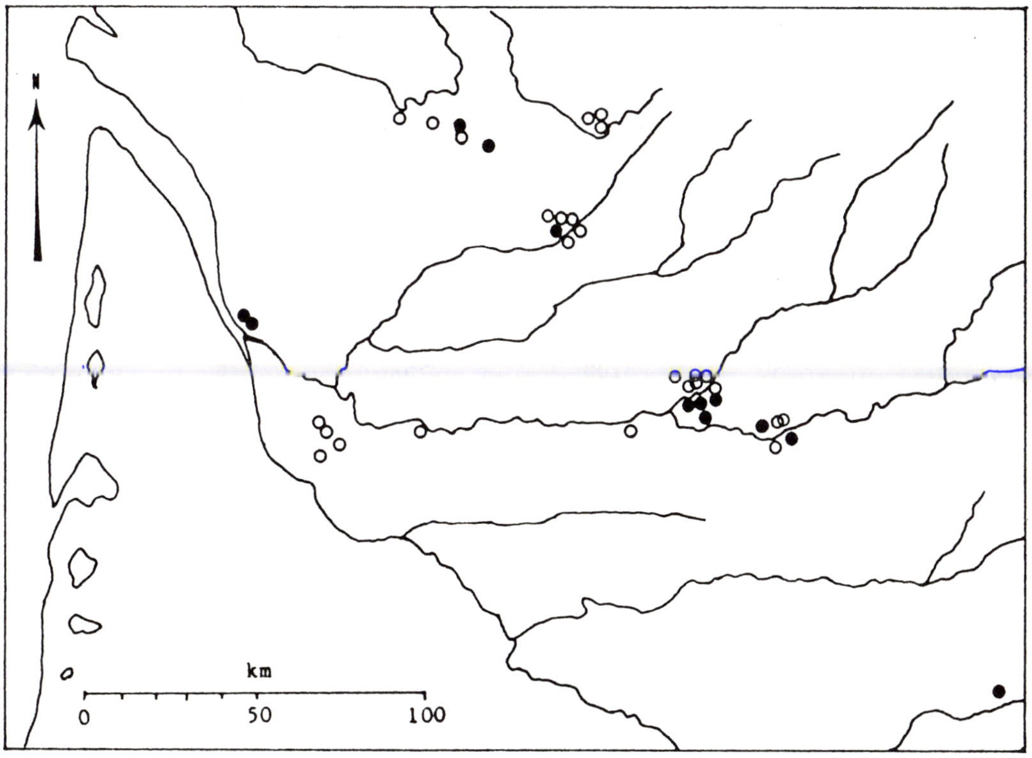

Maps showing the Distribution of Red Deer during the Aurignacian

Fig. 8.13

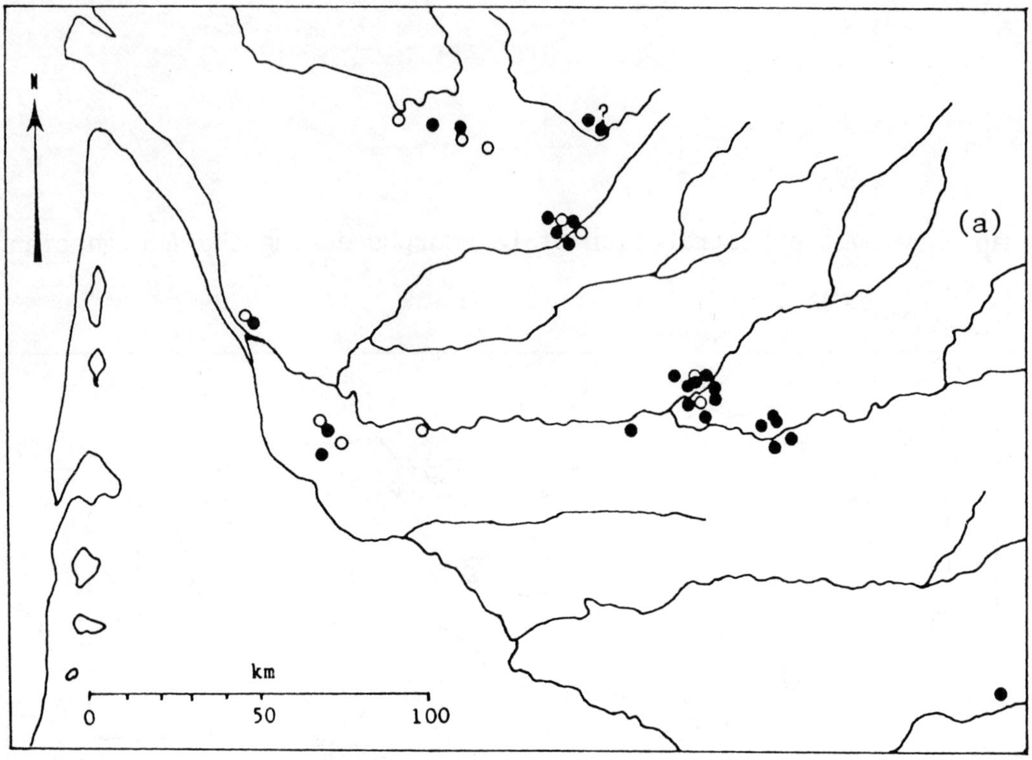

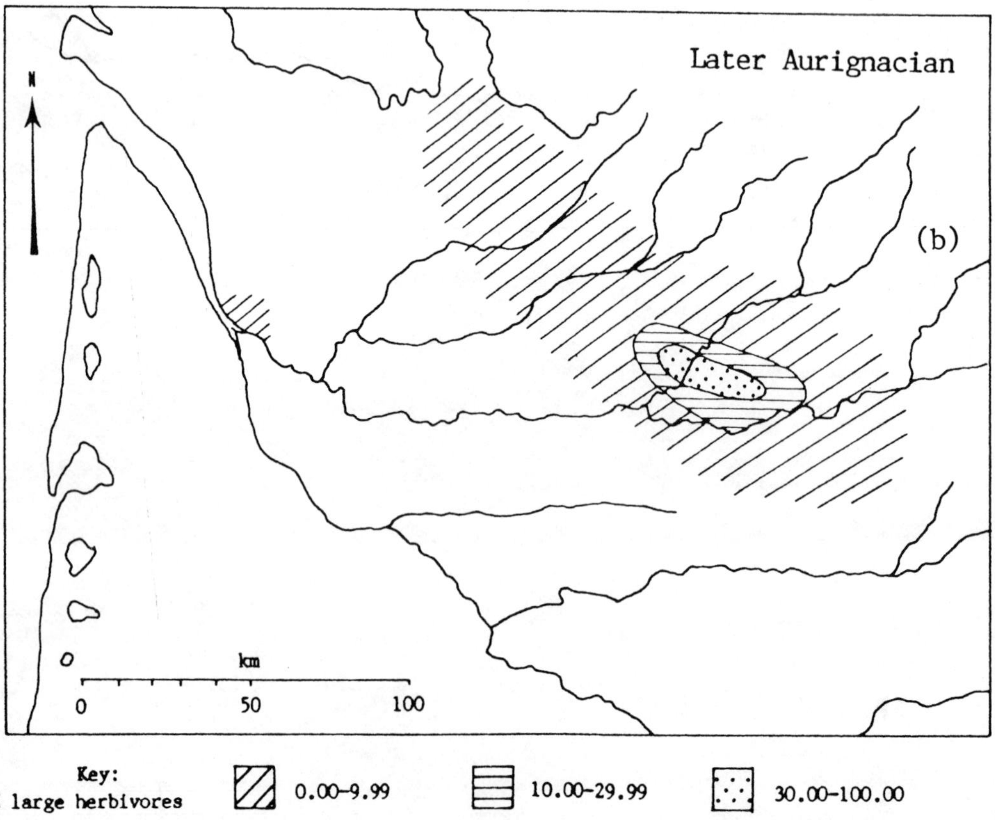

Maps showing the Distribution of Roe Deer and Boar during the Aurignacian

Fig. 8.14

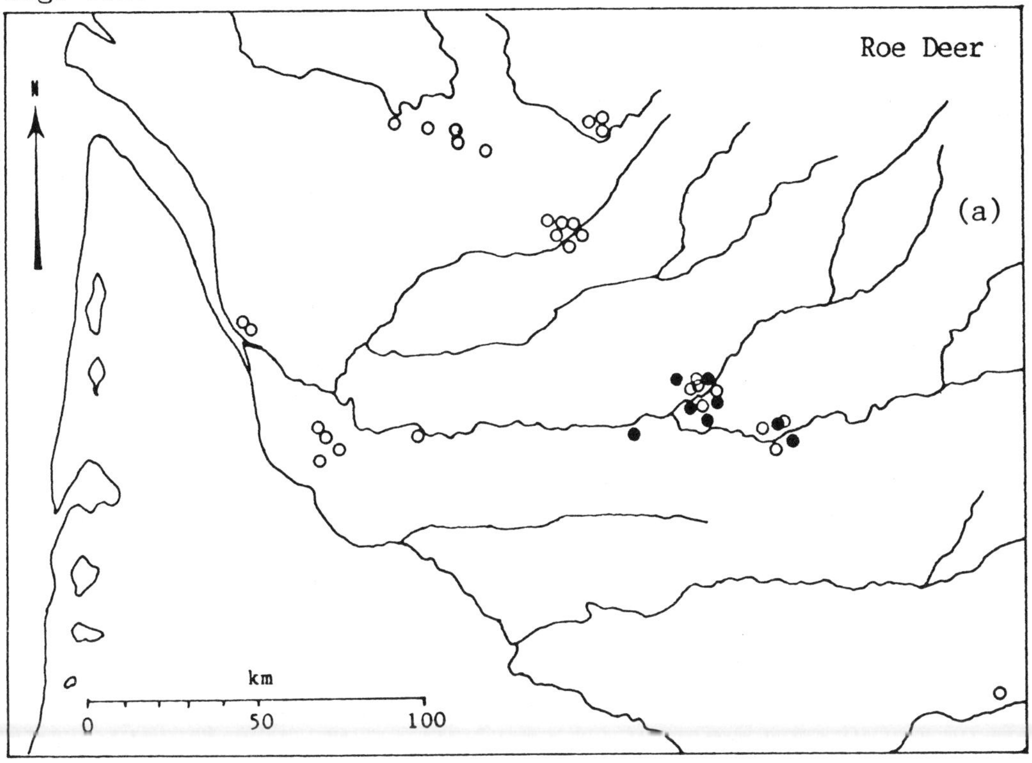

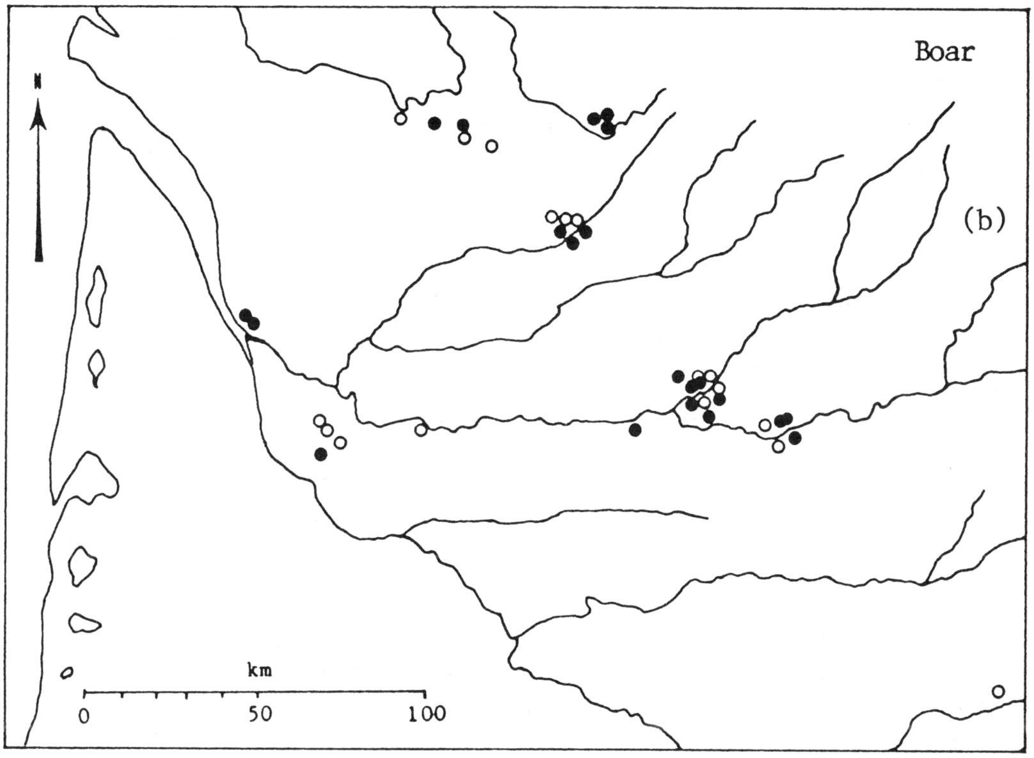

localised wooded conditions were available for exploitation (see fig. 8.14), once again especially in the Dordogne and Vezere.

Fig. 8.15 shows the distribution of dominant taxa during both Early and Later Aurignacian in South West France. The patterning observed largely reflects that suggested by the numerical data and presence/absence maps. However, the maps are of value in that they allow us to consider the quantitative data in association with those non-numerical data which provide some relative indication of abundance. Hence we increase our limited data base.

During the Aurignacian, species other than reindeer dominate in twenty levels, although there are others in which reindeer and alternative species are occasionally seen to co-dominate. Most of the levels in which reindeer is recorded as a secondary species occur during the Later Aurignacian, reindeer being outranked in seventeen separate levels. We see a roughly N.W.-S.E. orientation of the reindeer distribution, from the extreme south east (at Battuts) to the Charente, near Angouleme, although the overall abundance of the species declines slightly as we move north from the Vezere valley. Horse represents the most common alternative taxon, predominating at isolated localities, in areas largely characterised by wider valleys (eg: some reaches of the Dordogne and south of the Charente), open, flat terrain (in the Gironde) or interfluve surfaces (Vezere catchment). The red deer is seen to predominate or co-dominate five times in an area characterised by mosaic or patchy environments. It is noticeable that this occurs only during the Later Aurignacian, the area being more diverse in terms of the variety of ecological niches available and associated fauna. The suggestion is that we have pockets of woodland and similar conditions (in terms of 'cover') in the Lot, Vezere and Dordogne - areas in which local populations of other (non-reindeer) species were able to maintain viable reproductive levels and thus survive. Such areas are found on a much more limited scale in the north and west where bovids predominate only at La Chevre in the north during the Later Aurignacian, and fail to do so at all western sites during the Earlier period.

8.2 Upper Magdalenian.

Fig. 8.16(a) shows the distribution of Upper Magdalenian sites yielding reindeer. The species appears to be ubiquitous; it is absent at only two sites (Gironde) and occurs in all parts of the region. However its frequency shows marked regional variations. Frequencies exceed 90% at several locations in the area centred upon the Dordogne/Vezere confluence, gradually falling off towards the west as alternative resources appear to gain in importance (See fig. 8.16(b)). However reindeer does remain an important component of the fauna throughout the whole area. Unfortunately, in much of the northern zone, only qualitative data are available, hence the identification of a region in which the species is known to have occurred although often in unknown frequencies. In cases in which qualitative indication of presence or abundance is provided, some attempt is made to incorporate the site into the patterning. In some instances however, this is not feasible and here we can only ensure that the species' territory (or range) is seen to include the

Fig. 8.15

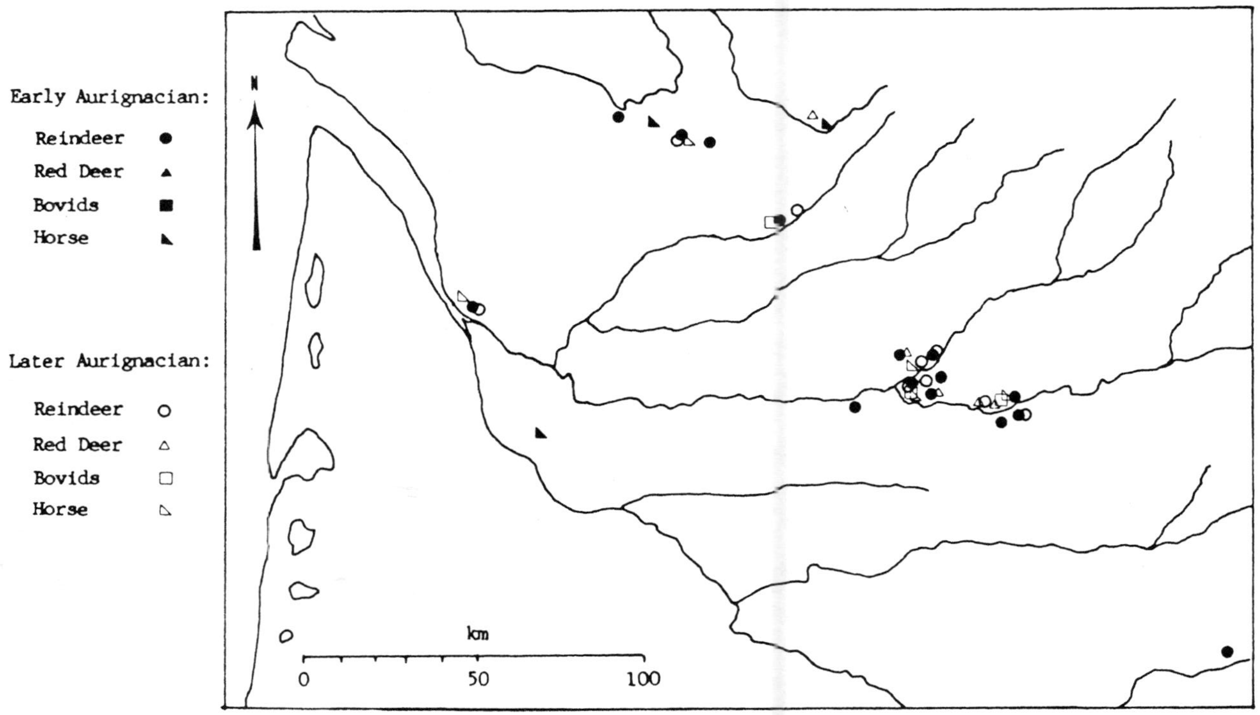

Map showing the Distribution of Dominant Species during the Aurignacian

Maps showing the Distribution of Reindeer during the Upper Magdalenian.
Fig. 8.16

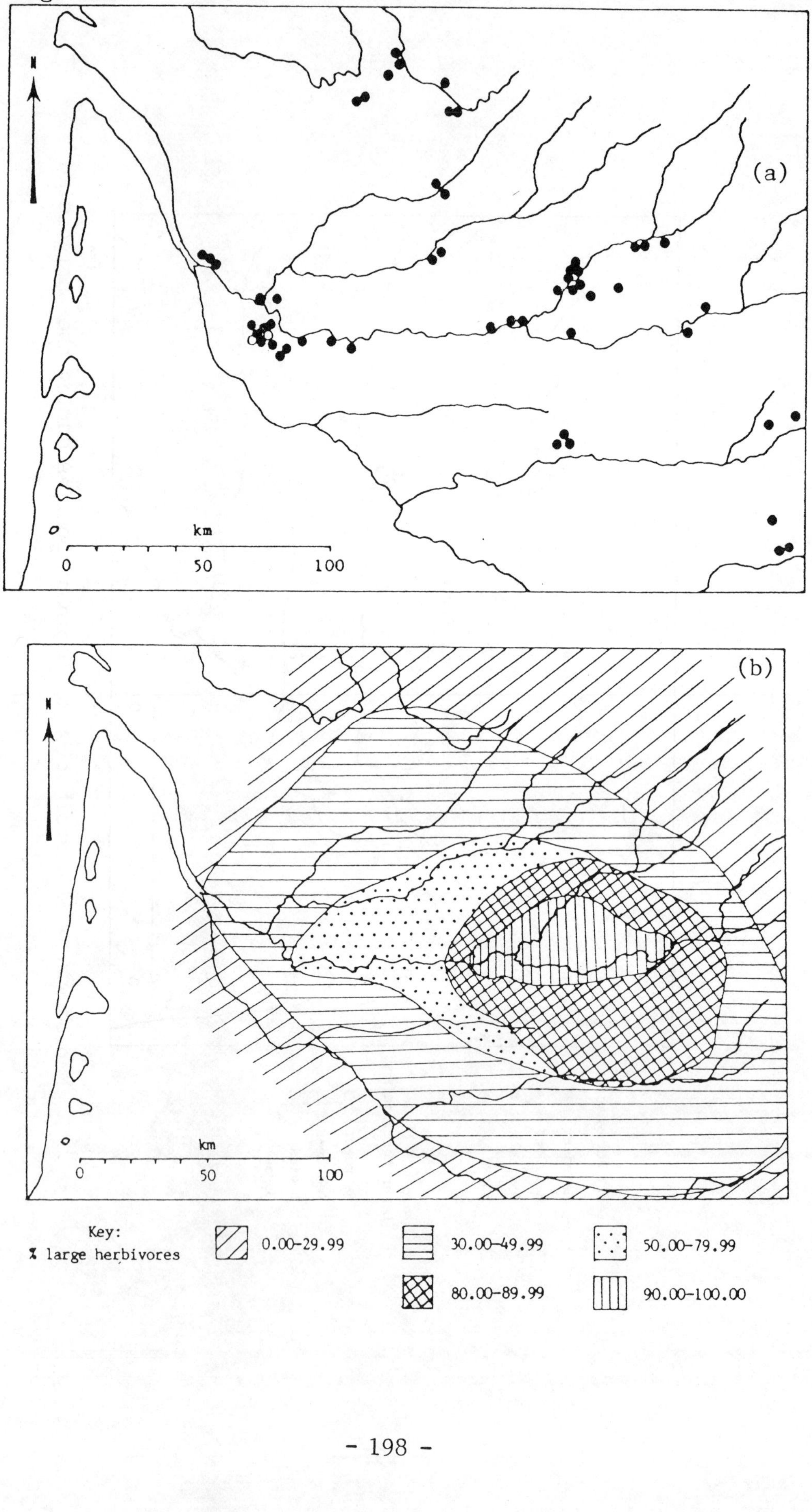

locations in question, eg: co-dominance of horse and bovids at La Roche de Castelmoron.

The decline in frequencies observed towards the east may be the result of harsh environmental conditions impinging from the Massif Central which, in addition to a change in altitude and topography may have prevented the expansion of the species's range towards the central ice cap. Veyret (1986) has recently attributed abundant snow and ice in the Massif Central during the Late-Glacial to the region's "oceanic" position, and thereby invoking conditions which may be deemed to have been unsuitable for the reindeer to the east of our area. To the west of the Perigord, conditions appear to be largely cold and dry with flat, open landscapes the norm, an area which may have offered less protection from the elements than did the valleys of the Perigord and Quercy and one in which herds may have dispersed making exploitation more difficult.

Examination of figure 8.17 shows that, as in the Aurignacian, the horse (which occurs at 76.8% of all sites) is probably of greater importance in the west of our region (78% of sites) than in the Perigord and Quercy (73% of sites). To the south of the Dordogne Equus caballus may be observed at such sites as La Roche de Castelmoron, where it co-dominates with bovids, accompanied by saiga and reindeer. The sites of the Dordogne and Vezere valleys show exceptionally low frequencies, especially around the confluence, indicating perhaps that the area was not sufficiently open. Only at La Madeleine do frequencies exceed 5.0%, the mean frequency in the area (excluding c.16-13 at La Madeleine) being 0.95%. The open terrain of the Gironde, Entre-Deux-Mers, Aquitaine plain and Isle valley well suits the horse, which occurs in relatively high frequencies at sites south of the confluence of the Isle and Dordogne rivers (Faustin and Fontarnaud). The species occurs rarely or in low frequencies to the north in the Charente, but appears again in relative abundance at Montmorillon (Vienne).

Upper Magdalenian bovids, found at 73% of sites in S.W. France (see fig. 8.18), also display a western distribution, reaching totals exceeding 80% at Fongaban (Gironde). In neighbouring Girondin areas, where 83% of sites yield bovids, frequencies range from 12% to 31%, totals gradually falling off towards the east, along the course of the Dordogne and Vezere river valleys where figures are low (3% at La Madeleine). Totals are also low towards the north, where only 50% of sites have bovids, reaching only 2-4% at Pont d'Ambon (Dronne). South of the Dordogne, bovid frequencies vary. In the Lot, at Sainte-Eulalie, frequencies average approximately 3%, while to the southwest, in the Agenais bovids are co-dominant at La Roche de Castelmoron, but absent at Le Martinet. Meanwhile, in the Vezere and Dordogne areas frequencies often fail to reach 1.0%, the exception being the site of Roc d'Abeilles.

The western distribution of bovids and horse is mirrored by that of saiga (see fig. 8.19). Saiga tatarica totals are of primary importance in the Gironde and surrounding, nearby areas, the species occurring at 89% of sites in the area. Frequencies fall off gradually towards the east and north as one moves along the Dordogne, Isle and Dronne valleys, the 200m contour being, according to Delpech (1983), a critical factor in the species' distribution. The major area of

Maps showing the Distribution of Horse during the Upper Magdalenian.
Fig. 8.17

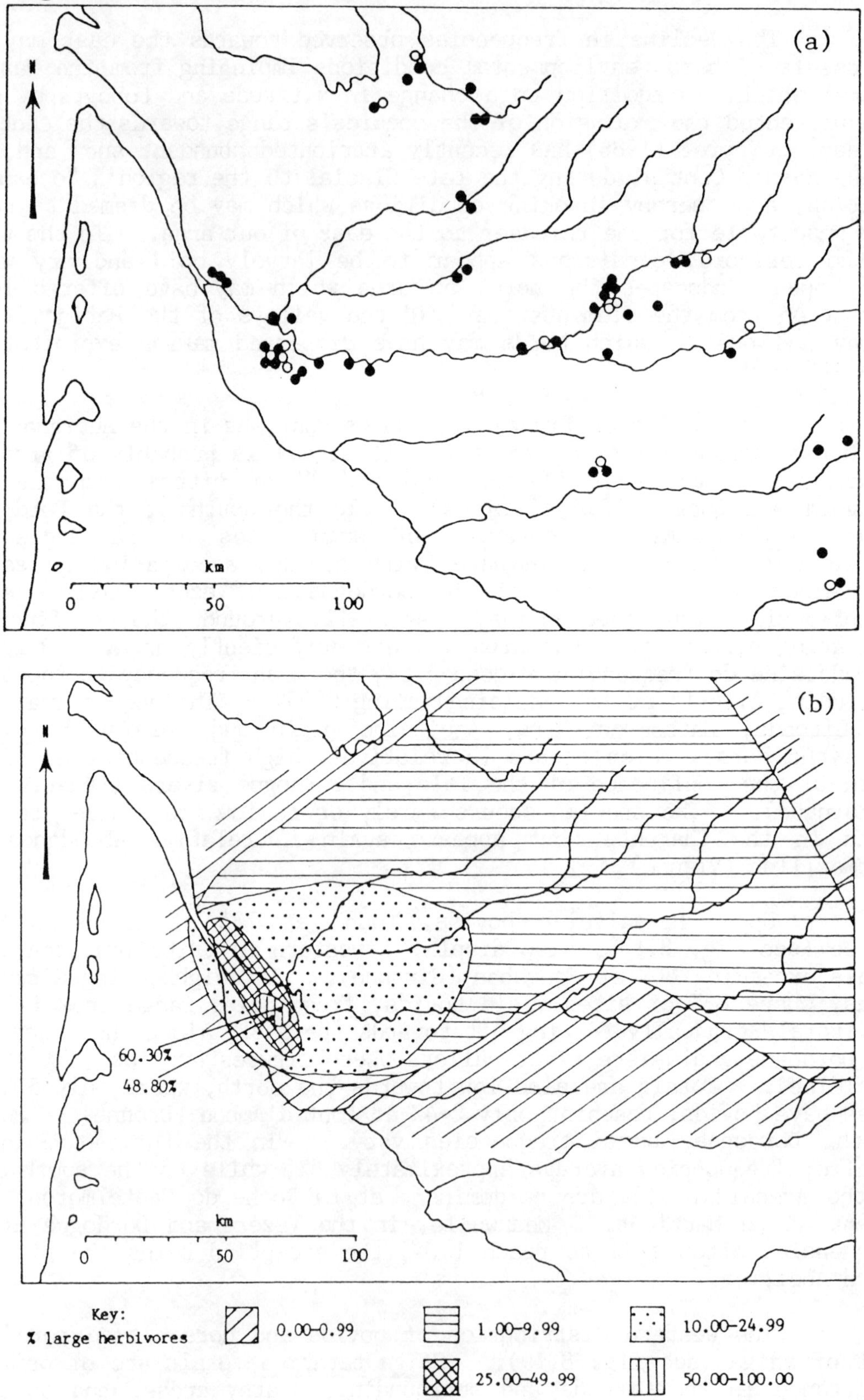

Maps showing the Distribution of Bovids during the Upper Magdalenian.

Fig. 8.18

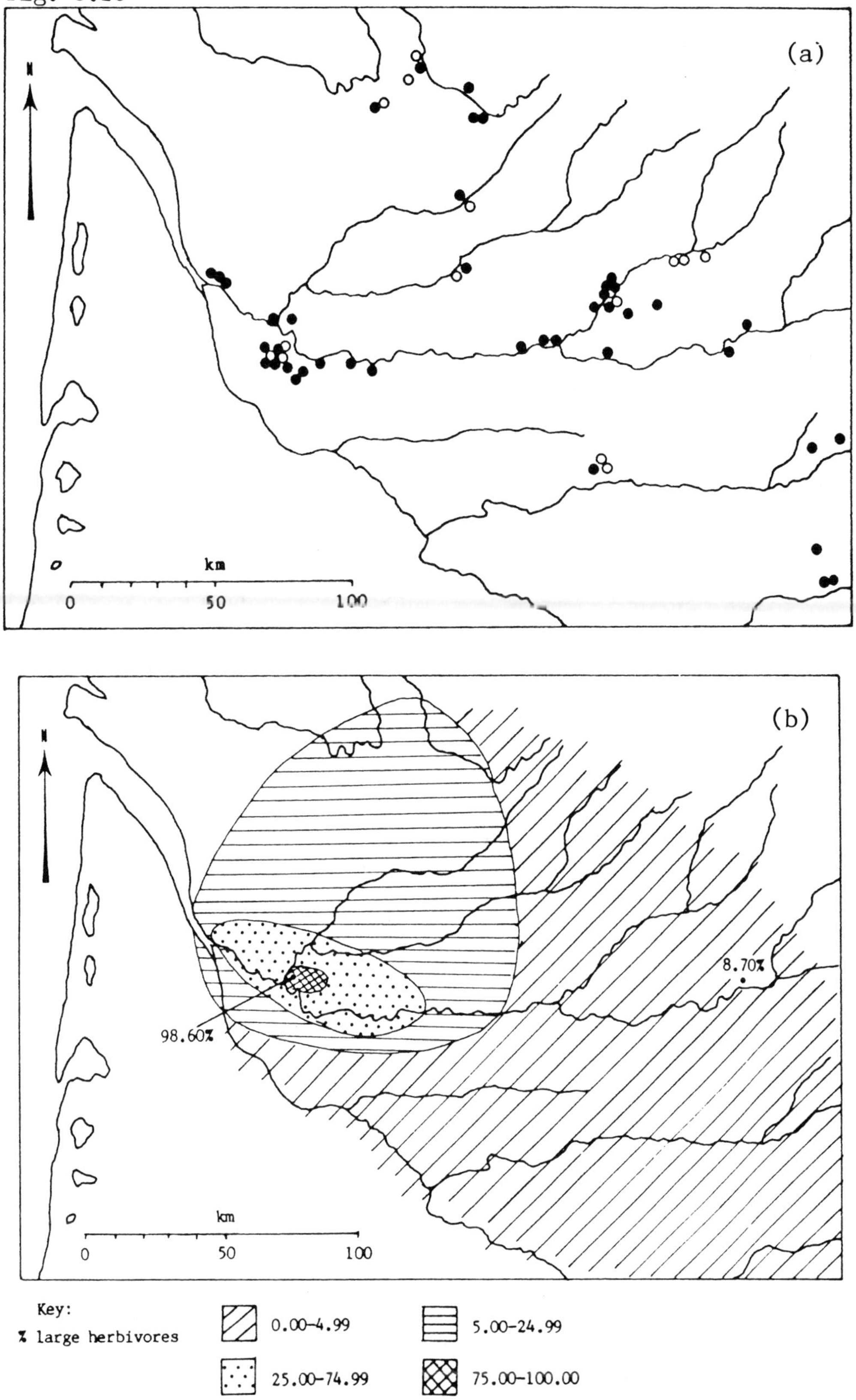

Maps showing the Distribution of Saiga during the Upper Magdalenian.

Fig. 8.19

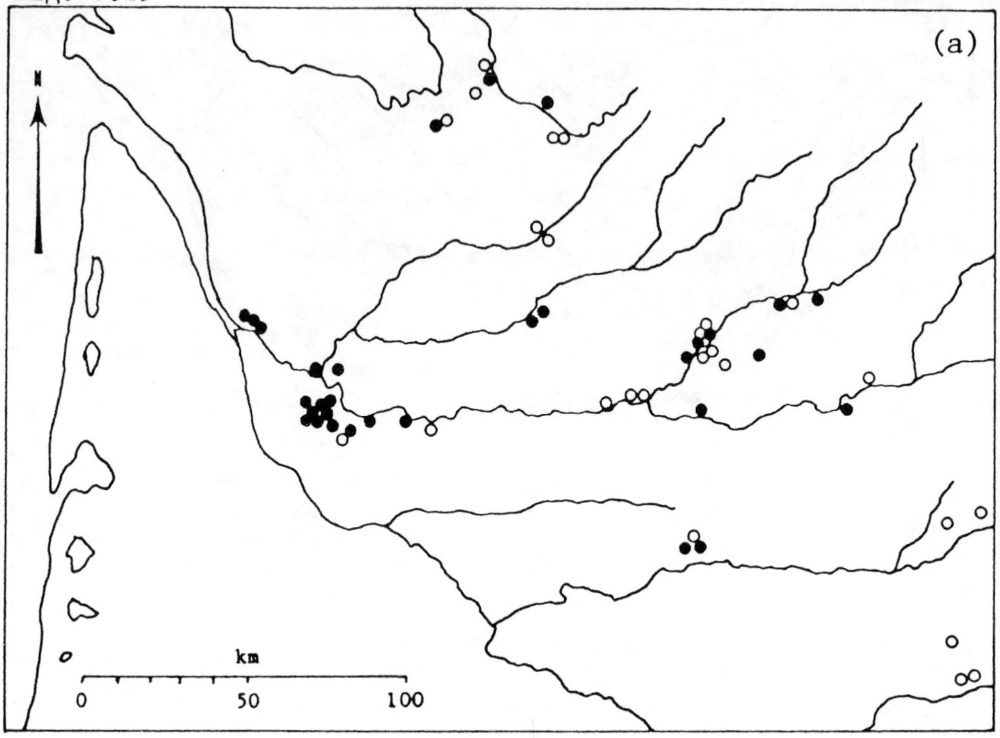

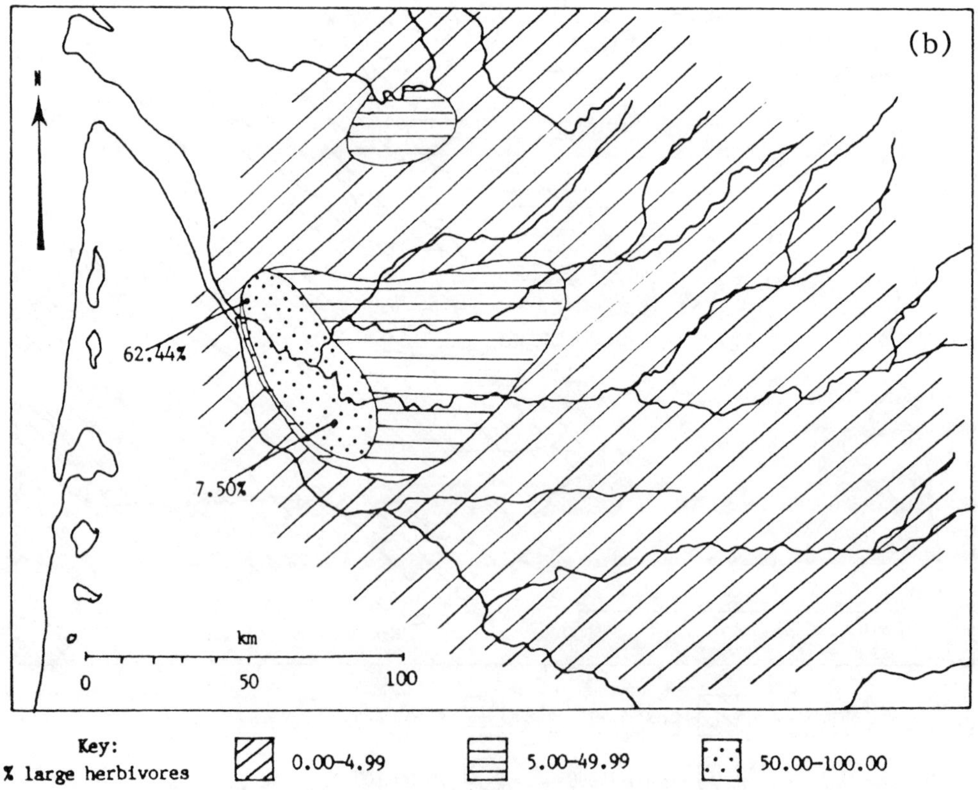

occurrences is centred on the Entre-Deux-Mers region, the lower course of the Dordogne and its confluence with the Isle. A subsidiary zone extends over the course of the Isle, the lower reaches of the Dordogne and nearby ranges. Meanwhile, frequencies are low in the valley and interfluve landscape of the Perigord, the species occurring at six sites, the mostly easterly being Esclauzur in the Brive Basin (Delpech 1983:210). To the south, in the Lot and Agenais, the species occurs at a few sites, but only at low frequencies for, preferring, as it does, flat, dry conditions, saiga shuns the rugged terrain of the Massif Central, Causses and, further to the south, the Pyrenean foothills.

The ibex (fig. 8.20) never attains high frequencies during the Upper Magdalenian, occurring at only 16 sites. As during the Aurignacian, the distribution of the species largely reflects the occurrence of river valley rock faces such as those to be seen in the Charente and Lot regions. Thus the species is 'absent' in the Gironde region, lower frequencies being observed at sites along the course of the Dordogne: at Gare de Couze, a single remain was uncovered, at La Madeleine, two bones (Delpech 1983:194).

Chamois figures are also consistently low, peaking in the Dordogne and Lot valleys (See fig. 8.21(a)), for, as in the case of the ibex, the distribution of the chamois largely reflects the occurrence of rockfaces and river gorges. The species may however reflect more humid conditions than the ibex (Delpech 1983:203) and, being more tolerant, is encountered in a larger portion of the region. Unlike the ibex, the chamois is observed in the Gironde at La Grotte des Fees and Roc de Marcamps where local topography is more suitable.

For the sake of discussion, the rabbit and hare are again considered together in the belief that similar hunting/trapping techniques would have been employed in their exploitation and that they would have occupied a similar niche in the human subsistence strategy (see fig. 8.22, table 8.1). The distribution of Lepus sp. and Oryctolagus sp. is largely determined by the nature of the geological substrate, each requiring soil conditions suitable for burrowing (eg: sand dunes) and preferring open meadow vegetation. They occur at 50% of western sites, the overall regional total falling to 42.8%. In general, frequencies are low reaching occasional peaks which may reflect special purpose sites situated in areas highly suitable for Lagomorph exploitation. For example, at Pont d'Ambon, where red deer comprises up to 90% of the large herbivore fauna, inclusion of lagomorphs reduces the red deer total to 8.10%, the rabbit comprising more than 89.40%. The species gain in importance as the Upper Magdalenian develops and gives way to the Azilian, especially in the south, in the Agenais. Here we see Terminal Magdalenian rabbit occurrences in addition to the Post-Glacial.

The rhinoceros occurs at only 5 Upper Magdalenian sites (8.9%) which are widely scattered. However 3 of the 5 sites are located in the north part of the region, with a fourth in the Gironde, a pattern similar to that observed during the Aurignacian (see fig. 8.22(b)).

Red deer is to be found at 59% of Upper Magdalenian sites (see fig. 8.23), frequencies reaching a peak at Pont d'Ambon (c.4) on

Map showing the Distribution of Ibex during the Upper Magdalenian.

Fig. 8.20

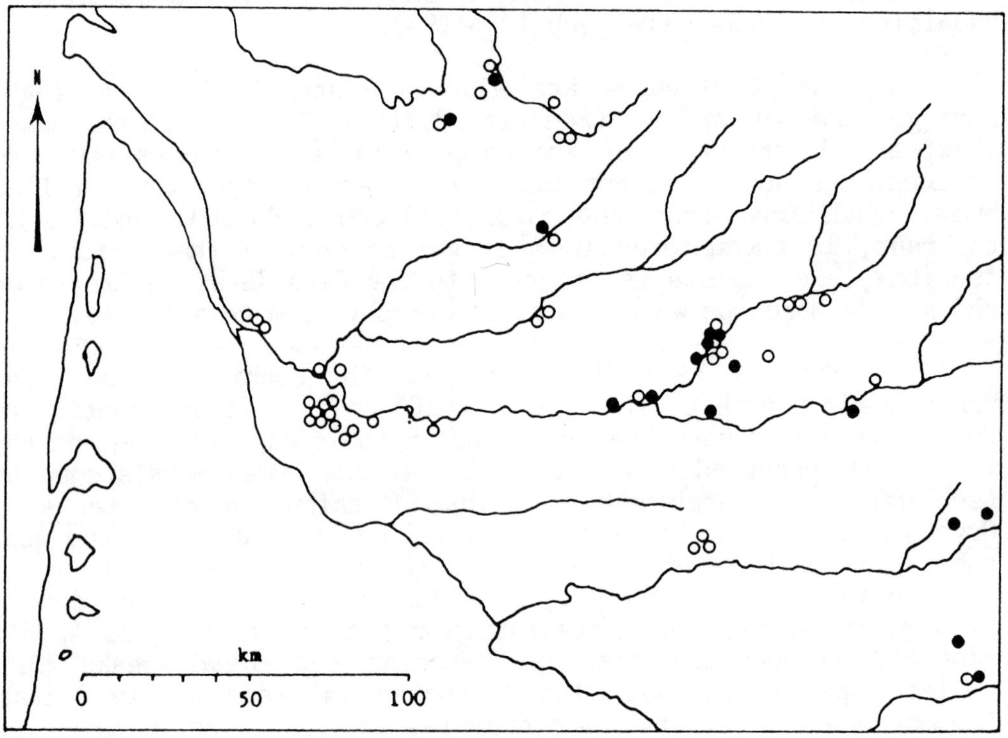

Maps showing the Distribution of Chamois during the Upper Magdalenian.

Fig. 8.21

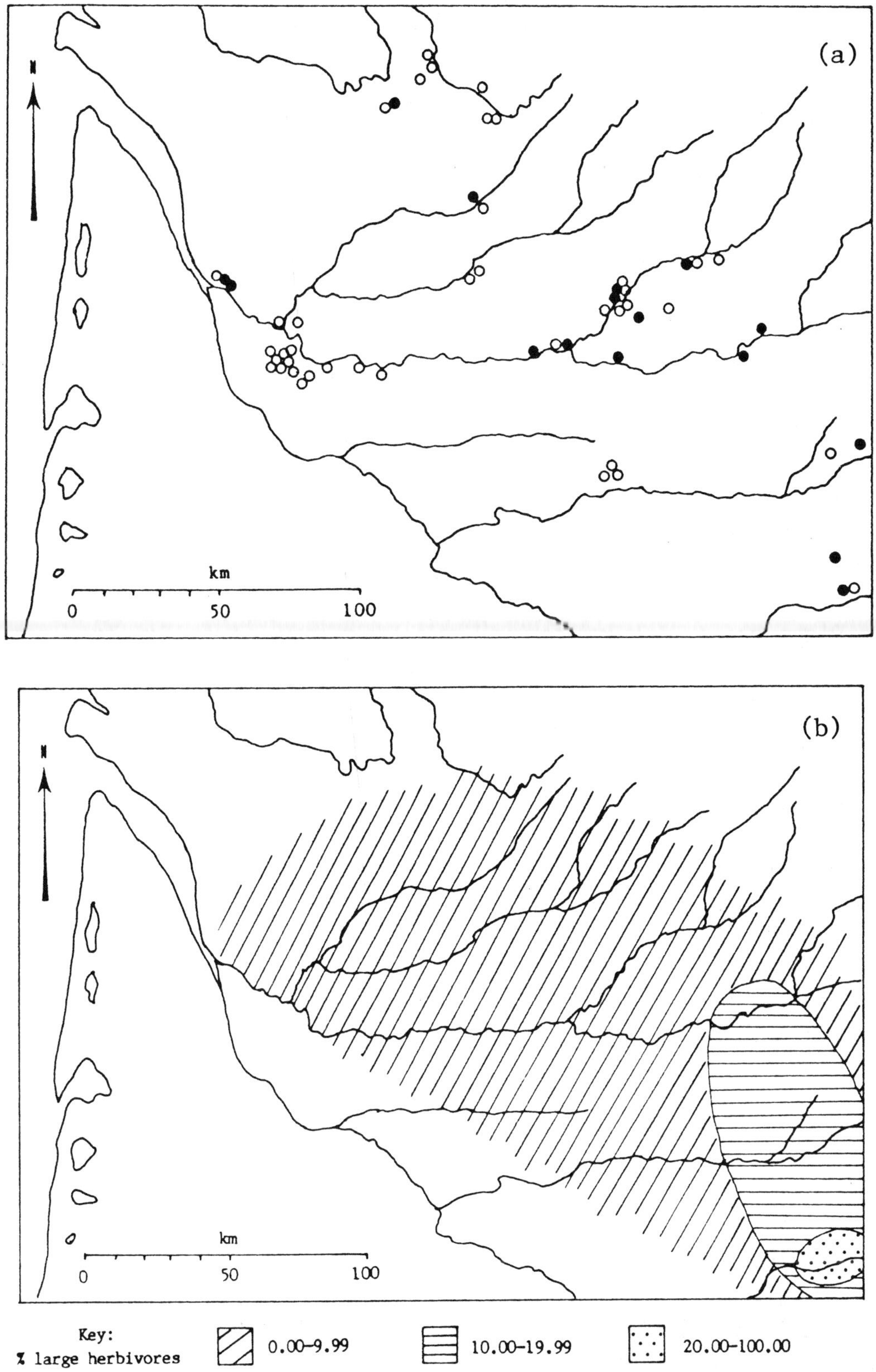

Table 8.1

Absolute Frequencies of Lagomorphs at Upper Palaeolithic Sites in S.W.France.

	N.I.S.P.
Roc de Combe	
Lepus sp.	69
Oryctolagus sp.	21
Maldidier	
Lepus sp.	2
Oryctolagus sp.	1
Le Flageolet I	
Lepus sp.	22
Oryctolagus sp.	1
Battuts	
Lepus sp.	8
Oryctolagus sp.	9
Laugerie Haute Est	
Lepus sp.	18
Laugerie Haute Ouest	
Lepus sp.	14
Le Piage	
Lepus sp.	1
Oryctolagus sp.	1
Sainte Eulalie	
Lepus sp.	2
La Madeleine	
Lepus sp.	119
Oryctolagus sp.	3
Combe Cullier	
Lepus sp.	5
Oryctolagus sp.	2
Le Flageolet II	
Oryctolagus sp.	10
Gare de Couze	
Lepus sp.	6
Oryctolagus sp.	437
Le Morin	
Lepus sp.	65
Oryctolagus sp.	535
Pont d'Ambon	
Lepus sp.	1
Oryctolagus sp.	6004+
Roc de Marcamps (Fees)	
Lepus sp.	30
Oryctolagus sp.	14

(Source: Delpech 1983; Beckouche 1981)

Maps showing the Distribution of Lagomorphs (a) and Woolly Rhinoceros (b) during the Upper Magdalenian.

Fig. 8.22

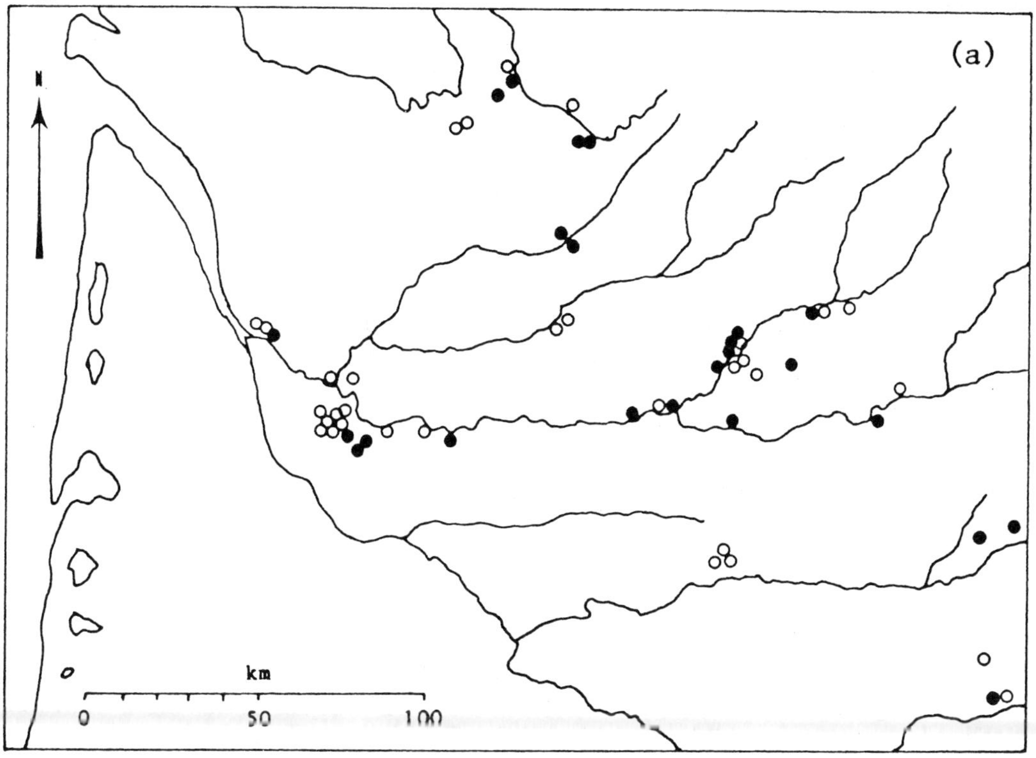

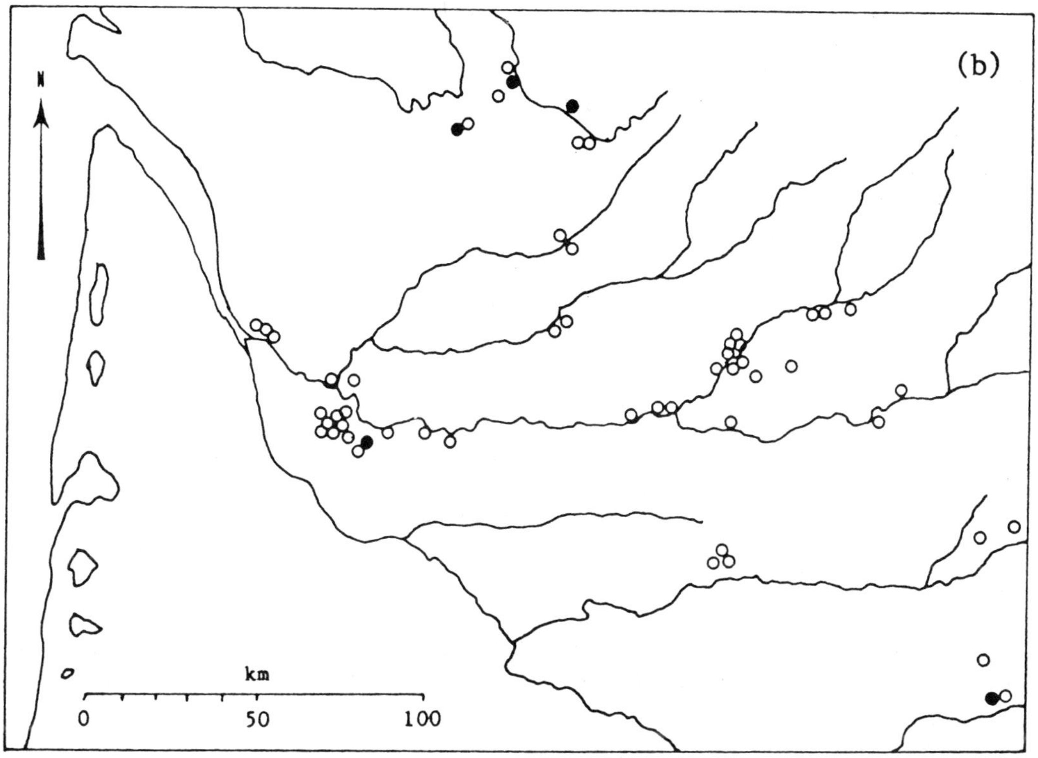

the Dronne, with a value of 89.47%, making up less than 10% of the large herbivore faunas further to the south. The high total associated with Terminal Magdalenian levels at Pont d'Ambon are typical of the later Azilian levels, and may in fact represent an 'azilianized' Magdalenian assemblage. The value of red deer at this site is therefore isolated in fig. 8.36 as an exception to the general patterns identified. Those sites located immediately on the Vezere and Dordogne yield low red deer frequencies (Gare de Couze, mean = 4.7%, Flageolet II = 2.45%, Combe Cullier = 1.58 - 3.65%, La Madeleine, mean = 0.0071%, the species occurring in only one level) while to the west of this area frequencies remain somewhat higher; at Abri Morin 10.0% of the ungulate total is represented by red deer, while at Abri Faustin the figure is 15.0%. To the south of the Dordogne the red deer disappears, absent in the Agenais region at such sites as La Roche de Castelmoron and Martinet dated, by C^{14}, to approximately 12 600 B.P. (LeTensorer 1981). In the south, it is only at the site of Sainte Eulalie (Lot) that we encounter red deer during the Upper Magdalenian. It is necessary to move to Azilian (post-glacial) levels in order to observe red deer again.

Roe deer, absent in the Charente (see figs. 8.24(a) & (b)) remains pitifully low throughout the course of the Upper Magdalenian. In essence the species has a western, scattered distribution occurring in all levels at the Abris Morin and Faustin. Further east the species appears in Magdalenian IV and V at Gare de Couze and La Madeleine respectively. In the north of our region, at Pont d'Ambon, its gradual rise in importance (c.5.0% of large herbivores) may be seen to herald the transition to early post-glacial (Azilian) wooded, more temperate conditions. Like the red deer, Capreolus capreolus is absent south of the Dordogne, in the Lot and Agenais.

Boar frequencies are very low throughout South West France (see figs. 8.25(a) & (b)), occurring, as it does, at 20 (35.7%) Upper Magdalenian sites widely scattered throughout the region. Of these only one occurs south of the Dordogne valley - in the Agenais, the landscape in this region being more suited to rock-face dwelling species. The distribution of boar may reflect, in part, the existence of pockets of woodland which later expand to form more extensive wooded conditions commonly associated with the post-last-glacial. The major occurrences are at Morin (Gironde) and in the north at Mege and Mairie (Dordogne).

To summarise, in terms of species dominance (see fig. 8.26), reindeer displays a marked concentration in the Vezere and Dordogne valleys, also dominating to the exclusion of all other taxa in the south east of the region (Lot). The species is the most widespread but is gradually replaced by bovids, horse and saiga in the west of the region, although such taxa never reach the phenomenal frequencies associated with the Perigord reindeer. While Rangifer tarandus continues to be exploited in the valley and interfluve landscape of the latter region, the other species are taken in the flatter, more open regions of the Gironde where cold steppe-like conditions prevailed.

Based upon the predominance and distribution of reindeer and the steppe element (horse, bovids and saiga), it is possible to divide South West France into two broadly defined regions:

Maps showing the Distribution of Red Deer during the Upper Magdalenian.

Fig. 8.23

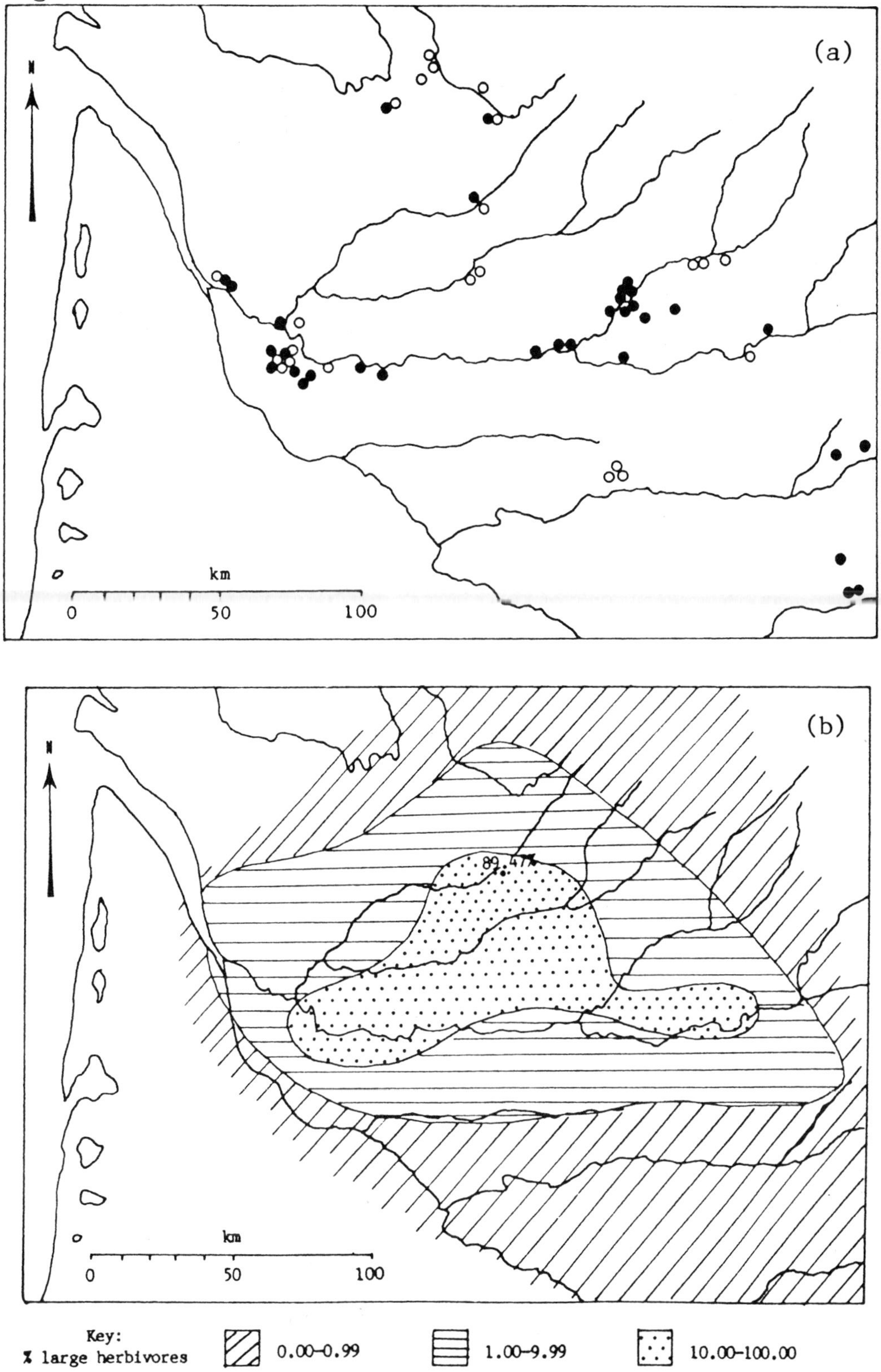

Maps showing the Distribution of Roe Deer during the Upper Magdalenian.

Fig. 8.24

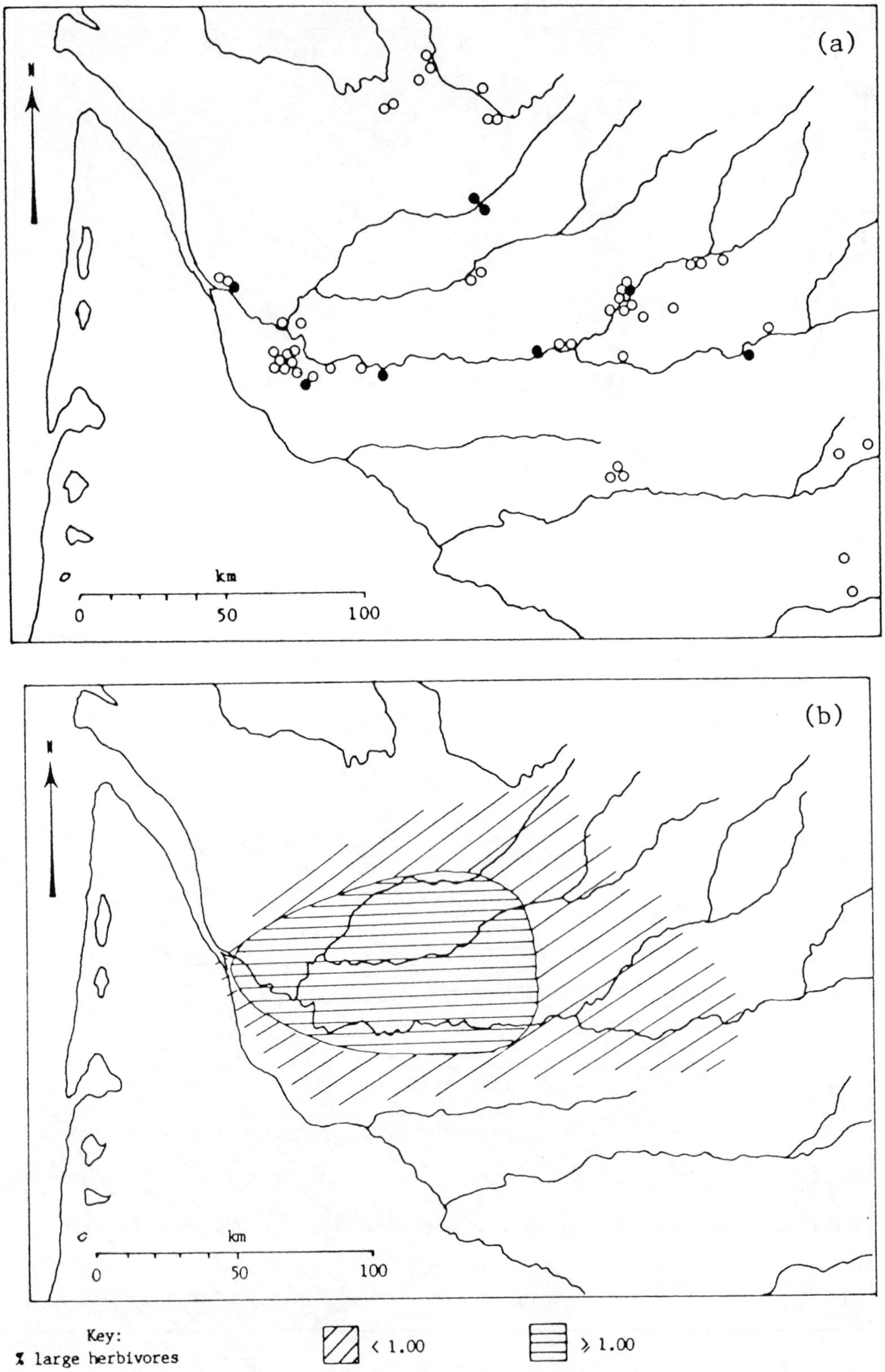

Maps showing the Distribution of Boar during the Upper Magdalenian.

Fig. 8.25

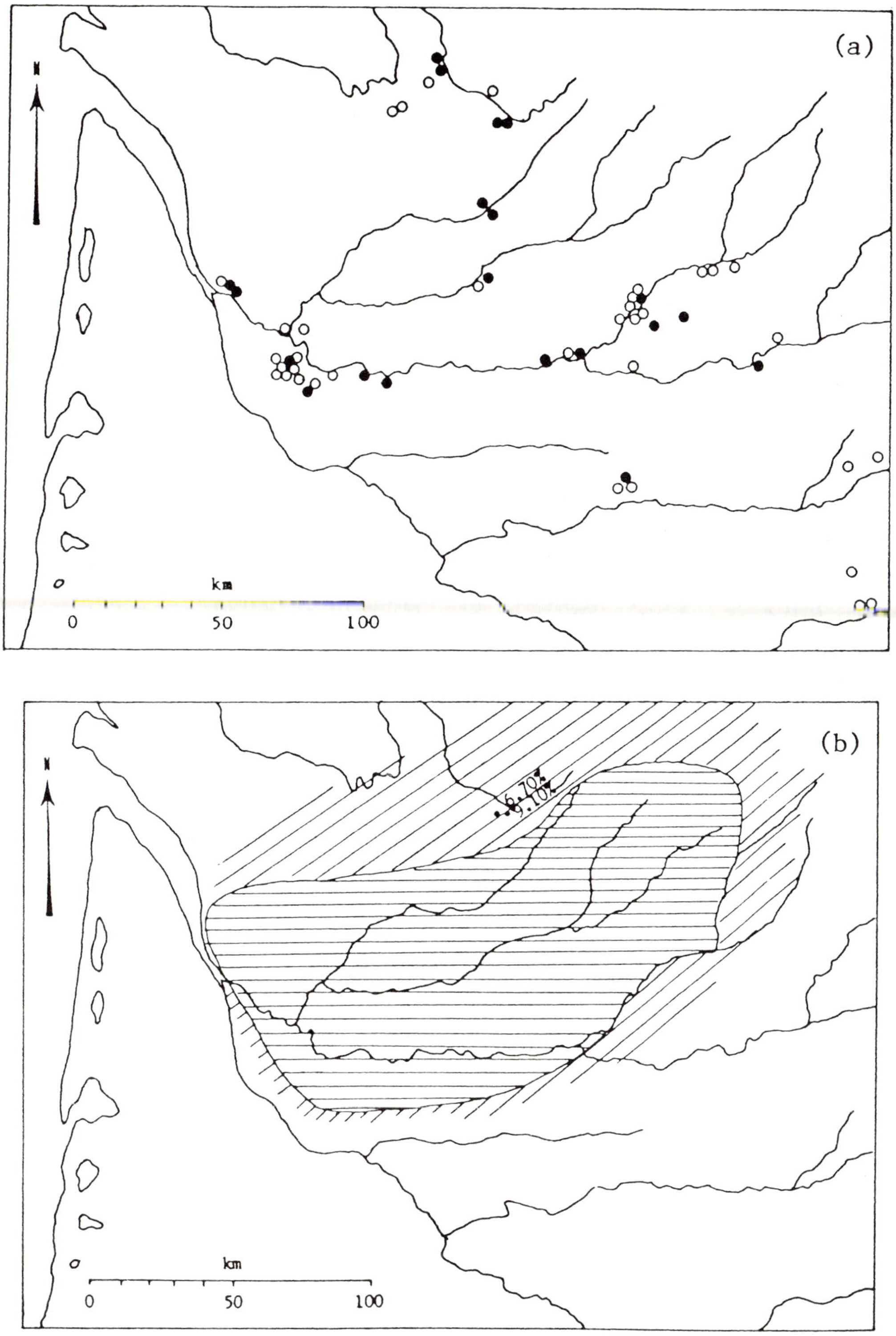

Key:
% large herbivores

< 1.00

≥ 1.00

Map Showing the Distribution of Dominant Species during the Upper Magdalenian

Fig. 8.26

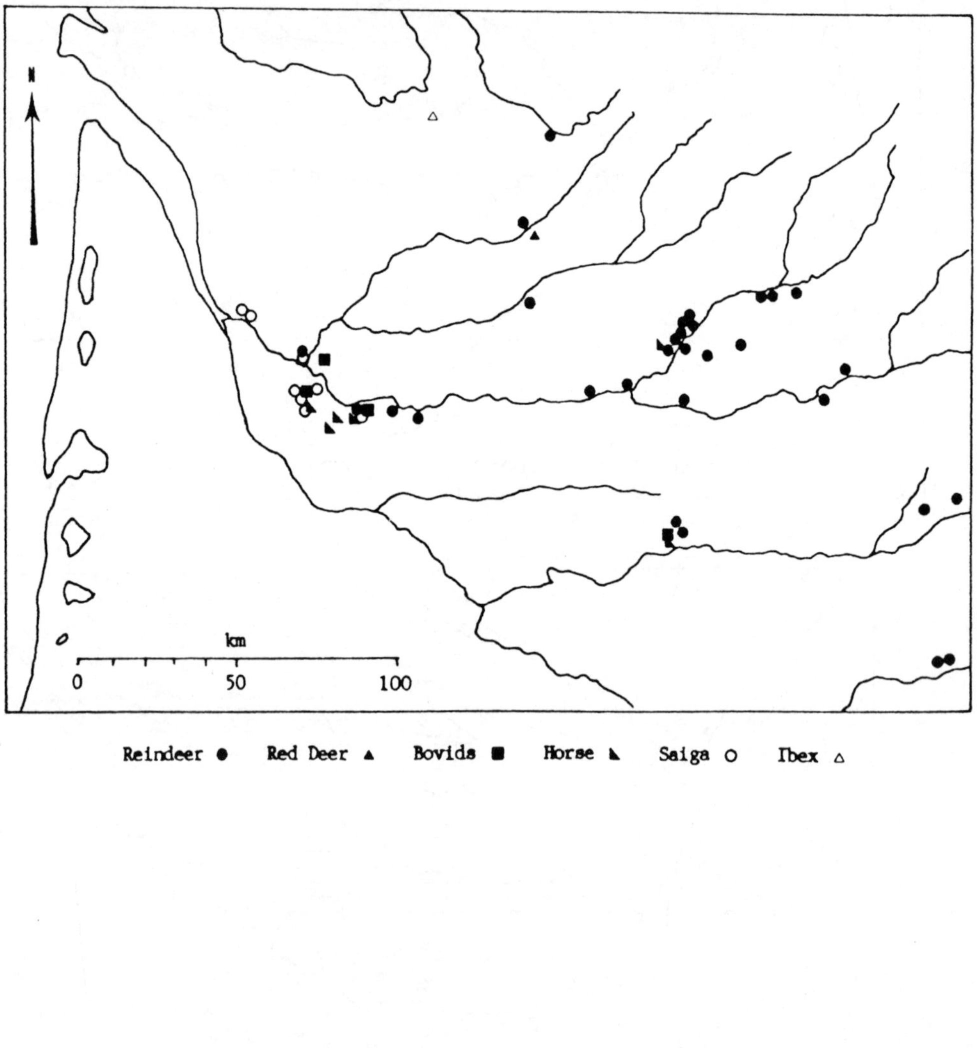

1 - The Dordogne, Vezere and eastern Lot, dominated by reindeer, and
2 - The Gironde and north west, where cold-steppe species also predominate.

These ranges appear to overlap somewhat although the distinctive element of the western component (ie. saiga) is restricted to the extreme west when dominance is considered. However, reindeer, horse and bovids (the last two representing the steppic range) do overlap in their distribution, each species displaying a tendency to register in the other's territory. Such may be explained in terms of gradually changing prevailing environmental and forage conditions along an east (Perigord / Massif Central) - west (Gironde / Charente Maritime) axis. To the west of a line running from the Charente towards the Lot assemblages appear to be relatively diverse in terms of the range of species encountered. Six species dominate at various sites - even the ibex on one occasion in the Charente. To the east of this line, only one occurrence of horse is noted, all other dominance records being reindeer. These reindeer-dominated assemblages which predominate until the end of the last glacial can only be considered as relatively specialised assemblages.

8.3 Geographic Variation in Taxonomic Richness and Diversity.

Having considered the spatial (and chronological) distribution of individual species abundance, an alternative measure of assemblage 'structure' was sought. This was done in the belief that more information could be derived from both the numerical and non-numerical data than had already been obtained. For example, the predominance of a single species often emerges, identified from percentage values ranging from 45.00 to 100.00. This represents a large range, the individual values being of varying significance depending upon the number of taxa included in the assemblage as a whole.

If, as an intuitive example, we compare fauna from levels 6 at Bois Ragot and D at Gare de Couze (both Late Magdalenian), the apparent importance of reindeer may be seen to be somewhat different, despite the similar percentage frequency of the species itself. At Bois Ragot, reindeer represents 91.30% of the large herbivore material, horse making up the remainder of the assemblage. Thus we have only two taxa. At Gare de Couze (D), reindeer comprises 91.60% of the material, the remaining 8.40% being composed of five species of varying importance, and therefore the degree of specialisation at Gare de Couze would appear to be the greater.

Ecologists, and palaeoecologists, have long recognized that there exist many potential causes of observed regional (and temporal) variations and similarities in species abundance. Given that the number of species is known to vary from one biome to another (Putman 1984), the total increasing with 'environmental complexity', attention was, in the present study, initially concentrated on this variable (N - the number of large herbivore species found in each level at each site). The use of N has the advantage, over other measures of taxonomic richness, of enabling us to look at both the quantitative and qualitative data available. However, the reader should be aware of the fact that a statistical relationship may be observed between sample size and the number of taxa encountered. This is a phenomenon which has received detailed consideration in much of the recent ecological literature (Giller 1984, Williamson 1988, Krebbs 1985, 1989), the potential implications, for archaeozoological studies, of this demonstrable relationship being discussed by Grayson (1984). Fig. 8.27 shows the expected, theoretical relationship between the total number of individuals, ie: sample size, and the number of species recorded (Southwood 1978:421). Studies such as those of Buzas et. al. (1982) and Koch (1987) show that a logarithmic series distribution may best predict the species occurrence frequency distribution in a wide range of palaeontological studies. By considering (N-1)/(log N) it may be possible to correct the index for sample size effects, and thus make it more suitable for further statistical manipulation. However, in the present study, while bearing in mind that any attempt to draw conclusions from the number of taxa yielded by a site or level necessitates caution, the decision was made to consider 'N'.

Fig. 8.27. <u>The Relationship between Sample Size and Number of Species (Taxonomic Richness).</u>
(After Southwood 1978:422)

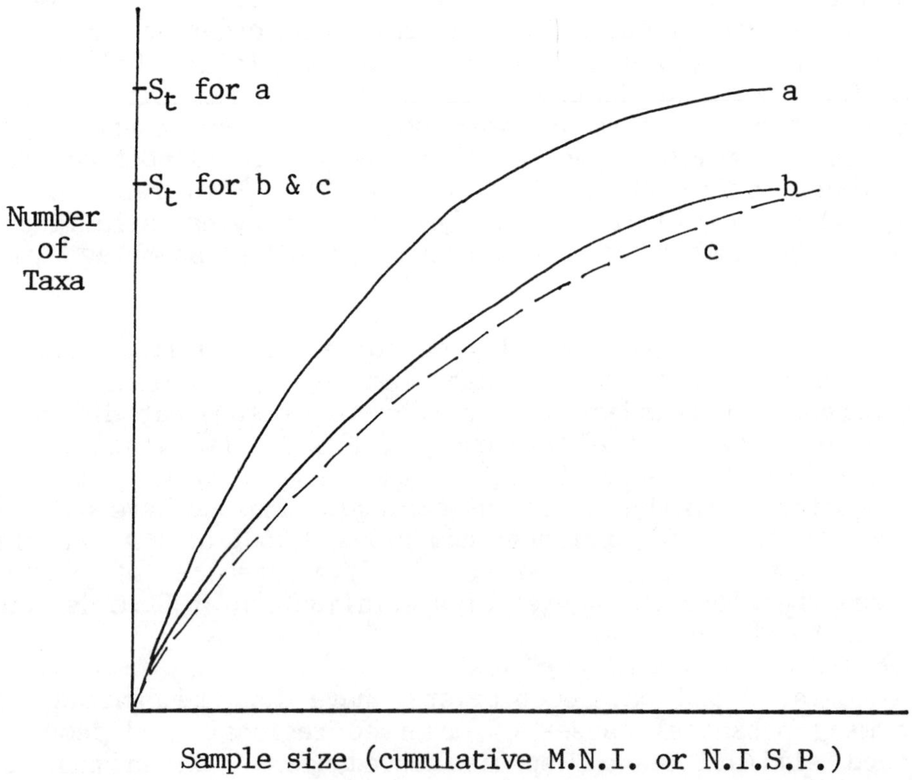

As a result of the problems encountered in considering N, an attempt was made to further examine variation in assemblage structure by turning to other measures. A brief survey of available ecological, palaeontological and archaeological literature revealed the existence of applicable "diversity indices" which measure the degree of heterogeneity of a sample (Krebs 1985:513), ie: take account of the distribution of each species. In the present study Simpson's Index of diversity was selected and is set out below:

$$C = \Sigma p_i^2$$

$$B = C^{-1}$$

where p_i = the proportion of species i.

> It is usual to determine the extent to which an assemblage is narrow (specialised) or broad (generalised) by considering B in relation to N. The value of B ranges from 1.0 to N, reaching N when the assemblage represents a perfectly 'general' distribution in which all species comprise the same proportion of the total fauna: a value of 3.8 at two sites, at one of which N = 4 and at the other 8, implies considerably different degrees of specialisation.

The index is based on the probability of randomly selecting two examples of the same species from a sample or assemblage. It depends upon the degree of concentration of species within the assemblage, reflecting the number of equally common taxa and considering only those species which are present and not those which are known to have occurred in the region but are absent at the site in question.

Both the number of taxa present and measured assemblage diversity (not ecological diversity, given the problems involved in extrapolating from archaeological data to environmental or ecological situation) are considered here, although detailed discussion of relevant theory is not provided. The reader is referred to Magurran (1988), Krebs (1985, 1989), MacArthur (1972), Pielou (1977) or Leonard & Jones (1989); the first provides a recent comprehensive summary of ecological diversity, its measurement and interpretation, while the last concerns the archaeological application thereof. Furthermore, introductions to ecology frequently include a chapter in which the subject is considered.

An increase in species richness and diversity is often, although not invariably, associated with a rise in primary productivity (Brown 1988:69) and environmental complexity and so, by examining the degree of taxonomic richness and diversity, we may, to some degree at least, be able to determine the extent of such environmental complexity. As species diversity declines, one may expect to see the rise in importance of a single, 'specialist' species, usually adapted to more extreme conditions. The survival of such specialists depends upon the relative constancy and stability of conditions. The disappearance of the reindeer at the end of the Upper Palaeolithic may reflect a change from relatively stable, severe, conditions of the late last glacial to post-glacial environments with marked seasonal fluctuations in temperature and precipitation unsuitable for the species. Its survival in small pockets in South West France reflects the distribution of areas in which suitable conditions continued to prevail - on a less predictable and stable basis.

Three established 'biocoenotic' principles, derived from zoogeographic studies concerning the number of species encountered (Illies 1974), apply to the above situation and should be borne in mind in the course of the following discussion:

> (1) The more variable the environmental conditions encountered in a biotope, the greater the number of associated species (Illies 1974:24), ie: complex, heterogeneous environments display more niches and more diverse associated flora and fauna (Putman and Wrattern 1984:332).
>
> (2) The more the environmental conditions encountered diverge from the normal or optimum for species, the less the species diversity and the greater the probability of the ecosystem being dominated by a single species (eg: the reindeer).
>
> (3) The greater the continuity of environmental conditions and the longer the biotope remains homogeneous, the more diverse and stable the associated species will become, ie: there is an increase in species diversity through ecological succession.

During the Aurignacian the number of species (N) varies from 2 to 10 showing a wide range of values which are summarised in fig. 8.28(a), on which the number of taxa at each site during the Early and Later Aurignacian is shown. The overall regional mean value (based on numbers shown in fig. 8.28(a)) is 5.673 (Early Aurignacian = 5.79, Later Aurignacian = 5.46). However, regional deviations are apparent.

Those sites to be found in the vicinity of the Dordogne and Vezere yield a mean value of 5.89 during the Early Aurignacian and 6.143 during the Later Aurignacian, those in the north 5.30 and 4.714 respectively, and in the west, where the site of Haurets dominates the sample (N=10), 6.75 and 4.0 respectively. This last region's sub-totals are small however, the Early Aurignacian totalling 4 sites, the Later Aurignacian 3. These values indicate that those assemblages from the second region are less diverse (poorer) than are those from the others - in terms of N. Such a contrast may indicate that the resource base was wider and more variable in the southern (central) and western regions, that in the north being more limited during the Early Aurignacian.

If similar figures are calculated employing the number of species which occur in each level at each site similar conclusions are reached, although the values concerned obviously change. The overall regional mean falls to 4.98 during the Aurignacian as a whole, while mean values in the north, south and west fall comparatively. In each

Maps showing the Distribution of the Number of Species during the Aurignacian and the Upper Magdalenian

Fig. 8.28

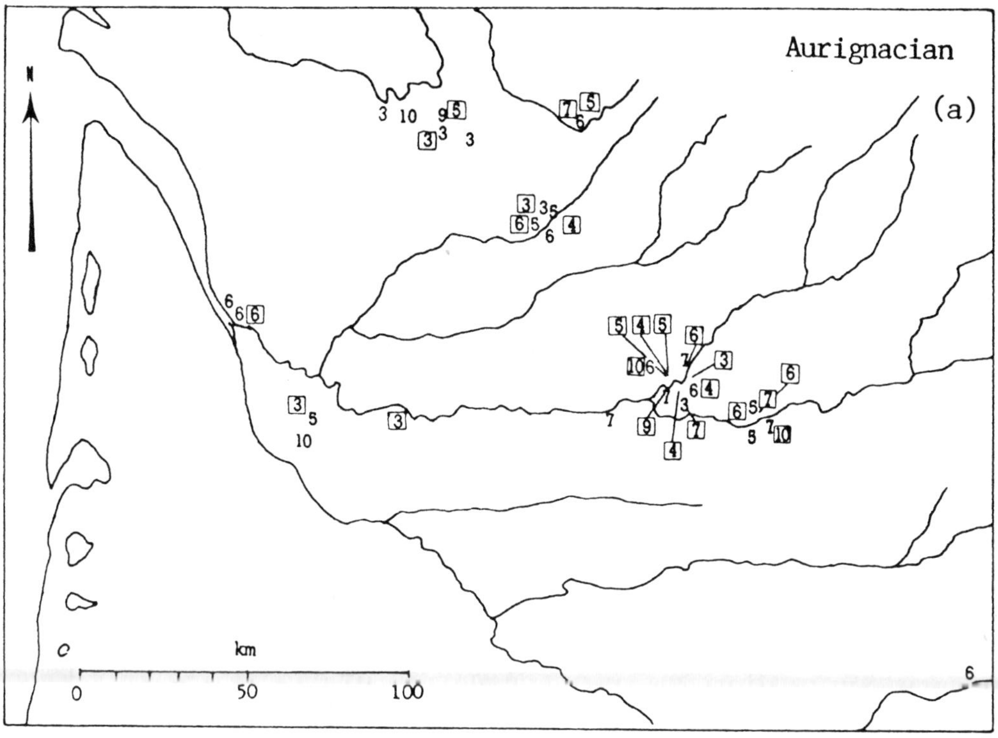

Key: N Early Aurignacian ☐ Later Aurignacian

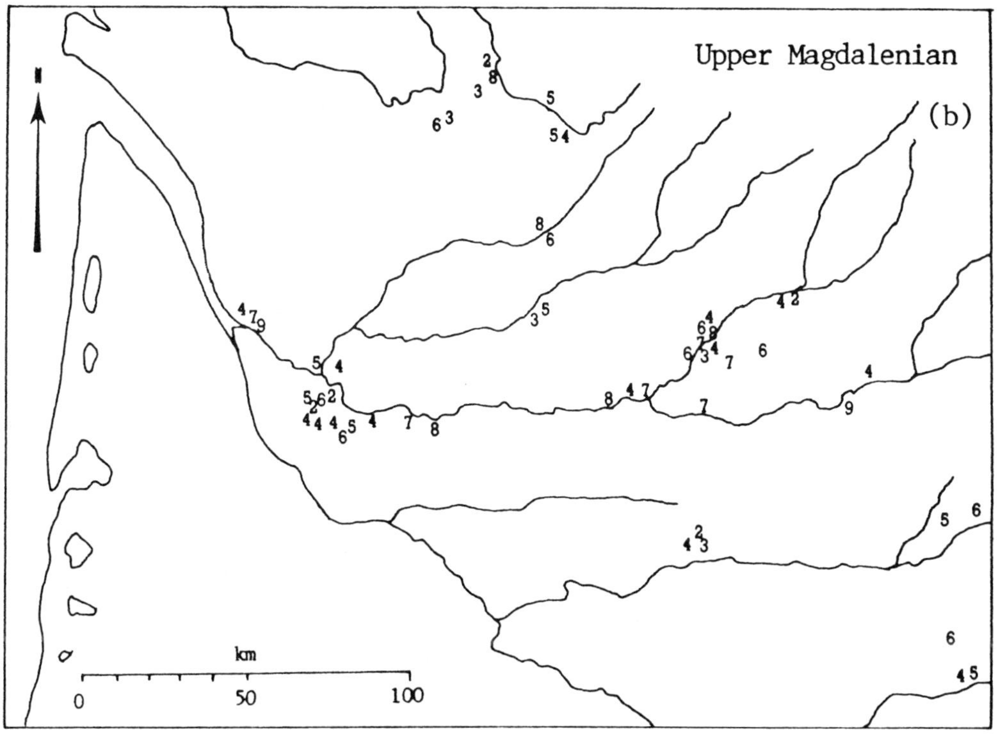

instance an average of 5 or 6 taxa are found in the Dordogne/Vezere (usually dominated by reindeer), 4 or 5 (usually 4) in the north and 5 or 6 in the west of the region.

During the Upper Magdalenian (see fig. 8.28(b)), when N ranges from 2 to 9 (the latter occurring only twice, at Roc d'Abeilles and Roc de Marcamps) the mean number of taxa occurring throughout the region is lower than that during the Aurignacian, namely 5.09. As during the earlier period, fewer species are regularly encountered in northern faunal assemblages than to the south, marked variations between north, south, east and west being seen.

On average 4.853 species make up the large herbivore prey component of assemblages in the north in contrast to 5.647 in the Dordogne and Vezere regions. In the Lot region a mean value of 4.375 is obtained, while in the west the figure rises to 5.059. Such regional differences may be invoked to explain, at least in part, the predominance of settlement in the Perigord and perhaps the existence of a second, important, cluster of sites (this time in the Gironde), for human communities depending upon naturally occurring resources (as do hunter-fisher-gatherers) are, given the choice, likely to congregate in regions of high resource abundance, diversity and predictability. In the Dordogne and Gironde regions, for example, we find areas which are characterised by high frequencies of a preferred species - reindeer in the Perigord, one of horse, saiga or bovids in the Gironde - in association with high taxonomic richness ie: a relatively wide range of alternative species which may be exploited when required.

The fact that the overall regional mean value of N is lower during the Upper Magdalenian than it is during the Aurignacian (Early, Later and combined) may imply that, over the region as a whole, fewer species were regularly exploited on a regular basis - perhaps not surprising given the overwhelming predominance of reindeer in many cases.

The individual regional means calculated for the north, west, Agenais and Dordogne/Vezere areas generally show a decrease from their Aurignacian counterparts similar to that seen for the overall means. In the west (Gironde) and north we see a rise in average number of species during the Upper Magdalenian from the Later Aurignacian, a rise which is associated with high frequencies of saiga and also abundant horse, bovids and reindeer. The Gironde may represent a region in which relatively high population levels of all four herbivores were encountered, at least seasonally, thus precluding the need to 'overexploit' one taxon at the expense of others.

As the number of species encountered during the Upper Magdalenian rises towards the south and east, so the frequency of reindeer increases. The predominance of Rangifer tarandus (exceeding 95% in places) over 5 or 6 other species implies that the reindeer was not culled in proportions directly reflecting the species' abundance in the environment. Given the presence of 5^+ species, random selection, on an encounter basis, should result in more diverse or generalised assemblages, such as those seen more regularly in the early post-glacial. The apparent increase in species number in the north, centre and south of our region may be seen to reflect a concomitant increase in the importance of those faunal resources associated with early post-last-glacial environments and decline in woolly rhinoceros, mammoth and other glacial forms.

Having considered the distribution of species numbers, it remains to take the discussion of diversity a stage further and examine the spatial representation of ecological diversity as measured by the modified form of Simpson's coefficient (B/N) (see figs. 8.28 & 8.30).

Diversity values (B/N) are highest, during the Early Aurignacian, in the west and north-west of the region - at Pair-non-Pair and Fontaury. While these sites yield assemblages which comprise a number of taxa which are relatively evenly represented, that of Bourgeois-Delauny is characterised by a more specialised assemblage. Here, as seen above (p. 187) horse predominates, occurring, it might be suggested, well inside the Aurignacian equid range.

Moving towards the Dordogne valley, lower diversity values are encountered (La Chevre 0.3084, La Ferrassie 0.2581), the most specialised assemblages being found at Roc de Combe (0.2236) and La Gravette (0.1581), in each of which the major taxon (reindeer) represents more than 90% of the large herbivore remains. Further to the south higher diversity is seen at Battuts (0.3276), where six taxa are encountered. Thus the reindeer-dominated region centred on the Dordogne valley is just that, namely an area in which, despite a wide range of available alternative resources, a single species was exploited from most sites in preference to others. Whether this was a seasonal concentration on migrating herds (supplemented by additional resources at other times), year-round culling of permanent reindeer or seasonal exploitation of seasonal resources by a seasonal population of hunter-gatherers can only be speculated upon until a large body of seasonality data for <u>all</u> taxa from Aurignacian sites is available.

During the Later Aurignacian assemblage diversity increases. Of the ten sites considered four display values of B/N in excess of 0.500, three in the Dordogne and one in the west, at Pair-non-Pair. Only two sites, Pataud and Roc de Combe, have values which indicate 'specialised' assemblages. At the latter site reindeer totals, although lower than those observed during the Early Aurignacian, remain high, dominating the nine other taxa which occur. At the Abri Pataud reindeer dominates eight other species. The more complex, heterogeneous assemblages (in terms of both N and B/N) encountered in the Dordogne are, in general, more diverse than their simpler counterparts observed in the Vezere.

An increase in species richness reflects, in general, a rise in spatial complexity which may itself be caused by growing species richness. A small region in a heterogeneous habitat is more likely to provide the necessary requirements for a wider variety of taxa than is an area in a more homogeneous zone. It has often been noted that herbivores occur more densely, and are therefore more readily available for exploitation, in areas where numerous small patches of differentially-structured vegetation exist in close association. As Moss et. al. (1982:74) say - 'more edge, more herbivores'. During the Later Aurignacian the Dordogne may be seen as a region of 'edge'.

During the Upper Magdalenian assemblages are often more specialised than their Aurignacian counterparts. Indeed, as was

Maps showing the Distribution of Diversity Values (B/N) during the Aurignacian

Fig. 8.29

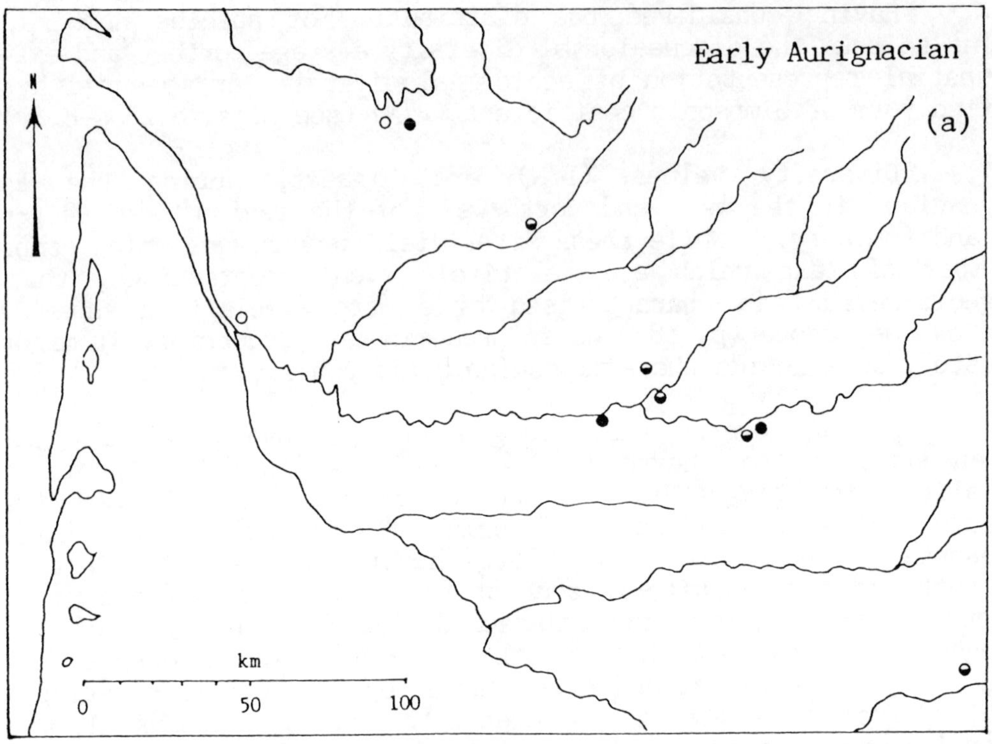

Key: Diversity values (B/N) ● 0.00–0.249 ◒ 0.25–0.499 ○ 0.50–0.749 ⊘ 0.75–0.999

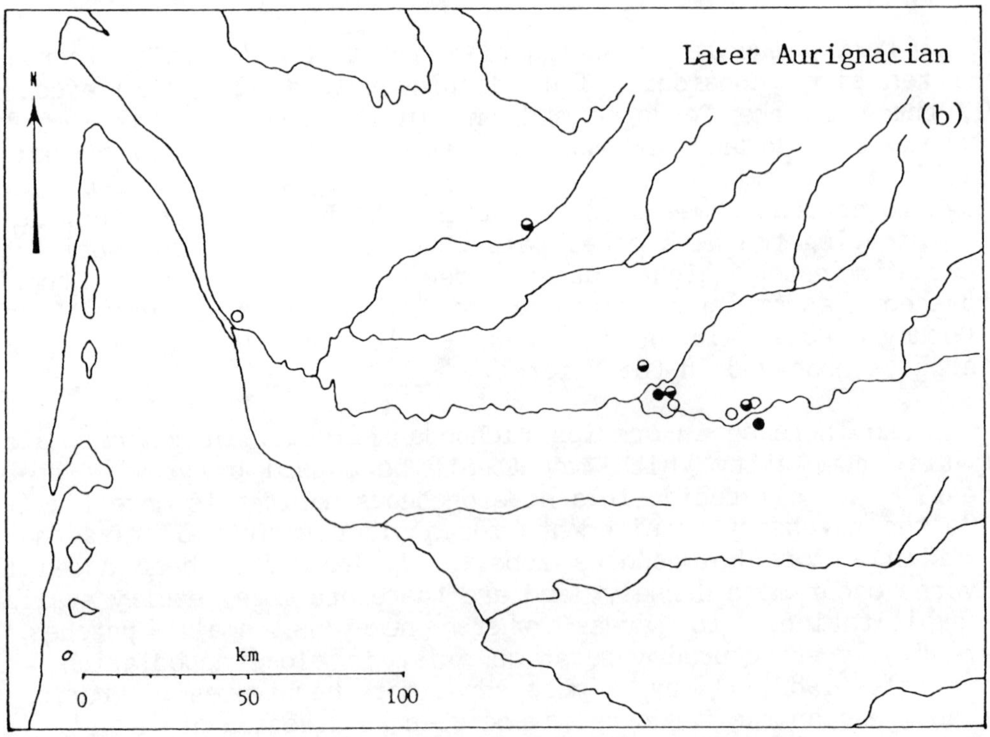

Key: Diversity values (B/N) ● 0.00–0.249 ◒ 0.25–0.499 ○ 0.50–0.749 ⊘ 0.75–0.999

Map showing the Distribution of Diversity Values (B/N) during the Upper Magdalenian

Fig. 8.30

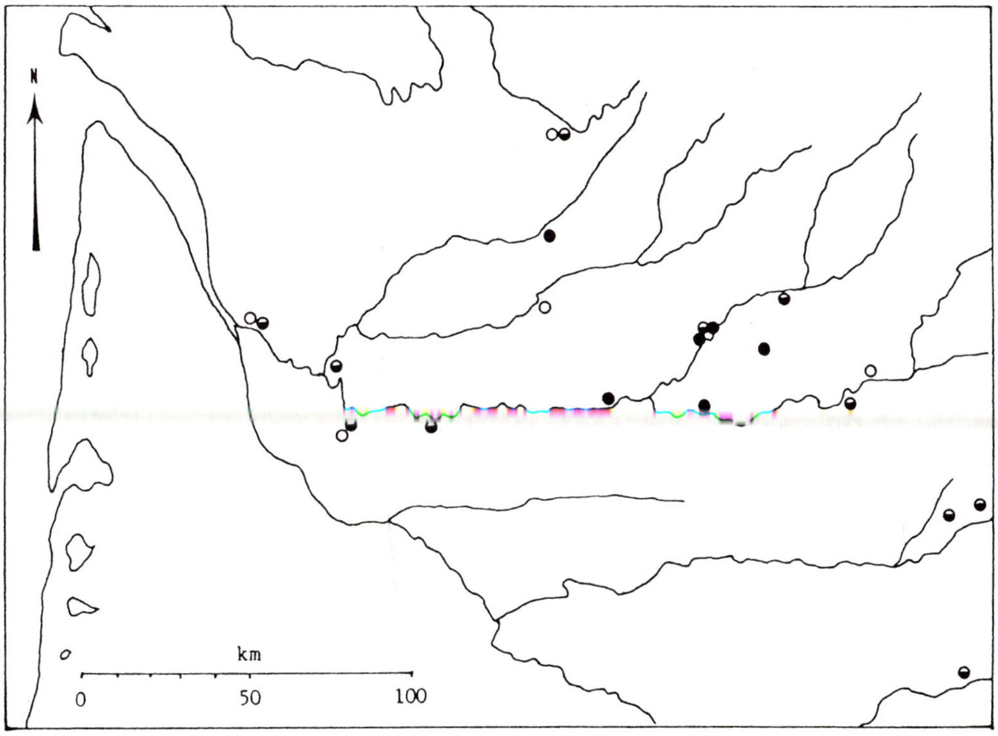

Key: Diversity values (B/N)

● 0.00–0.249 ◓ 0.25–0.499

○ 0.50–0.749 ⊘ 0.75–0.999

observed during the Early Aurignacian, sites in the Dordogne/Vezere region yield more specialised assemblages than elsewhere, with one exception towards the north (again non-reindeer). This time however, lower diversity, or greater specialisation, is to be seen in the Vezere suggesting either the exploitation of less diverse environments than was the case earlier on, or substantially more specialised exploitation strategies.

Sites to the south are more generalised, those species exploited being taken in more even quantities. This might imply that the herd animals (usually reindeer) which were exploited elsewhere on a scale to result in low values of B/N were not available to the same extent. The same may be said of western assemblages where Fongaban represents the most specialised site, values of B/N being 0.320, 0.257 and 0.541 in levels 2, 3 and 4-6 respectively, for those assemblages derived from Girondin contexts yield higher diversity indications, being, as they are, located in an area characterised by 'abundant' horse, bovids and saiga herds during at least a part of the year. Two of the six assemblages considered have diversity values exceeding 0.500, while the remaining four all fall into the 0.25-0.499 category.

Multiple occupation sites display relatively consistent diversity measures, eg: Gare de Couze and La Madeleine. Of the 23 levels encountered at these two sites only two (both at la Madeleine) display values exceeding 0.36. At no point do we find highly specialised assemblages immediately following diverse ones, or vice versa. This may be seen to imply that the environment and / or the relevant foraging strategy were consistent. Such may lead us to infer the repeated seasonal occupation of a site in order to exploit the known periodic abundance of a resource, for the low diversity figures lead us to expect specialisation as a feature of the subsistence strategy.

In the present study reindeer forms this targeted species although to the west, it failed to impinge upon the subsistence strategy to the same extent. If we accept Gordon's (1986a) conclusions, that both the Gironde (west) and Dordogne - Vezere (Central) regions are areas of winter occupation, we are left to wonder why specialisation on reindeer occurs to a much greater extent in the east than the west. Likewise, if we assume that hunting technology and 'know-how' were alike 100 km apart we can not infer behavioural causes of the differences observed. Such may be assumed based upon the conclusion of several authors concerning the seasonal movements of Late Upper Palaeolithic human groups and projected 'trade and information exchange' systems (Gamble 1982, Bahn 1982). We are forced to conclude that environmental conditions were different in the west and to the east, the change becoming apparent as we approach the valley-interfluve landscape of the Perigord and upland of the Massif Central.

8.4 Discussion.

During the Aurignacian two clusters of sites yielding faunal data are apparent - in the centre of the region (in the Dordogne and Vezere valleys) and in the north (Charente and Dronne). These assemblages indicate that reindeer was exploited throughout the whole region on a regular basis, as were horse and, to a lesser extent, bovids. Montane species were generally of greater importance in the Dordogne and Vezere area than elsewhere, although peak ibex values are seen at Battuts (in the south) during the Early Aurignacian. Although the chamois, which is less widespread than the ibex, is known to occur to the north-west and south of the Dordogne/Vezere confluence, it rises in terms of frequency towards the Massif Central.

More temperate species occur more frequently in the southern part of the region as a whole. Roe deer is restricted to the Vezere/Dordogne, red deer is more common there than elsewhere (failing to occur at only two sites), while boar clusters in the Vezere with scattered pockets to the north and west. Meanwhile, lagomorphs, which are dispersed throughout much of the region, are more often found in the southern part of the Dordogne and Vezere where half the occurrences are located.

Finally, the woolly rhinoceros represents a species which is widely scattered, although it is not recorded in the far south and occurs only twice in the Dordogne/Vezere. The presence of a species of this size, involving substantial risk to the hunter-gatherer may imply the use of traps and co-operative hunting strategies or the scavenging of 'natural' deaths. The same would apply for the mammoth, which largely occurs at the same sites as the woolly rhinoceros. The megaherbivores encountered do not represent a major component of the Upper Palaeolithic resource base. The presence of several sites with several species and the marked predominance of a single taxon (reindeer) suggests that the 'other' species formed an alternative available resource, should the preferred prey fail.

During the Upper Magdalenian we see, in the west of the region the predominance of those species which are associated with relatively steppic conditions (bison, horse and saiga). A relatively cold, open environment is thus implied, although, given the observed distribution of more temperate, woodland taxa, scattered pockets of woodland or light forest may have occurred resulting in a 'forest-steppe' ecotone. In the Agenais reindeer and horse are the major species, the former largely predominating. Saiga and bovids comprise and important secondary group of species in an area in which a woodland component is rare but may be associated with pockets of rich, river-valley vegetation. Meanwhile, montane species are regularly observed - again in the Agenais, Lot and Dordogne, associated with river rockfaces and steep valley slopes.

In the region centred on the Dordogne/Vezere confluence, where reindeer predominates over all other taxa, a large proportion of the total potential resource spectrum was regularly exploited - albeit in limited quantities. The area is one in which a relatively high number of taxa is available, although, with reindeer representing in excess of 80% of the large herbivore remains at many of the sites yielding suitable data, diversity values in the Dordogne/Vezere are low. However, it may be that the existence of a wide range of 'other' taxa at Upper Magdalenian sites indicates that sudden failure of the normally stable, relatively predictable reindeer could be compensated for, in this area, by increased exploitation of a wide range of alternative resources which includes other large herbivores, lagomorphs, birds and fish, in addition to an unknown plant resource spectrum.

As during the Aurignacian, highest frequencies of reindeer are usually found to occur, during the Upper Magdalenian, at sites located on the main river valley routes, with lower percentages found further away in the subsidiary valleys, suggesting that the major valleys formed important herd migration routes. Given the propensity of reindeer or caribou today to course natural features (eg: rivers) prior to crossing them, following the contours of the local topography and thereby using the line of least resistance (see chapter 4), it is hardly surprising to find abundant reindeer material at river-side sites. Such animal behaviour could be quite easily exploited, while alternative, perhaps non-migratory, species could be taken elsewhere. Support is leant to this by the fact that assemblage diversity (B/N) rises with increasing site altitude and, to a lesser extent, with distance from major river valleys (altitudes and distances being taken from White 1985).

In the northern zone, we see a resource base which is essentially similar in nature to that encountered immediately to the south. Reindeer and horse are wide ranging, bovids occupying a subsidiary position. Saiga occurs in the west of the area, occupying cold, dry, relatively flat and open terrain. To the east we see a slight increase in the importance of woodland species, implying a more closed, perhaps humid landscape.

In summary, several environmental types are suggested by the spatial faunal associations observed. These compare favourably with those identified by statistical analysis.

1. <u>INTERFLUVE</u>.

(a) <u>Steppe</u>: Horse, bison, saiga (Western cold-dry steppe). Open cover of grass, interspersed with isolated trees (Forest steppe).
(b) <u>Tundra</u>: Reindeer, hare, arctic fox, lemming, bison?. (Arctic or alpine meadow cover. Moist oceanic tundra)

2. <u>WOODLAND</u>.

(a) <u>Taiga</u>: Tundra species plus elk and red deer. Dense arboreal (coniferous) cover.
(b) <u>Deciduous woodland</u>: retains taiga faunal elements with addition of species with low resistance to cold and aridity, eg: boar, roe deer, <u>Bos</u> sp.

3. <u>WETLAND</u>. Boar, roe deer. River bank and marsh/bog.

4. <u>SKELETAL</u>. Montane species (ibex, chamois) on rock-faces etc.

The postulated distribution of the above environmental zones is shown, for the Late Upper Palaeolithic, in figure 8.31. It largely reflects the occurrence of topographic features, ie: the valley and

Landscape Types in South West France during the Upper Palaeolithic

Fig. 8.31

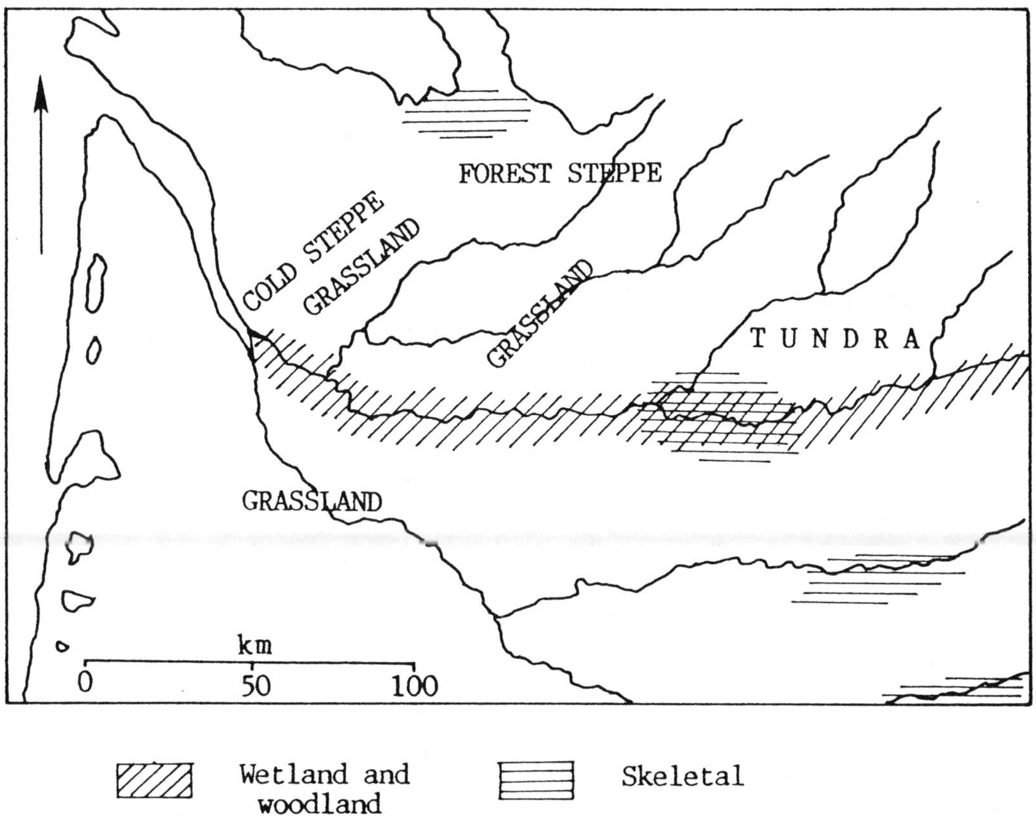

interfluve landscape of the Vezere and Dordogne, the rockfaces of the Quercy and the flat terrain of the Gironde. However, changes in the spatial distribution of faunas are seen as we progress through the Upper Palaeolithic, the Middle and Upper Magdalenian being considered here.

During Magdalenian III northern South West France is divided into three faunal regions:

(1) A forest-steppe and tundra (mixed) region, where biotopes meet. This occurs in the north, centred on assemblages with high reindeer frequencies and a broad subsidiary resource base, including relatively abundant saiga and horse.

(2) A reindeer dominated, moist (oceanic) tundra zone centred on the Vezere/Dordogne confluence and extending southwards into the Lot, where, at a local scale montane communities gain in importance. This 'mosaic' environment is characterised by a range of taxa, including regularly occurring woodland/forest species.

(3) An open, cold-steppe region, centred on the Entre-Deux-Mers region (Gironde). Here we find abundant saiga and horse, associated with important reindeer and bovids. The area is one in which large populations of seasonally-varying herbivores were available for exploitation - probably on a year-round basis.

In addition to these regions, one must, of course, bear in mind the potential of the land mass encountered westward of the present day coastline - due to lower sea-levels. It is almost certain that coastal exploitation took place in some form, although the nature of resource procurement, and indeed the resources themselves, remain totally unknown to this day. Whether it was simply 'marine' resource exploitation and/or a continuation of the Gironde system is unknown. Suffice it to say that shells from the Atlantic coastline are known from inland sites.

By Magdalenian IV (see fig. 8.32) the distribution has changed somewhat as the late glacial ecosystem develops. By this time saiga frequencies have reached their peak in the Gironde and are beginning to decline, although horse continues to be important. Both taxa are to be found slightly to the north, saiga being of considerable importance in the southern Charente catchment. Reindeer remains of paramount importance in the east of the region where woodland species fluctuate in numbers. Low levels of steppic species occur, mainly saiga, as this species spreads eastwards along the Dordogne, Isle and Vezere river valleys. However, the local topography is not suitable for saiga and percentages remain low.

Once again we can postulate the existence of a significant potential zone of coastal exploitation, although, given a rise in sea-

Maps showing changing Zoogeographic Patterning in South West France between 15 000 and 10 000 B.P.

Fig. 8.32

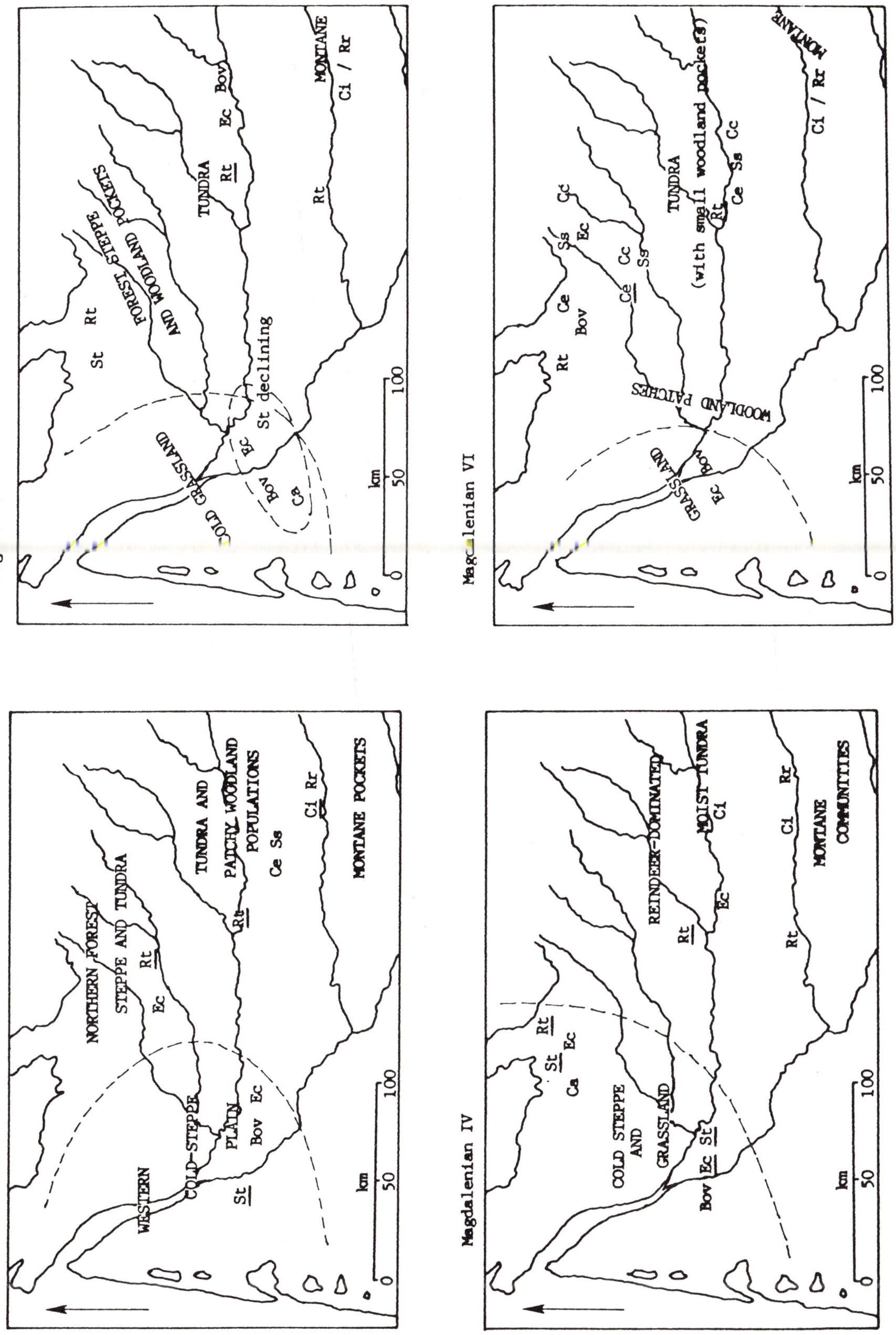

levels in the range of 30m, the region exposed will have become smaller.

During Later Magdalenian (V) reindeer continues to dominate assemblages in the Perigord where cold, moist tundra conditions still prevail. Ibex is encountered in close association with the reindeer to the south east of our immediate region (ie: in the Lot and Tarn), while to the west, steppic conditions remain, although the saiga is now to be found in lower frequencies than was the case earlier on (eg: at Fontarnaud, a site where one of the final occurrences of woolly rhinoceros is observed). This may be explained by an increase in woodland cover in the Gironde. As sea-levels continue to rise, by this time to ca.-70m, so the area to the immediate east of the coastline will become smaller and more oceanic or humid, precipitation increasing. Dry, open conditions which the saiga requires in order to maintain high population levels gradually would begin to disappear, to be replaced by woodland - a biotope to which saiga is not adapted.

During Magdalenian VI a change in faunal assemblage composition is apparent. Despite the remaining dominance by reindeer of Perigordian assemblages, more temperate, mild, woodland species are gaining in ascendancy. Thus, in the Dordogne we see reindeer in association with red deer, bovids, horse, boar and some roe deer. Ibex populations are encountered further to the east with sporadic occurrences in the Vezere catchment. Woodland conditions are also encountered in the Garonne, although to the north and west herd ungulates predominate, in the form of horse and bovid populations. By this point sea-level has risen to -50m and saiga is on the point of local extinction, which, together with the more widespread extinction of woolly rhinoceros and mammoth, heralds the beginning of early post-glacial faunas.

Chapter Nine

BUTCHERY CARCASE MANAGEMENT.

For some years now it has been apparent that a working knowledge of both modern and prehistoric butchery and carcase disarticulation processes is of value to the archaeologist. Much of the work has centred on the question of natural disarticulation, due to the problems, highlighted during the past fifteen years, of distinguishing human (cultural) from carnivore or natural (geological) deposits (Hill 1979; Brain 1980, 1981; Binford 1981).

Recent years however have seen an increase in the study of human butchering procedures - both modern and prehistoric - for a knowledge of the remains left by contemporary hunter-gatherers may provide insights into the structure of the archaeozoological record. The more valuable parts of carcases, (usually the meat-bearing ribs, vertebrae and proximal long bones) may be found at what might have been base camps. Field camps, temporarily occupied, whether on only one occasion or repetitively, may be characterised by abundant, discarded parts of lower value (eg: metapodia, phalanges, skull and mandible), butchery and procurement being the major tasks undertaken. At more long-term occupation or residential sites a mixture of parts is most likely seen, a variety of activities being reflected.

Research reported by Binford and Bertram (1977), describes the differential survival of anatomical elements in terms of the density of each element type, for it is worth noting that those elements which are commonly considered to be of high value (largely in terms of food-yield) are in fact of low density, as measured by Binford & Bertram (1977), and therefore may have less chance of surviving. Survival (or failure to survive) may be due to several factors. The part concerned may have been intensively processed. Alternatively it may have succumbed to post-burial destructive taphonomic processes. Similarly, low value is frequently associated with high density and thus high preservation potential (Lyman 1985). Thus the frequency of high utility/low density values may perhaps, be argued to be of greater significance than abundance of low utility/high density pieces.

The work of Binford and Bertram was followed up by Binford (1978) in "Nunamiut Ethnoarchaeology". Binford's main aim here was the determination of subsistence practices among the Nunamiut through the identification of procurement, processing and storage techniques. Employing utility indices, he argued, it should be possible to detect butchering and other decisions. For example, in a situation in which meat was sought rather than marrow the proximal or distal femur (meat index 100) should be the most commonly chosen element. Whether this part represents 100% (maximum M.N.I.) or is absent remains a question of whether the assemblage results from the collection of parts used or the discard of unwanted elements. Despite the inevitable problems encountered when employing procedures devised for the analysis of one specific case-study to other, different situations, procedures are described which, Binford suggests, should allow us to examine patterning observed in the archaeozoological record. He provides various 'indices' (although Chase (1985) has recently stated that the large majority of these are predictive models rather than indices) which may allow us to evaluate the significance of element representation in terms of meat, marrow and white grease. These indices provide a "reference dimension" (Metcalfe & Jones 1988) with which the archaeological record may be compared, the indices predicting, to some extent the anatomical structure of assemblages. They are combined to form what Binford (1978:72) terms the General Utility Index and then modified to form the "Modified General Utility Index", both of which Chase (1985) considers models rather than indices.

The indices described by Binford (1978) fall into two broad categories, of which only one will be considered here - the Modified General Utility Index, Meat Index, Marrow Index etc. The other category is developed later by Binford (1978) and is considered too situation-specific to be employed here.

The Modified General Utility Index (henceforth M.G.U.I.) was based upon the premise that animals are cut up as units of bones as opposed to individual elements, and that low utility parts are often found to be attached to those of higher value. Thus back-feet may 'ride in' with hind limbs. The M.G.U.I. takes account of this by raising the value of lower utility elements according to the higher values of the parts to which they are attached, although neither G.U.I. nor M.G.U.I. incorporate a measure of transportability per se, which might most conveniently be considered in terms of mean element and associated meat weight.

In the present chapter all the species considered are examined in terms of reindeer/caribou utility values. This is done in the belief that results should be comparable. The use of different indices might prevent this, despite the fact that they measure the same things and are correlated (see fig. 9.1). Only the Grotte des Eglises, where ibex is considered, presents a problem. Although the sheep index may be deemed more relevant, the size difference between the sheep used by Binford (1978) - total live weight 98.991bs (44.905kg) - and the average ibex weight is such as to suggest that the larger, reindeer/caribou indices are more suitable. Furthermore, Pleistocene ibex are generally considered to have been larger than their modern counterparts, and thus the size differential increases.

The problems discussed by Chase (1985) need to be borne in mind during the discussion, although his apparent belief that a linear regression line (a straight line, Y=a+bX) may be expected when predictive models are employed archaeologically would seem to be somewhat optimistic. We can not expect a 'perfect fit' when dealing with the archaeological record, ie: material which has been biased by post-kill processing, preservation conditions, collection and even analytical procedures. Furthermore, as the results which follow show, much of the patterning observed seems to agree closely with results presented by Binford (1978) and Speth (1983). In addition, those plots showing simply the percentage of the maximum M.N.I. value observed (M.N.I. being calculated simply by dividing the number of an element type by the number of that element in the body) are presented and show close similarity to profiles described by Binford for known season and activity. The methodology, devised as it was for the study of

The Relationship between M.G.U.I. and Meat Index Values for Caribou and Sheep.

Fig. 9.1

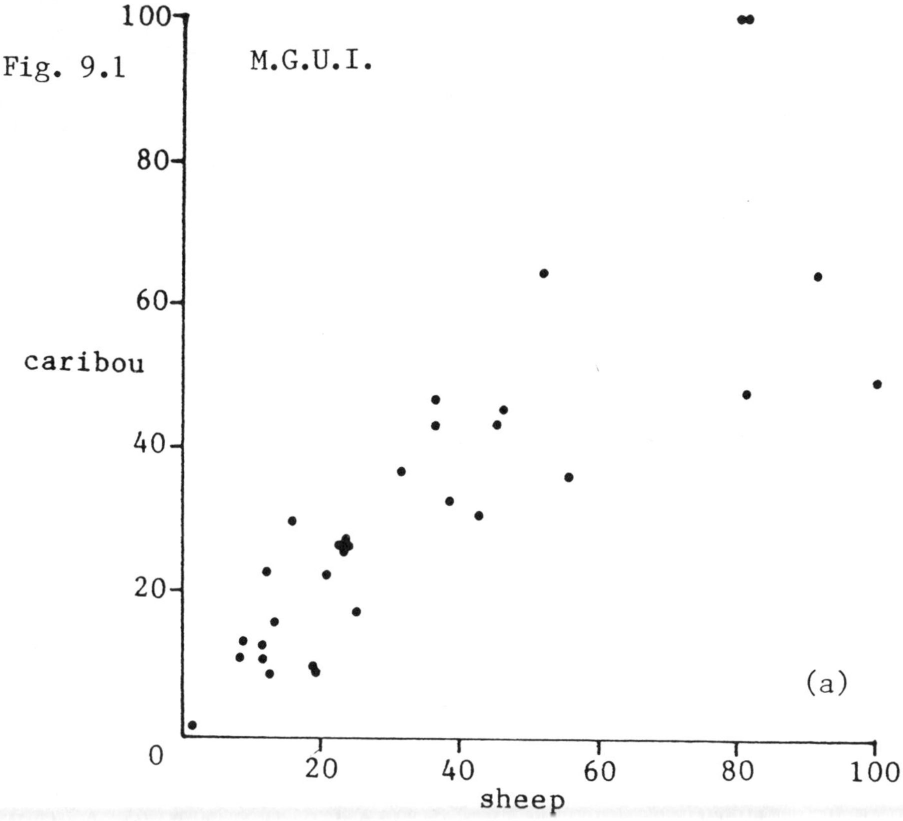

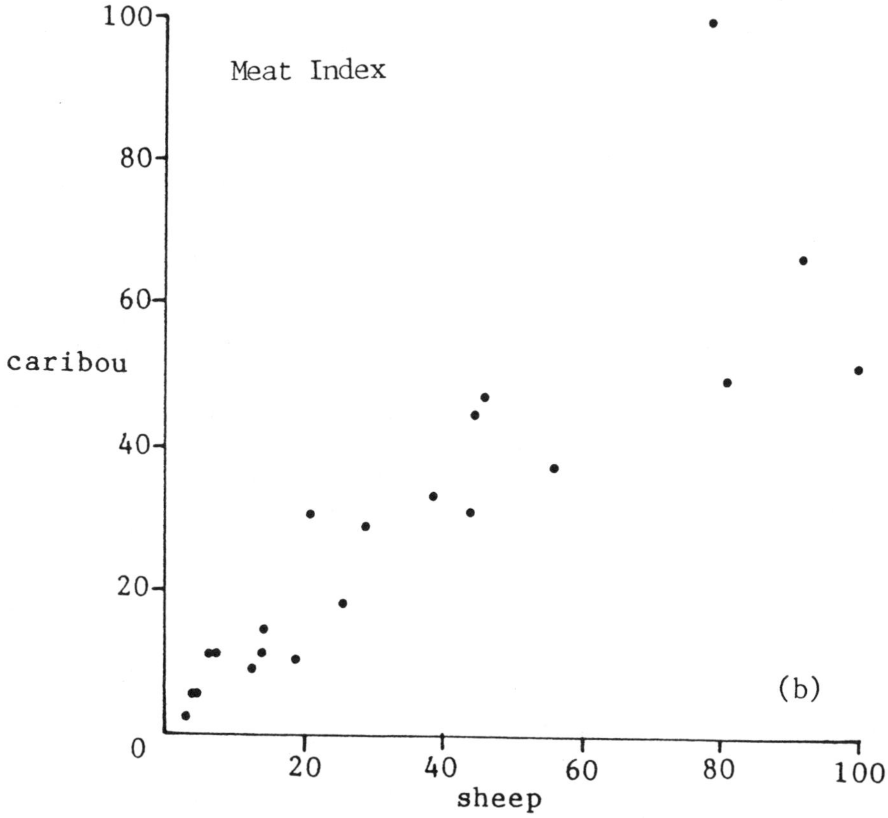

Nunamiut (Alaskan caribou-dependent hunters) subsistence strategies, was originally adopted here due to the broad similarities in faunal assemblage structure in Upper Palaeolithic South West France and the Alaskan archaeozoological record, ie: assemblages in which between 75% and 95% of the material is reindeer lend themselves to analysis using data derived from caribou material. Subsequent analysis of other assemblages resulted when it became apparent that variations in results could be, at least to some extent, related to fluctuating relative frequency of species.

9.1 Methodology.

In the present study the approach adopted is that advocated by Binford (1978) and modified by Speth (1983), although the results of earlier research reported by the likes of Lacorre (1939), Capitan and Bouyssonie (1924), Bouchud (1961) and Delpech & Rigaud (1974), will be incorporated where possible in order to take account of the extent of fragmentation and bone surface modification as well as element type representation.

For the sites considered (limited by the scarcity of suitable, largely published data), the %fM.N.I. value for each element was calculated in the way suggested by Binford (1978): having identified element types (fractional) M.N.I. counts were calculated as described above. Adopting the highest fM.N.I. value as equivalent to 100 %, the proportion each other element represents was computed, eg: given a maximum of 12.5 fM.N.I. from distal metatarsals, an fM.N.I. count of 7.5 from proximal humeri represents 7.5/12.5 or 60%fM.N.I.. The %fM.N.I. value of each element type was then plotted (y-axis) against its M.G.U.I. value (x-axis) (see table 9.1). Where a choice of M.G.U.I. values is provided (e.g.: mandible with/without tongue) the two values are shown linked by a solid line. In addition, comparative percentages for body parts remaining at various sites are shown.

When considering fM.N.I. counts we are not forced to assume that animal units were in fact complete individuals. In the traditional manner of calculating M.N.I. figures there is no way in which we can account for the introduction of parts of animals. In many cases among hunter-gatherers, animal units or meat parcels are introduced on to sites since the effort involved in transporting the complete carcase from the kill site outweighs the benefits of so doing. Binford (1978:70) suggests that the M.N.I. count should be obtained by dividing the frequency of a certain element in an assemblage by the frequency of that element in the animal. In this way we obtain 'fractional' frequencies. Field experience taught Binford that little attention is paid to left vs. right hand side of carcases when butchered and, this being the case, he claims that the lack of attention to this aspect of the assemblage is unimportant. Body size and nutritional state (some carcases are discarded as being of too poor quality) are the more important factors in determining carcase usage.

Finally, before turning our attention to results of analysis, it should be noted that in the present study no antlers are included. The decision to omit these was made due to the fact that published sources often fail to distinguish between shed and unshed antlers.

Table 9.1

M.G.U.I., Marrow, Meat and Density Values.

Anatomical Part	M.G.U.I.	Marrow Index	Meat Index	Density
Antler	1.02	1.00		
Skull	7.49	1.00	9.05	
Mandible	30.26	5.74	31.10	1.55
Atlas	9.79	1.00	10.10	1.45
Axis	9.79	1.00	10.10	1.38
Cervical vertebrae	35.71	1.00	37.00	1.28
Thoracic vertebrae	45.53	1.00	47.20	1.26
Lumbar vertebrae	32.05	1.00	33.20	1.37
Pelvis	47.89	7.85	49.30	1.52
Ribs	49.77	1.00	51.60	1.07
Sternum	64.13	1.00	66.50	0.76
Scapula	43.47	6.40	44.70	1.40
Humerus (p)	43.47	29.69	28.90	1.01
Humerus (d)	36.52	28.33	28.90	1.41
Radio-cubitus (p)	26.64	43.64	14.70	1.33
Radio-cubitus (d)	22.23	66.11	14.70	1.36
Carpals	15.53	1.00	5.20	1.19
Metacarpals (p)	12.18	61.68	5.20	1.28
Metacarpals (d)	10.50	67.08	5.20	1.25
Femur (p)	100.00	33.51	100.00	1.29
Femur (d)	100.00	49.41	100.00	1.41
Tibia (p)	64.73	43.78	25.50	1.38
Tibia (d)	47.09	92.90	25.50	1.46
Tarsals	31.66	1.00	11.20	1.29
Astragalus	31.66	1.00	11.20	1.28
Calcaneus	31.66	21.19	11.20	1.28
Metatarsal (p)	29.93	81.74	11.20	1.33
Metatarsal (d)	23.93	100.00	11.20	1.20
Phalanx I		30.00	1.70	0.90
II	13.72	22.15	1.70	0.81
III		1.00	1.70	0.76

(Source: after Binford 1978, 1981)

Only unshed antlers should be included in a study such as this. Shed antlers do not indicate that any part, however small, of a carcase was present at the site. Such antlers merely point to the accumulation, intentional or 'accidental' (natural) of antlers. As a source of 'raw material' they are significant. As a source of food or 'subsistence material', they are not.

9.2 Jaurens: a 'natural' deposit.

Before examining the degree of element survival and representation at a selection of Upper Palaeolithic sites in South West France, a 'natural deposit' of appropriate age will be considered.

The site of Jaurens lies in the Correze, near Nespouls. A series of radiocarbon dates has been obtained, placing the deposit well within the range of the Early Upper Palaeolithic.

Ly-359 (secteur median)	29 300 +/- 1 400 B.P.
Ly-892 (secteur median)	30 350 + 3 000 / - 1 900 B.P.
Ly-1938 (secteur 20)	32 630 + 2 900 / - 2 100 B.P.
Ly-1939 (secteur 1)	29 700 + 1 500 / - 1 300 B.P.

Jaurens therefore forms a useful control against which we may compare subsequent analytical results. Animals are generally believed to have arrived at the sites as carcases (Guerin et. al. 1979), little of the material reported bearing evidence of carnivore (ie: hyaena) gnawing (see below). Although some water transport is proposed, albeit of short duration, several elements are found still in articulation, a fact which suggests that post deposition modification is minimal. However, according to Debard (1979) the geological matrix comprises alternating levels of clay and sand, an estimated 98% of material being of allochthonous nature, transported by water. Debard attributes this to the age of the deposit and the climatic conditions under which it formed. Associating it with the mild wet climate and short, dry spells of Wurm III, Perigord III-V, the Jaurens deposit would appear to be contemporaneous with the Early and Intermediate Aurignacian at the Abri Pataud (levels 11-7), the Evolved Aurignacian at Roc de Combe, Aurignacian II-III at La Ferrassie and Aurignacian II at Le Facteur.

Bone material is well preserved at Jaurens, sufficiently so in fact to yield the skeletal remains of four foetuses of horse - the major species. However, as Mourer-Chauvire (1980) points out, not all elements are equally well preserved, the vertebrae and ribs being much rarer than other parts. The presence of foetal bones is used by Mourer-Chauvire to infer the season of death, an event she places at the end of winter or early spring. This makes the material which she describes even more suitable for consideration here, as the most commonly cited season of death (and therefore the commonly presumed season of human exploitation) falls within the late autumn, winter and early spring (Spiess 1979:201; Gordon 1988).

Teeth represent the most commonly recovered element at Jaurens, a total of 427 being ascribed to the horse. Based solely on the lower teeth available, the M.N.I. value obtained is 42 (34 adults and 8 juveniles), while Levine (1979) reports an M.N.I. value of 22. Mourer-Chauvire however pushes the total still higher by redividing the juvenile material by 'secteurs' of the site. She thus adds 12 juveniles to the 34 adults over the age of 4 years recognised. By ignoring the teeth, as we shall do in the present study, we reduce the M.N.I. count to 19.5, a figure which represents only 42.39% of the alternative maximum derived from dental material. The foetal material is excluded completely from subsequent discussion.

The horse bones recovered at Jaurens show little sign of post-mortem modification. Possible human activity is indicated by surface marks on one skull fragment, while only a small proportion of limb bones (humerus, radio-cubitus, femur and tibia) show signs of hyaena gnawing. Humerus shafts show more evidence of gnawing than do other elements, distal epiphyses far outnumbering the total proximal ends. Mourer-Chauvire (1980) attributes this to the fact that the distal portions are harder and less rich in spongy tissue and thus they are more likely to survive. Furthermore, according to Binford & Bertram (1977) and Binford (1981) the density value for the distal humerus exceeds that for the proximal end. Meanwhile, the ratio of distal to proximal tibiae places the deposit in the "zone of destruction" as defined by Binford (1981:219), although the case of the humerus is less clear-cut.

Of the 32 femurs only 3 are complete, while 14 shaft fragments display tooth marks. Meanwhile, a total of 34 tibiae are recorded. Of these 24 are distal, 2 proximal and 7 complete. Of the 24 distal segments, 12 bear gnawing marks. Mourer-Chauvire suggests that carcases were exploited, to a limited extent, before being transported to their final position in the gallery (1980:42). She infers that hyaena activity was slight based on the fact that only one identifiable coprolite was uncovered while no suggestion that the site was used as a hyaena den is made by the author.

Given the above information, derived almost solely from reports in the Nouvelles Archives du Musee d'Histoire Naturelle de Lyon for 1979 and 1980, it simply remains here to present the data provided by Mourer-Chauvire in a manner which is comparable with that obtained from other Upper Palaeolithic sites - all of which are believed, on the basis of close association of hearths, stone tools and other artefacts, to be due to human activity.

Figure 9.2(d) shows the percentage fM.N.I. values in the form of an element representation profile, while in figure 9.2(a) to (c) we see the proportional values plotted against density, meat index and M.G.U.I. respectively. Density figures are employed in order to gauge the extent to which survival is directly related to the hardness (density) of the bones. The values used are those of Binford and Bertram (1977:109) and subsequent modifications recorded by Binford (1981:218). The curve labelled S.P. represents the Survival Potential of bones obtained adopting the following equation, detailed by Binford (1981:217).

$$S.P. = -352.78 + 1050.4(d) - 1008.69(d^2) + 332.882(d^3)$$

From fig. 9.2(a) we can see that it is not simply density which governs preservation in the case of Jaurens.

Fig. 9.2 Jaurens

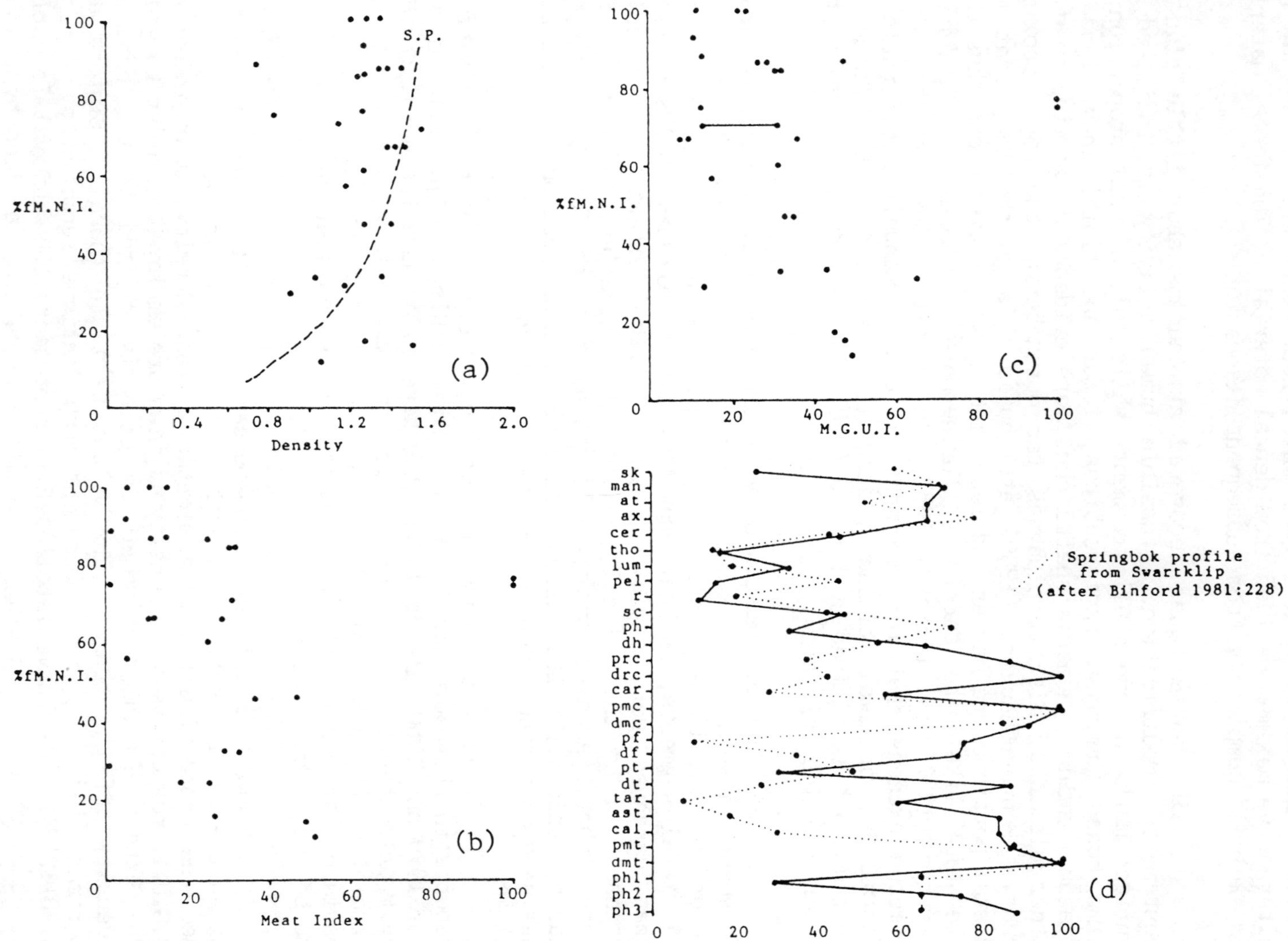

Given that the deposit is associated with some hyaena activity it was decided to compare element representation with the meat index, on the basis that it is the meat which would have been sought by the hyaena. With the exception of the femur (with a meat index value of 100), bone representation broadly declines with increasing meat index values. In general the relationship is not good, although it may be used to imply that those bones which were removed, either before or after transportation to the site itself, were those with higher meat utility values.

The plot shown in figure 9.2(c), in which %fM.N.I. values are compared with M.G.U.I. values, once again points to the apparent over-representation of the femur. As with the meat index, percentages fall as M.G.U.I. values rise, although not to the extent to be seen with the sites described below.

Jaurens forms a useful control sample against which to compare other assemblages, if only in that it shows us that the pattern observed in the presence of human activity is not one which is due to chance or 'natural' survival patterns.

9.3 Results.

In the following discussion, those sites from which we have reindeer data will be considered first, due to the fact that the methodology employed was initially devised employing North American caribou material. For this reason the approach should be more suitable for the analysis of reindeer assemblages than those comprising other species. Although the work reported by Speth (1983) concerning the Garnsey bison kill-site has shown the utility index to be of relevance to alternative species, it does reinforce the need for circumspection when considering other taxonomic groups, especially the perissodactyla such as horse and mammoth. Thus a site somewhat similar (in terms of archaeozoological material) to the Garnsey kill-site, namely Fongaban, is considered. Discussion then turns to sites where element representation data for more than one taxon are available: La Gravette, Le Roc de Marcamps and Reignac.

Employing the data presented by Poplin (1977), four element representation profiles may be drawn for the Abri Pataud: Perigordian IV, V, VI and Protomagdalenian. The decision was made to consider Poplin's (1977) data, largely taken from Bouchud's figures but showing some light variation, as they were deemed to be more suitable for the analyses in question. Bouchud (1975) divides long bones into three parts: A - proximal articulation, B - shaft, and C - the distal articulation, a procedure which is not particularly useful in the present discussion, when considering the division of parts into only distal and proximal portions (considered to relate to the relevant 'half' of the bone). By ignoring category B, as we are 'forced' to do, even using Poplin's data, we are, of course, severely reducing (and 'biasing') our data, but the only alternative is to attempt to assign undesignated shafts to epiphyses - a risky undertaking!

The profiles shown in figs. 9.3 and 9.4 fall into two groups: Perigordian IV and V ($M.N.I._{max}$ = 280.5 & 306.5 respectively) and Perigordian VI and Protomagdalenian ($M.N.I._{max}$ = 66.0 & 39.0). In

Fig. 9.3

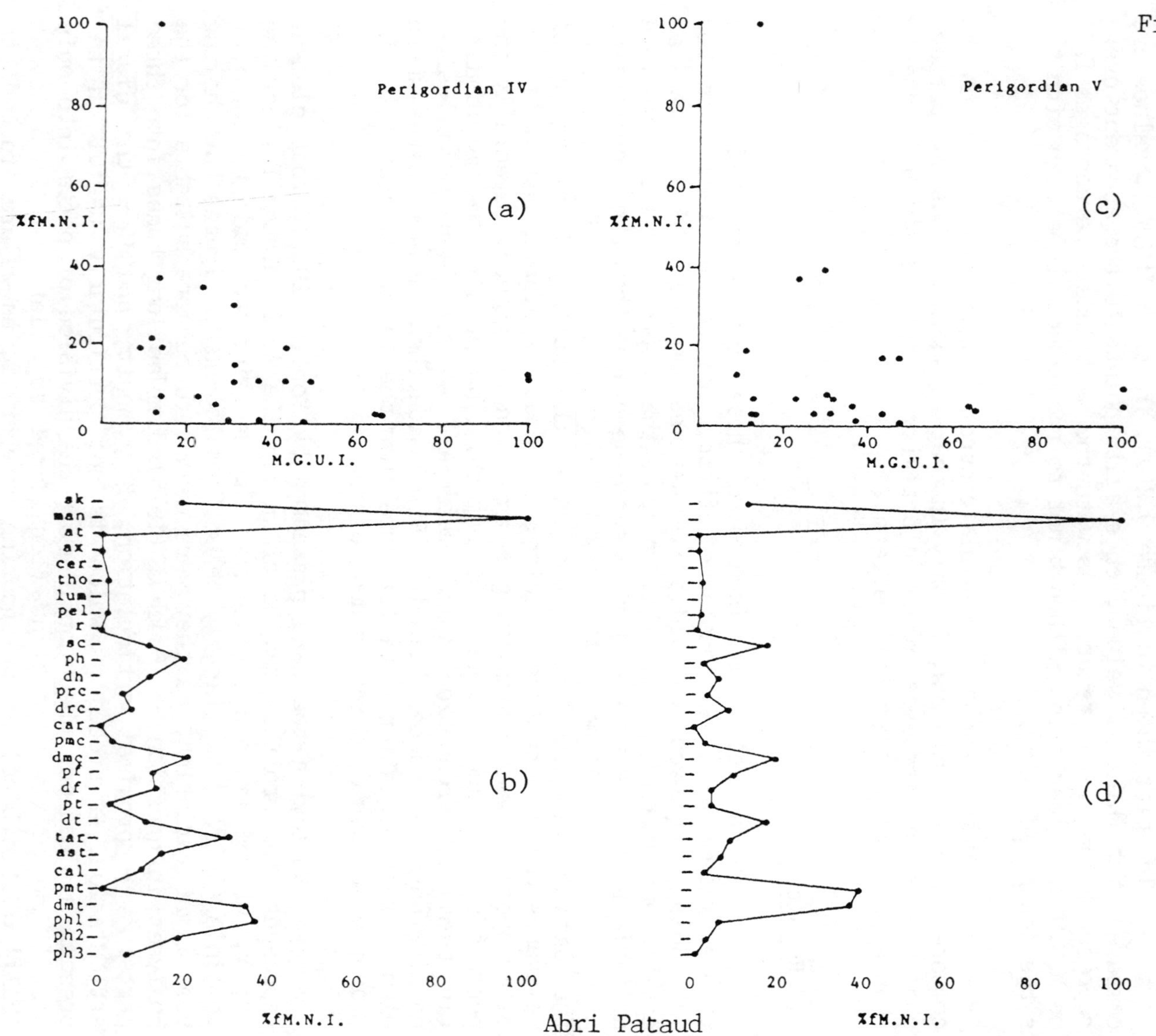

Abri Pataud

Fig. 9.4

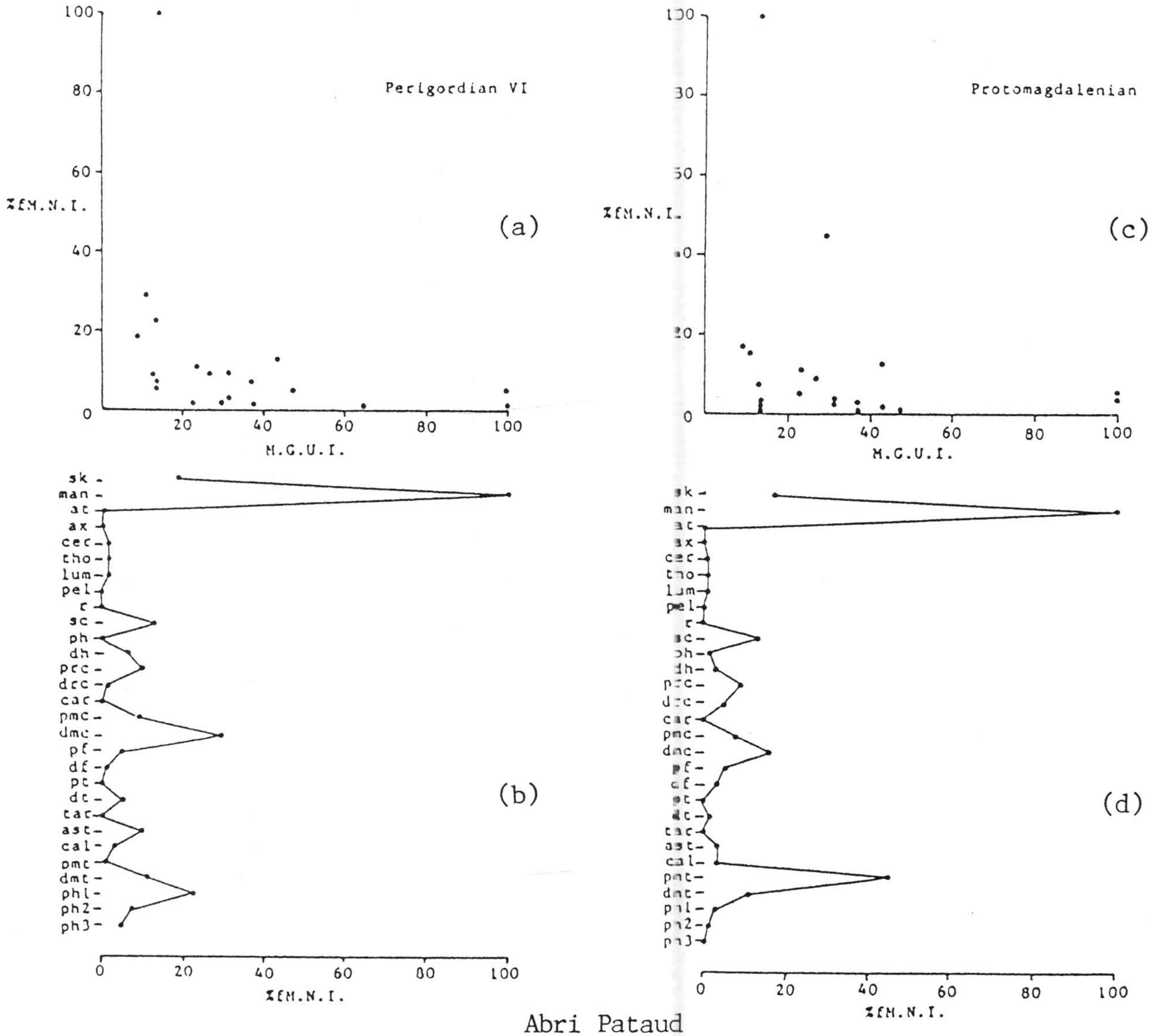

Abri Pataud

the first two assemblages the range of %fM.N.I. values represented by each element is substantially greater than in the second, with the notable exception of the metatarsal, although the frequency of the majority of elements falls within a 20% (0-20) range in all four samples. In all levels the fM.N.I.$_{max}$ is yielded by the mandible total. The rise in the frequency of elements (amongst which distal ends outnumber proximal parts in all but the Protomagdalenian: 21:50 during Perigordian IV, 28:35 in Perigordian V, 16:35 in Perigordian VI and 53:29 during the Protomagdalenian) occurs at a lower M.G.U.I. value in the later levels, suggesting perhaps more intensive post-kill processing. Perhaps we have evidence of a change in human behaviour for while the data from Perigordian IV and V levels yield a curve somewhat similar to that of Binford's (1978) gourmet strategy, Perigordian VI and Protomagdalenian assemblages show a slight shift towards the bulk curve, perhaps reflecting greater availability of game during the earlier periods. However, given that the assemblages each reflect, in all probability, more than one season of occupation, such hypotheses must be treated with care.

Bouchud (1975) maintains that during the course of the occupation of the Abri Pataud, all carcases, both of reindeer and other species (with the exception of the mammoth), were transported to the site complete. This belief is maintained based on the fact that all parts of the carcase are seen to be present (there are in fact a few element-types lacking). No account is taken of the frequency of each body part. However, if we examine the indications he provides which relate to element frequency, we can draw further conclusions. Those parts preferentially represented are those of the long bones, normally the lower hind limb (eg: Bovids in Perigordian VI, Red deer in Perigordian Vc). These and the skull are heavily broken, suggesting perhaps marrow and brain extraction, although plotting %fM.N.I. against standardised marrow indices (Binford 1978:27) fails to show the relatively clear-cut (expected) pattern observed at Le Flageolet (see below). Detailed sex and ageing data are required for each element type for the full implications of our analyses to be considered. In addition, detailed taphonomic data should be obtained in order to ascertain the extent of fragmentation on this 'consumption' site (Gamble 1986), to which it is suggested that carcases were brought already dismembered.

Consideration of the Abri Mege percentage profile shown in figure 9.5(b) reveals a trend somewhat reminiscent of the spring hunting stands of the Nunamiut identified by Binford (1978:186). The data are taken from Capitan et. al.'s (1906) site report which includes details of reindeer element frequencies (largely numerical). However, the number of carpals and tarsals recovered is recorded only as "nombreux". For this reason their presence has been documented in fig. 9.5(b), but no estimation of their frequency made. They are necessarily omitted from fig. 9.5(a).

Abri Mege is small, being only 70 m^2 in area, south facing, 8m above river level, and on the edge of a 350m wide valley. It is ideally located for the interception of migrating (reindeer) herds as they perhaps move towards the nearest major river, 20 km away - in much the same way as may be suggested for several of the sites in the Sergeac valley, where all the engravings of moving herd animals point towards the Vezere.

Fig. 9.5

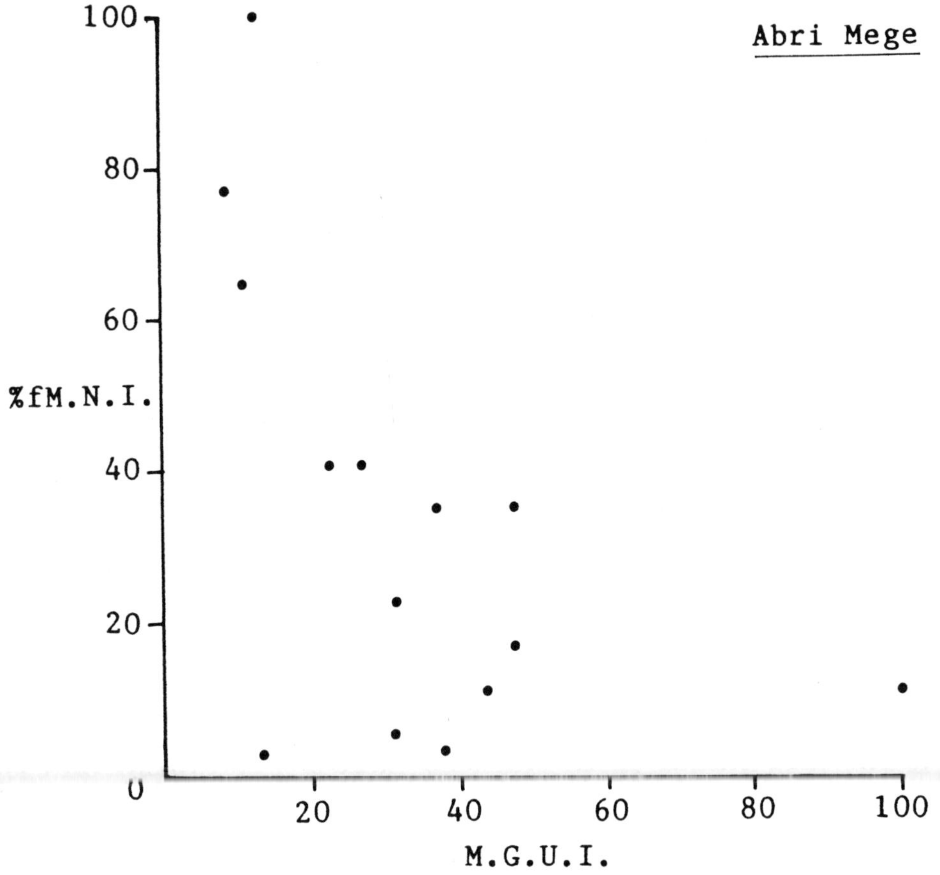
Abri Mege
100
80
60
%fM.N.I.
40
20
0
20
40
60
80
100
M.G.U.I.

(a)

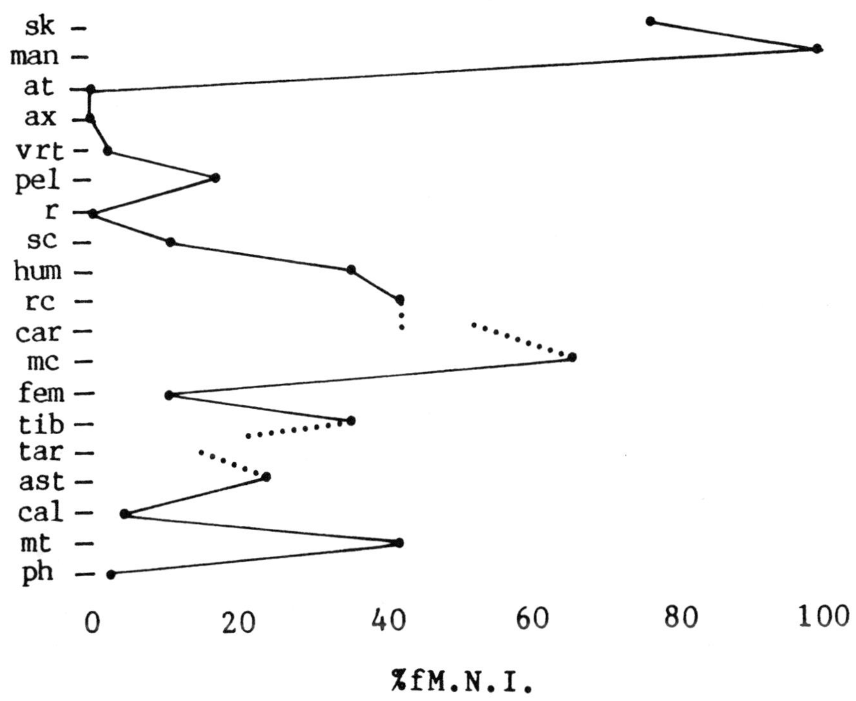
sk
man
at
ax
vrt
pel
r
sc
hum
rc
car
mc
fem
tib
tar
ast
cal
mt
ph
0
20
40
60
80
100
%fM.N.I.

(b)

Figure 9.5(b) shows the relationship between %fM.N.I. and M.G.U.I. values at the Abri Mege, for an fM.N.I. of 8.5 reindeer. A negative, curvilinear relationship is observed, the range of %fM.N.I. values in which most elements fall being relatively wide (5.88%-41.18%), covering a similar range of M.G.U.I. values (8.74-47.1). The fact that elements with an M.G.U.I. value exceeding 20 are relatively well represented may be seen to imply that medium utility parts were neither removed from the site, and transported elsewhere, nor very well processed, and certainly they were not all processed to the point of destruction, suggesting that the site was not occupied for a long period of time as a base or residential camp. Meanwhile, high utility elements were removed from this site, a observation which is often associated with 'hunting stands', where low utility elements are discarded in substantially greater numbers than are their higher-utility counterparts. At Abri Mege we therefore have a 'bulk curve' (as defined by Binford 1978:81), resulting from selective procurement and processing of high and moderate utility elements, and the discard (at an increasing rate) of lower value parts.

As indicated above, although the data from Abri Mege do yield a broadly bulk curve, the apparent 'over-representation' of a few elements of medium utility (35.00-50.00) suggests that the reindeer, which represents the predominant taxon, was not exploited to its full potential. This is a phenomenon which we will come to expect in association with the assumed major prey species and which has already been noted at the Abri Pataud where reindeer exceeds 85% in many of the levels. Thus gourmet curves are found with the predominant taxa and bulk curves with the majority (but not all) of the secondary species. The implication is that species which were exploited less regularly were more intensively processed than were the others.

At the Abri Mege, four elements are considerably more common than might be expected: radius, metatarsal, humerus, and tibia. Two main butchery units are suggested by these parts, both of which carry a relatively large quantity of meat: forelimb (humerus, radius) and lower hind limb (tibia, metatarsal). However, the ribs, sternum and cervical vertebrae, all of them with high or medium meat yields, are absent, while other vertebrae comprise a very small proportion of the total. Speth discusses the differential treatment of various elements in terms of sexual and seasonal factors, the implications of which may be of significance here. However, the relevant data are lacking in the present study and thus we can take the study no further than this, although it is important to be aware of the problems caused by such factors as seasonal and sexual variation in the quality of meat and hides.

At Flageolet II, a SSE-facing rockshelter 80m above the Dordogne river valley, we encounter a late winter reindeer kill (Gordon 1986)(see fig. 9.6(a)). Once again those parts of this predominant taxon with higher M.G.U.I. values are represented by fewer elements. The frequency of the majority of elements (all but 3) falls between 0% and 25%, over a wide range of utility index values (c.8.0-47.0). There is a marked rise in frequencies below M.G.U.I. = 30, suggesting some similarity with the 'Gourmet curve' identified by Binford (1978) in which low and moderate utility elements are discarded.

Fig. 9.6

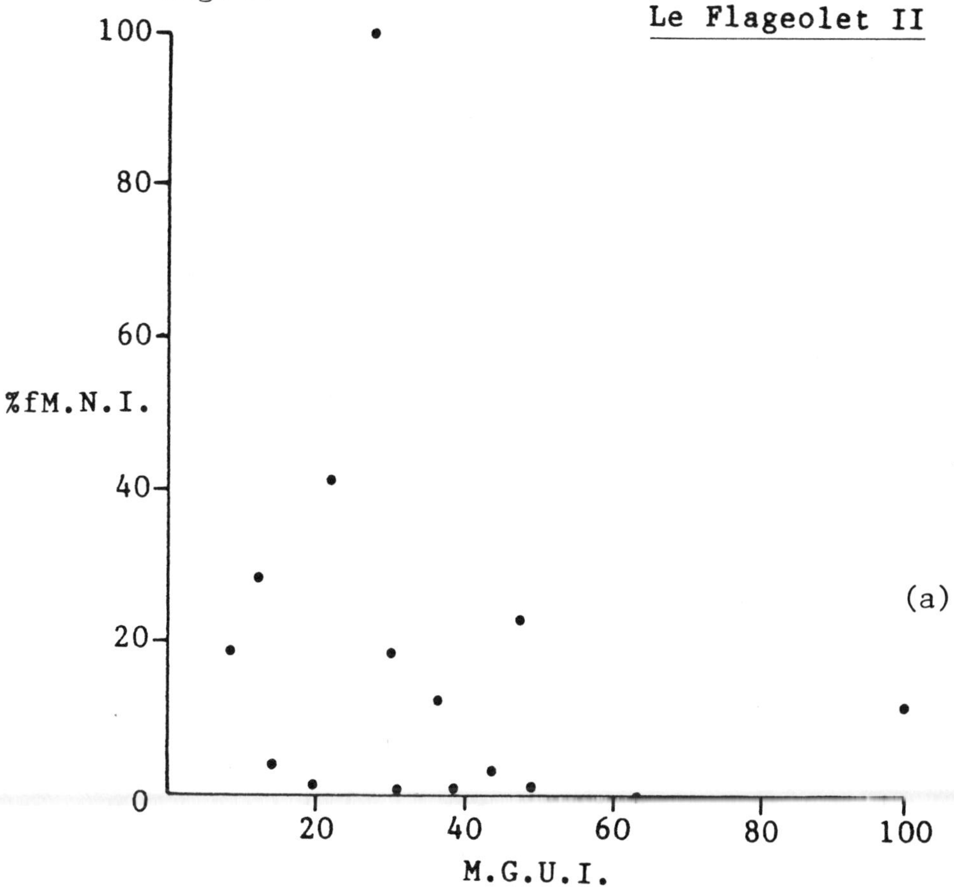
Le Flageolet II
100
80
60
%fM.N.I.
40
20
0
20
40
60
80
100
M.G.U.I.
(a)

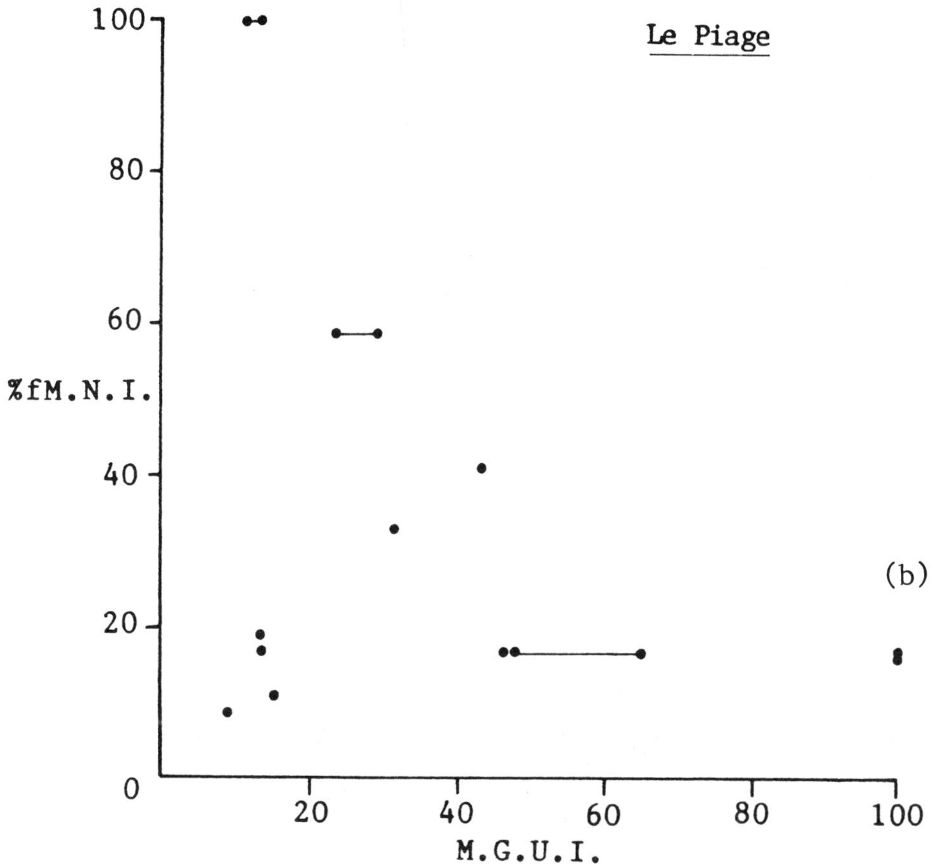
Le Piage
100
80
60
%fM.N.I.
40
20
0
20
40
60
80
100
M.G.U.I.
(b)

Fig. 9.7

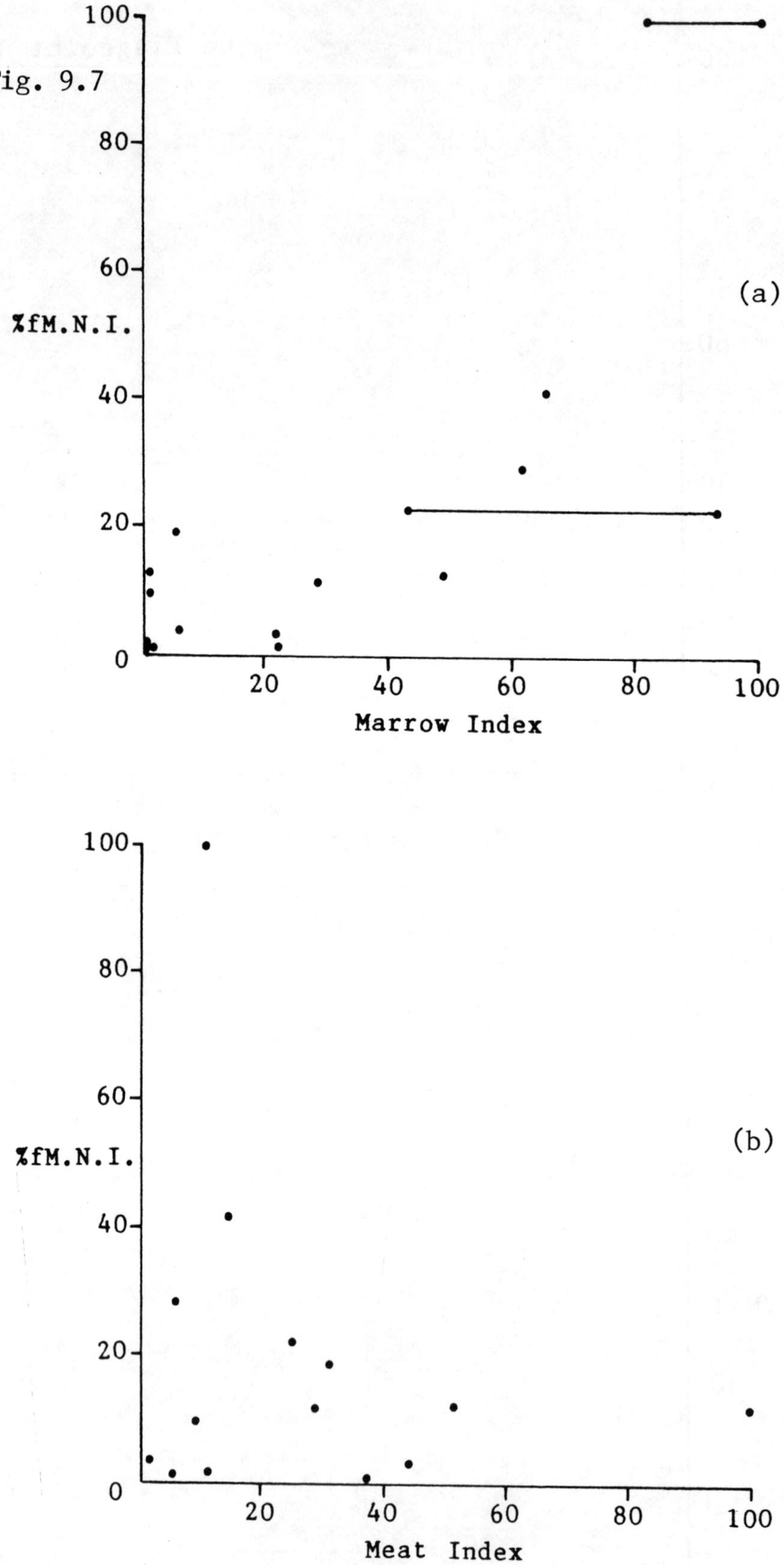

Relationship between Marrow and Meat Index Values and %fM.N.I. at Le Flageolet II

The very low frequencies of the axial skeleton (vertebrae and ribs) and scapula suggest that carcases were initially dismembered away from the site. Those parts of the carcase with high marrow values are relatively well represented while, in comparison, high meat yielding elements are not common and a curve similar to that observed for the M.G.U.I. is obtained - not surprising given that 'Meat Index' and M.G.U.I. values are highly correlated (r=0.9238). These curves are shown in fig. 9.7. It must be remembered, however, that such elements are particularly susceptible to loss through decay, carnivore activity or fluvial sorting (Speth 1983:85). A strategy of primary dismemberment at a kill site (in all probability in the valley somewhere below the site itself) is likely to have been adopted at such sites of relatively high altitude, thereby avoiding the need to transport carcases a long distance. By removing the bone from pieces of high meat yield, and perhaps low marrow or grease, and transporting only the flesh from these bones to camp, more usable food can be obtained. Five metacarpals and eight metatarsals show signs of butchery, in the form of artificial lines running perpendicular to the axis of the bone (Delpech 1972). Two large cuneiform bones also show evidence of butchery. Delpech (1972) claims that such marks may be the result of severing the carpal/tarsal - metapodia articulations in order to obtain tendons, allowing the discard of the feet. We may also be seeing evidence of the use of skins as containers (perhaps for meat parcels?), to which lower limb bones remain attached and used as handles in the way described by Binford (1978).

Fig. 9.8 The Relationship between Ranked Marrow Utility and Element Representation at Le Flageolet II

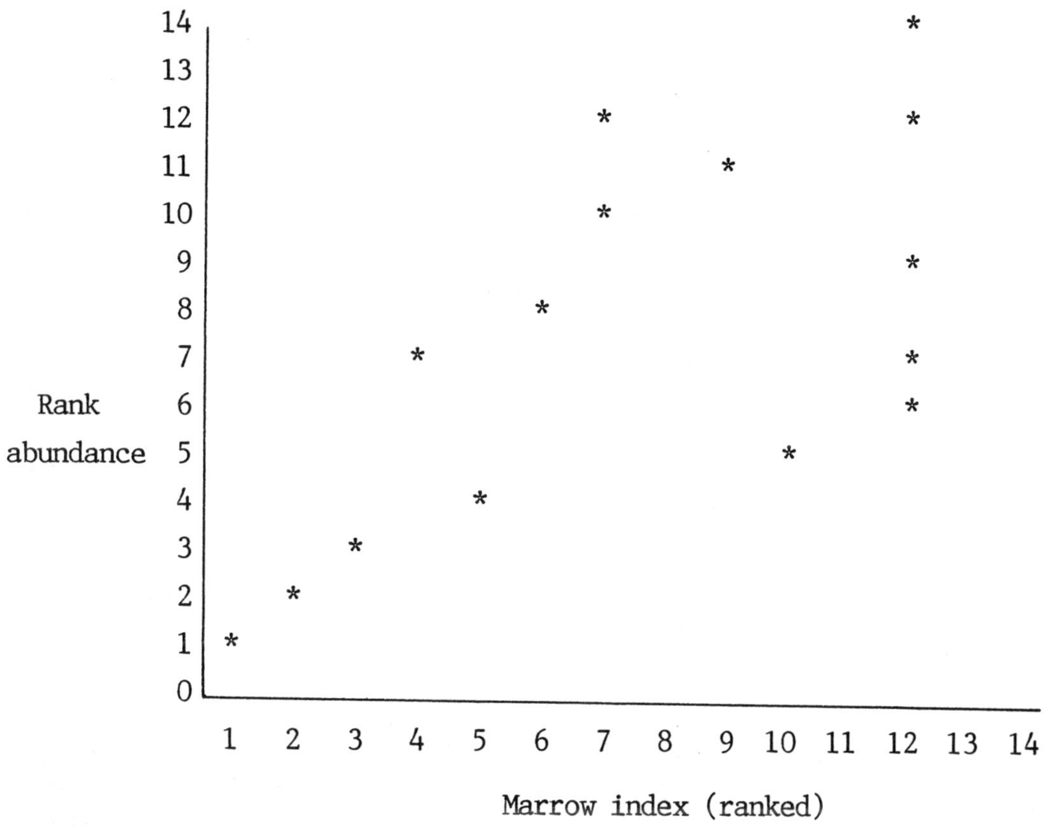

In 1974 Delpech and Rigaud presented their conclusions concerning the mode of fragmentation of bones from the Noaillian level VII at Le Flageolet I. They conclude that the extraction of marrow and grease was a primary factor, basing their argument upon the relative abundance of marrow-bearing elements and comparing this with Nunamiut preferences. A necessary assumption is that this ethnographic material forms a comparable database. However, statistical comparison by Spearman Rank Correlation of Nunamiut preference and rank order of elements (as detailed by Delpech and Rigaud (1974)) does show this (r_s=0.853, n=10, level of significance 0.01), while a similar conclusion may be reached for Le Flageolet II. By ranking body part representation at the site and among the Nunamiut (adopting the ranks shown in Binford 1978, fig. 1.10) and again calculating the Spearman Rank Correlation we find ourselves able to reject a null hypothesis (H_0) of no mathematical association between marrow preference and element rank. r_s=0.849, which, when n=9, means that we can reject H_0 at a confidence level of 0.01. The relationship between rank fM.N.I. and marrow utility rank is shown above in fig. 9.8, where r_s reaches only 0.708. However, this too is significant at 99%.

Le Piage (Plaine d'Auniac, Lot) Protomagdalenian reindeer element representation is shown in figure 9.6(b). Only eleven types of element are present, dispersed over a wide range of % M.N.I. and M.G.U.I.. However, the negative, curvilinear relationship expected in the case of the predominant taxon is observed, element frequencies rising at M.G.U.I. values of approximately 30 - 44, metacarpals (M.G.U.I. = 10.50-12.18) representing the $M.N.I._{max}$. There is some similarity to the gourmet curve, although the relative lack of data necessitates caution when attempting to reach conclusions given the available material.

A gourmet curve (see fig. 9.9) can be observed for the major species, reindeer, at Gare de Couze, although only a relatively small number of different element types is actually recorded by Prat (1962) - despite the high species total given by Delpech (1983). In the lower level frequencies rise steadily between M.G.U.I. values of approximately 30 and 36, while a similar pattern is evident in the upper level where the femur and proximal tibia are completely absent. The latter, interestingly, is particularly susceptible to marrow-fat depletion in the spring, one of the seasons of occupation suggested by Gordon (1988). Higher value parts were either removed from, or never brought to, the site or were reduced to 'unidentifiable' fragments. In the **couche superieure** however, frequencies rise at approximately M.G.U.I. = 40.00. Element representation profiles are shown in fig. 9.10, reflecting the late winter/spring occupation of both this south-facing, cave, rockshelter and open-air site and the similar, nearby site of Sainte-Capraise, which according to Gordon (1988), was occupied during the winter.

Levels III and I (Magdalenian III and VI respectively) at the primarily winter-occupied Lot site of Sainte Eulalie (Gordon 1988) present gourmet-type curves (see fig.9.11), the proximal metatarsal providing the $M.N.I._{max}$ in each case. The plots are similar to those observed at Gare de Couze. Couche I yields no high utility elements (M.G.U.I.<50) implying that such parts of the carcase were removed from

Fig. 9.9

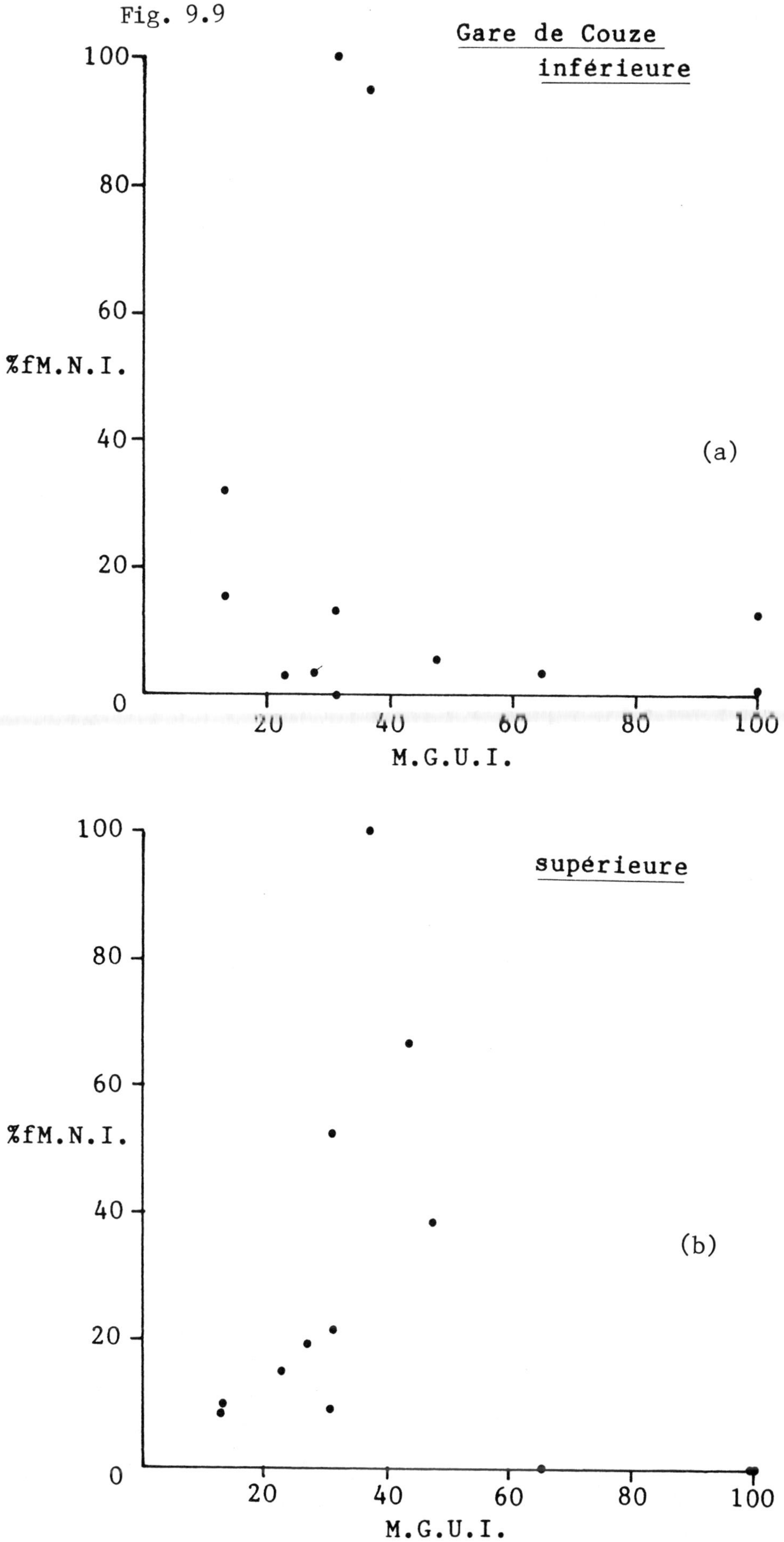
Gare de Couze
inférieure
100
80
60
%fM.N.I.
40
(a)
20
0
20
40
60
80
100
M.G.U.I.
100
supérieure
80
60
%fM.N.I.
40
(b)
20
0
20
40
60
80
100
M.G.U.I.

Fig. 9.10

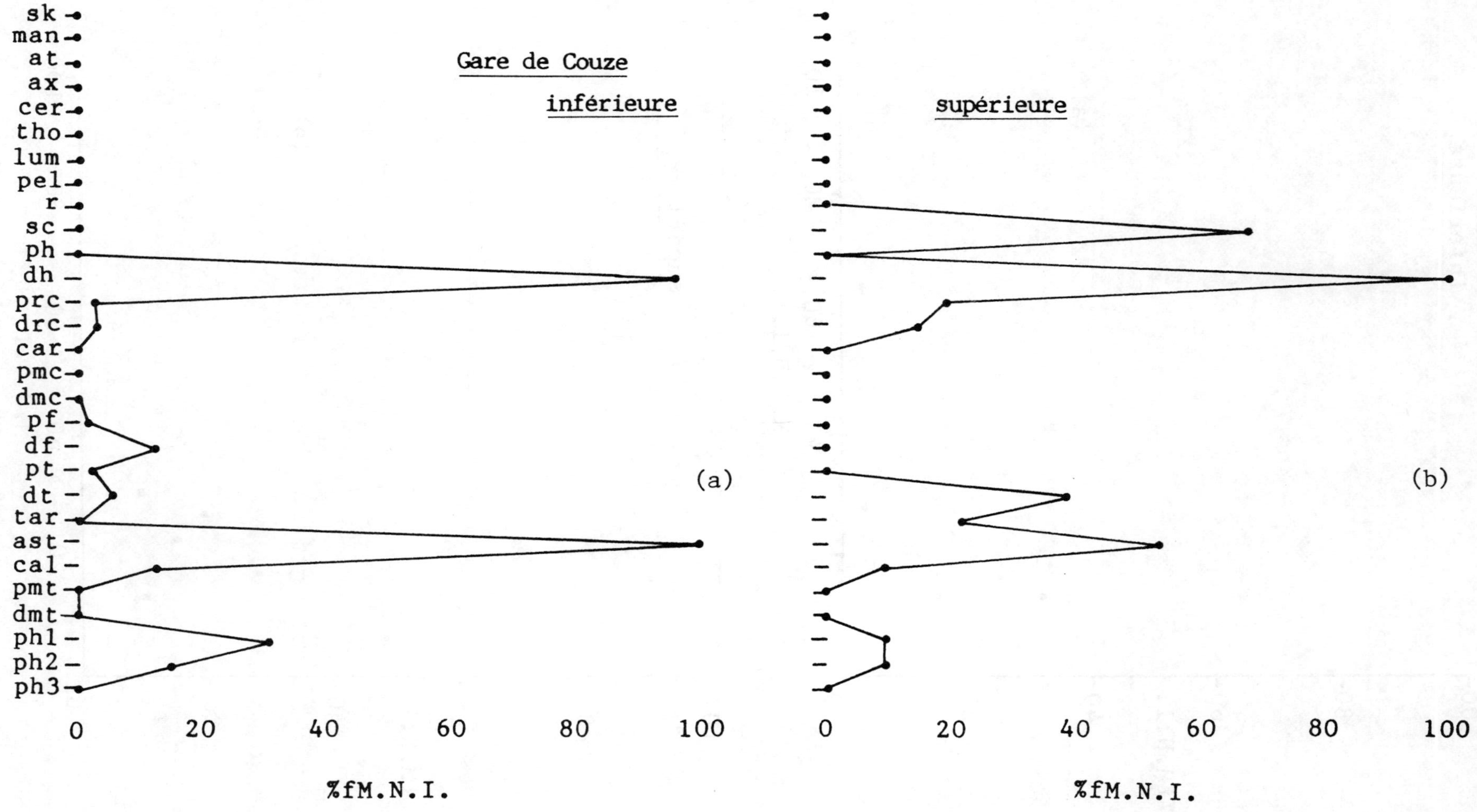
Gare de Couze
inférieure
supérieure
(a)
(b)
sk
man
at
ax
cer
tho
lum
pel
r
sc
ph
dh
prc
drc
car
pmc
dmc
pf
df
pt
dt
tar
ast
cal
pmt
dmt
ph1
ph2
ph3
0
20
40
60
80
100
%fM.N.I.
%fM.N.I.

Fig. 9.11

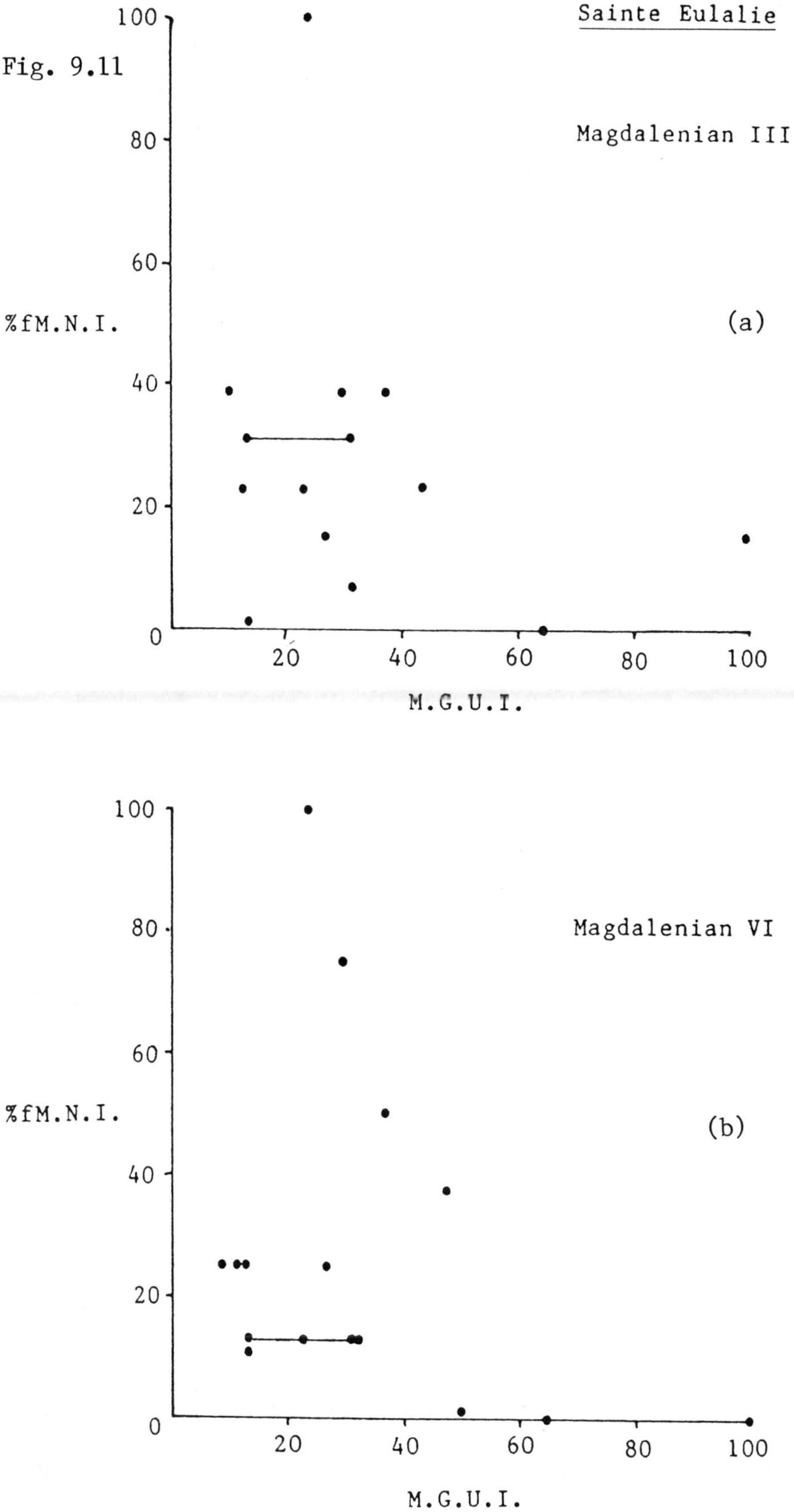

Sainte Eulalie
Magdalenian III
%fM.N.I.
(a)
0
20
40
60
80
100
M.G.U.I.
Magdalenian VI
%fM.N.I.
(b)
0
20
40
60
80
100
M.G.U.I.

the site or processed to such an extent that they are now unrecognizable. Such proves to be a recurrent problem, requiring an examination of fragment size distribution in order to determine the extent and role of fragmentation. The use of published site reports, as in this case, often precludes this.

The major difference between the two levels at Sainte Eulalie lies in the extent to which points, on the graphs, are dispersed, there being a more clearly defined cluster in the earlier level. What is of interest is that it is in this level that the femur (high M.G.U.I.) is observed and together with the other parts present, suggests that reindeer were brought to the site complete or near-complete.

The reindeer data recently presented by Olsen (1987) derived from Magdalenian VI deposits at the Grotte des Eyzies are shown graphically in figure 9.12. Unfortunately Olsen fails to distinguish between distal and proximal parts of each element. Thus, like several of the older site reports, we have totals for undivided tibia, femora etc. Instead she meticulously divides the material into axial, left and right elements, and it is in this way that her data are presented in Appendix I, together with her M.N.I. calculations. Nevertheless it was felt to be of value to consider the available data, calculating our own fM.N.I. frequencies. It is the fM.N.I. values which are employed in the present example. The maximum value is yielded by the calcaneus, the rise in fM.N.I. occurring at a high M.G.U.I. value. Here we see a distinct gourmet curve, to be expected given that reindeer represents 95% of the large herbivore bone fragment total. Based simply upon the fact that most parts of the carcase were recovered from the deposits, Olsen (1986) concludes that reindeer were brought to the site complete for subsequent processing, a premise which is largely supported by the present analysis.

The bovid data available from the large rockshelter of Fongaban are shown in figure 9.13. At this site we once again see a gourmet curve, this time associated with bovids. Bovids, however, represent the major species at Fongaban which may be interpreted as a special purpose kill-site, perhaps only temporarily (seasonally ?) occupied. The majority of elements occurring fall within the M.G.U.I. range of 30.00 to 50.00. However, with the exception of a small number of femur and the tibia, elements yielding a large amount of meat (femur, ribs, scapula) are poorly represented, and may have been removed to be consumed elsewhere. Alternatively such elements may once again have been intensively processed and reduced to small fragments. Given the high frequency of tibia and common femur, it would seem unlikely that Fongaban represents a single purpose hunting stand. Rather, it may be suggested that we have here a deposit representing some form of base camp, suitably positioned for the exploitation of migrating herds of bovids. An alternative explanation may be the discrimination against calving females during the spring, the proximal tibia being particularly prone to marrow-fat depletion at this time of year (Speth 1983:91-92). The abundance of distal femur may also be explained in this way. However, if such is to be given serious consideration sexing data are required, in addition to bovid-derived seasonality indication which unfortunately are not yet available. Inferences have to be based on the very small quantity of reindeer material analysed.

Fig. 9.12

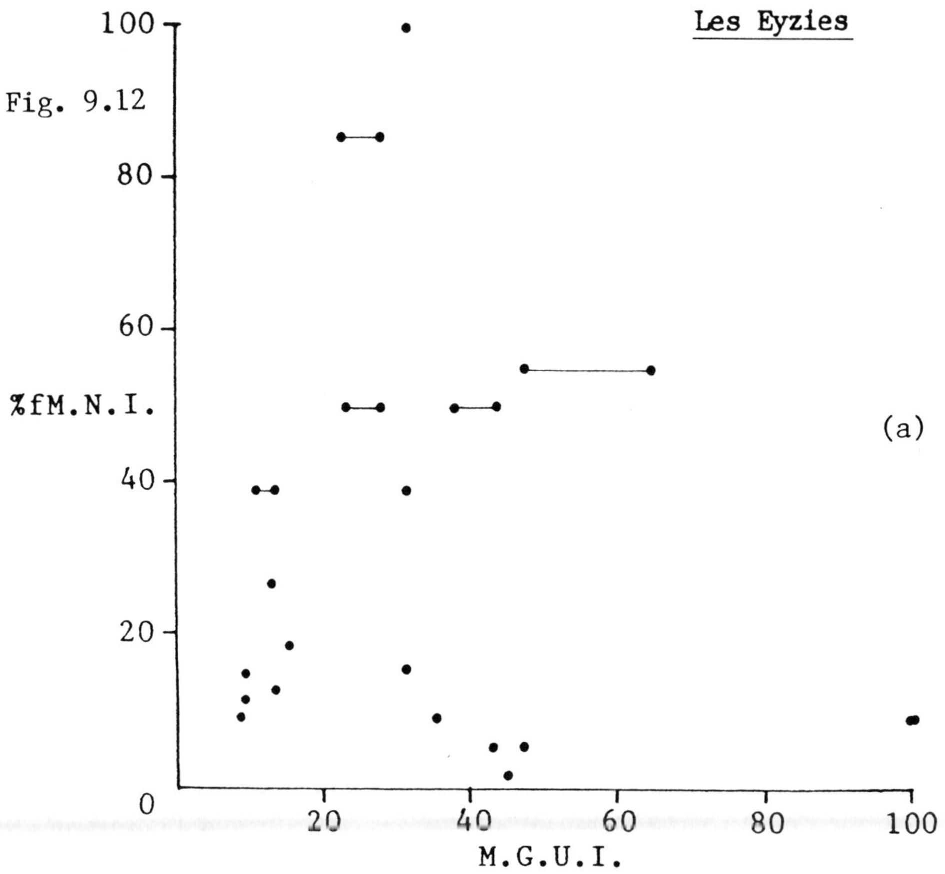

Les Eyzies
100
80
60
%fM.N.I.
40
20
0
20
40
60
80
100
M.G.U.I.
(a)

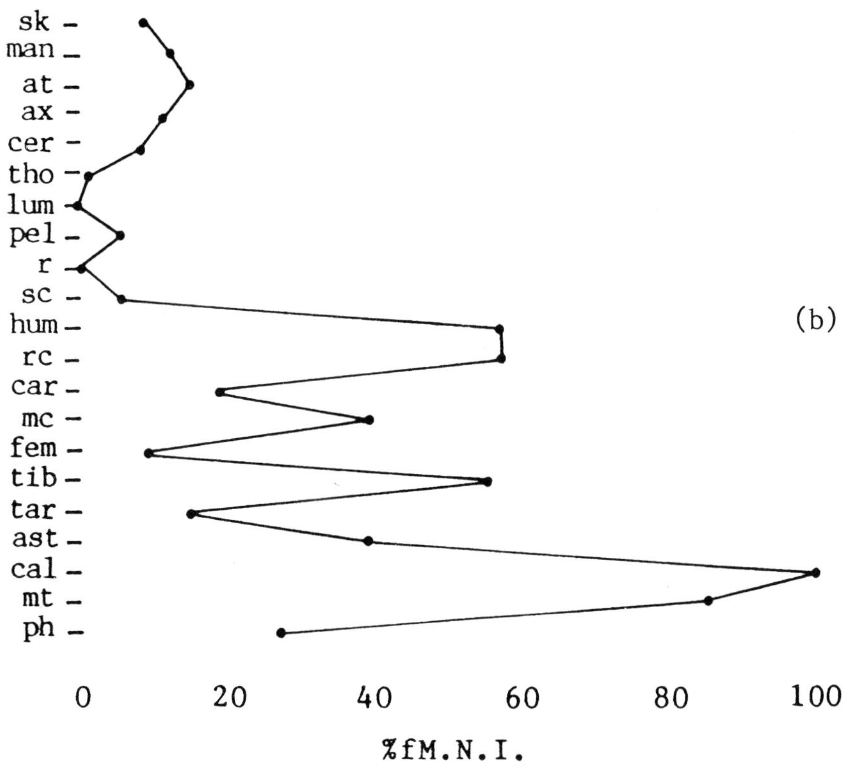

sk
man
at
ax
cer
tho
lum
pel
r
sc
hum
rc
car
mc
fem
tib
tar
ast
cal
mt
ph
(b)
0
20
40
60
80
100
%fM.N.I.

Fig. 9.13

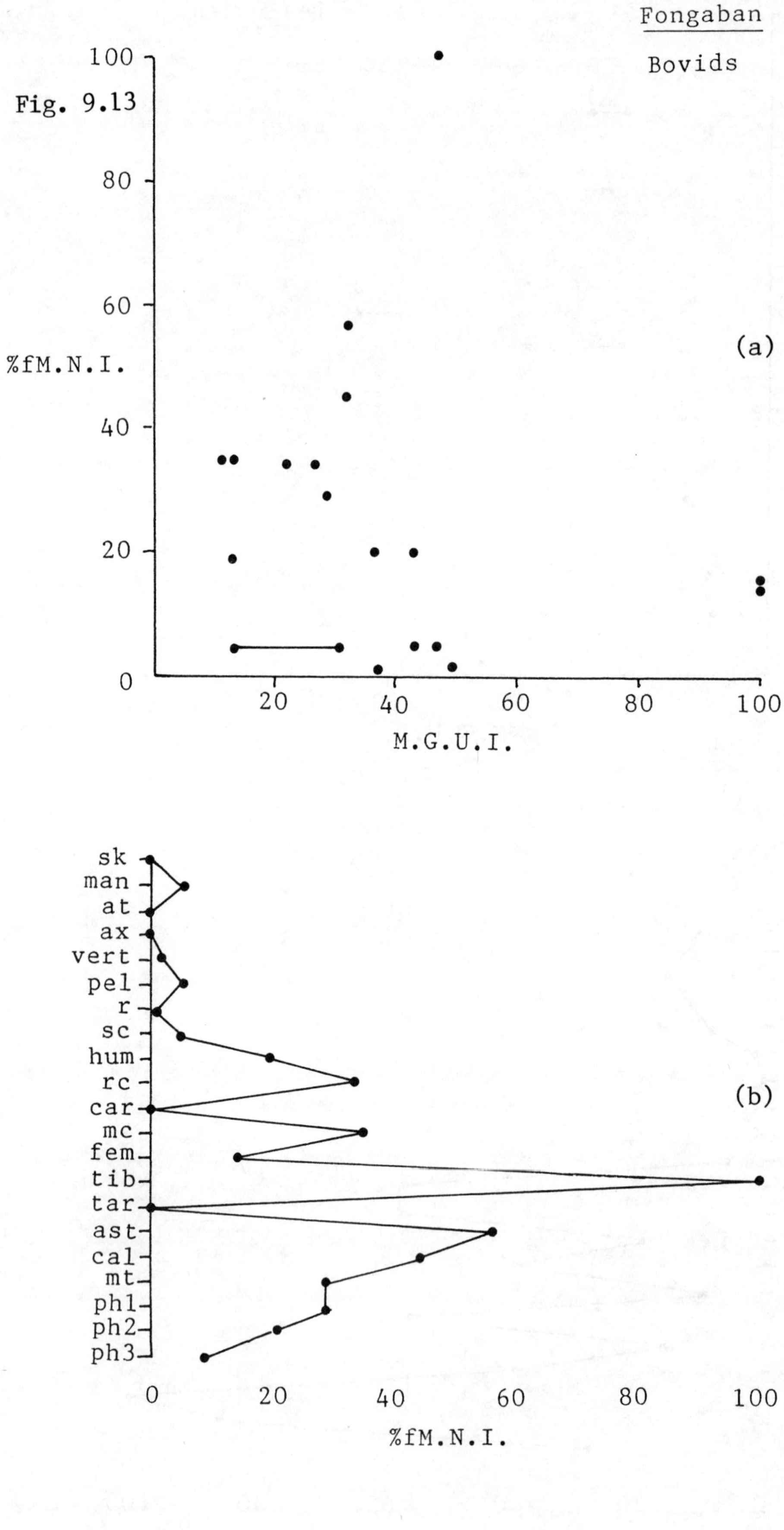

Fongaban
Bovids
100
80
60
40
20
0
%fM.N.I.
(a)
20
40
60
80
100
M.G.U.I.
sk
man
at
ax
vert
pel
r
sc
hum
rc
car
mc
fem
tib
tar
ast
cal
mt
ph1
ph2
ph3
(b)
0
20
40
60
80
100
%fM.N.I.

According to Gordon (1988:181), Fongaban is likely to have been occupied predominantly during the late winter and early spring, a time at which meat from calving females shows evidence of nutritional stress. His conclusions are based on winter and early spring predominance at nearby sites such as Saint-Germain-la-Riviere, Faustin, Le Morin, Fontarnaud, Fauroux, Vidon and Moulin-Neuf, although the reindeer material at Fongaban itself points to summer occupation. The establishment of season of death of the bovids at Fongaban is of paramount importance if we are to be able to consider more fully its role in the settlement-subsistence system of the northern Gironde region. Here we may have a camp, occupied during the summer while reindeer were relatively unavailable - for either reasons of human choice (nutritional) or environmental necessity (absence of herds from the region). The example of Fongaban shows up a weakness inherent in Binford's utility indices - the failure to account for seasonal fluctuations in meat and marrow quality. Ideally a series of indices is required, each of them species- and season-orientated.

Two sites, strictly speaking outside our immediate area of study, yield contrasting patterns. The Salle des Morts (Enlene) and Grotte des Eglises, both in the Ariege, show markedly different configurations. At the former (see fig. 9.14(a)), a north-facing, spring occupied site (Bahn 1979, Gordon 1988), we see a gourmet curve associated with the predominant species.

At the high altitude site of Grotte des Eglises (see figs. 9.14(b) and 9.15) (elevation 350m (Straus 1987)), we find a situation so far unencountered where the femur totals make up a high percentage of the total fM.N.I.counts in both couche 4 and couche 6. However, in c.4, despite the frequency of the femur, a gourmet-type curve can be discerned, a rise in frequencies occurring at an M.G.U.I. value between approximately 30.00 and 40.00. In level 6 (see fig. 9.15) we see a situation in which almost the entire carcase is represented. Only the pelvis and proximal humerus are completely absent, while most of the high meat-yielding axial skeleton is relatively rare. However, high %fM.N.I. values of low and medium M.G.U.I. value parts (foot bones, tarsals, calcaneus, astragalus and distal tibia) are seen and the gourmet curve is once again encountered with the predominant taxon (in this case ibex). Based both on the fact that both sexes are almost equally well represented and Delpech & Le Gall's seasonality indications (1983), Straus (1987) concludes that Les Eglises was often occupied during the rut, now occurring in late autumn. At this time, when males are at their fattest, ibex was taken, (in the form of boneless meat parcels?) together with grouse and salmon, for removal to camps elsewhere (Straus 1987: 173). As Binford remarks in the case of the pre-firearm Nunamiut, fall-hunting in the mountains may be characterised by intensive carcase processing at or near the kill in order to "reduce the bulk relative to the usable or consumable meat" (Binford 1978:417). It is thus suggested that the Grotte des Eglises might be considered as a specialised ibex butchering location which, as Straus (1987) points out, was probably situated downslope from the site of the kill. The extent of marrow processing (as suggested by Straus 1987), given the seasonal quality of the meat might perhaps be questioned. When fat content of meat is 'high', marrow processing is not 'necessary'. Alternative causes of bone fragmentation need to be considered.

Fig. 9.14

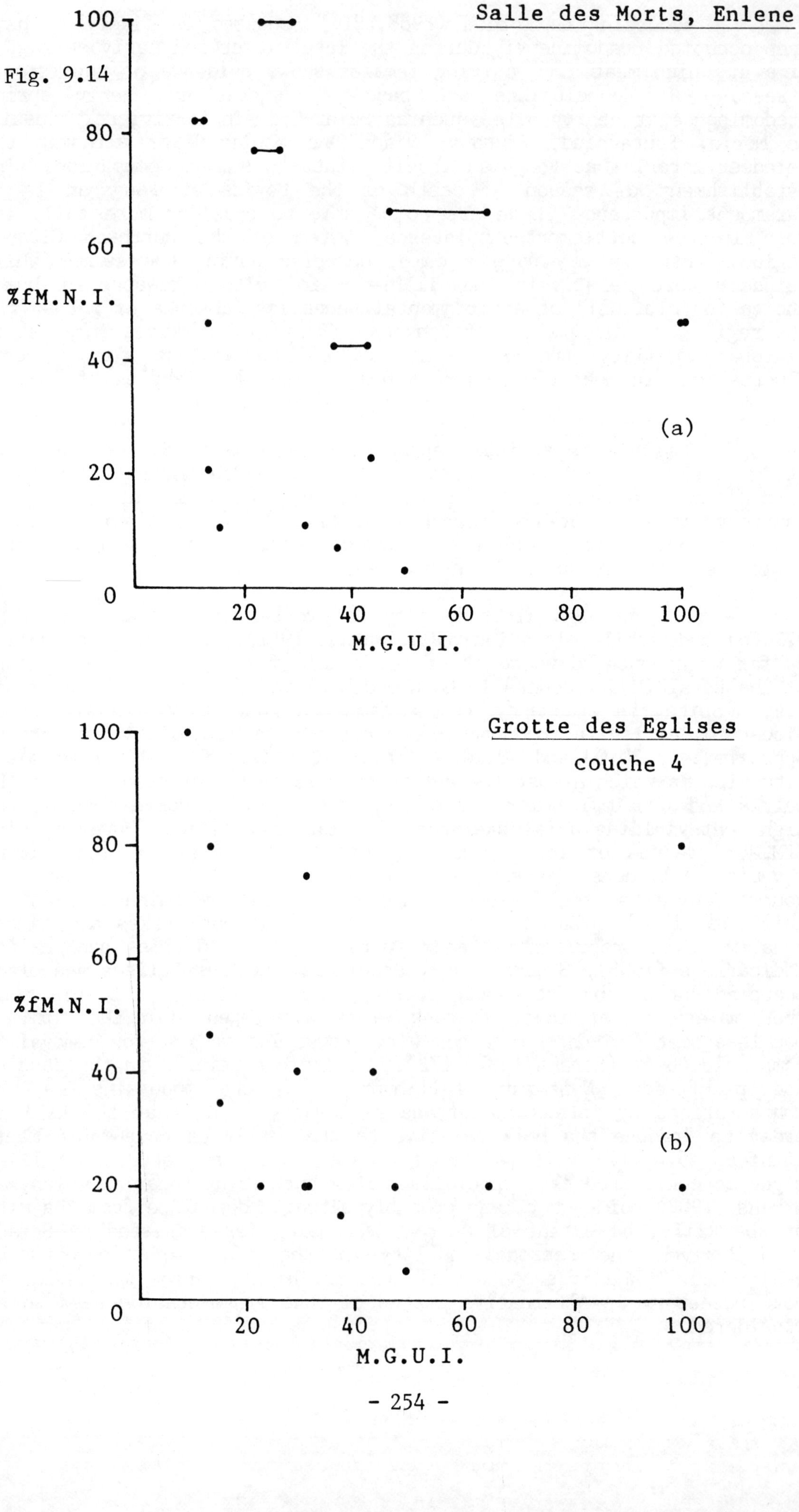

Salle des Morts, Enlene
100
80
60
%fM.N.I.
40
20
0
20
40
60
80
100
M.G.U.I.
(a)
Grotte des Eglises
couche 4
100
80
60
%fM.N.I.
40
20
0
20
40
60
80
100
M.G.U.I.
(b)

Fig. 9.15

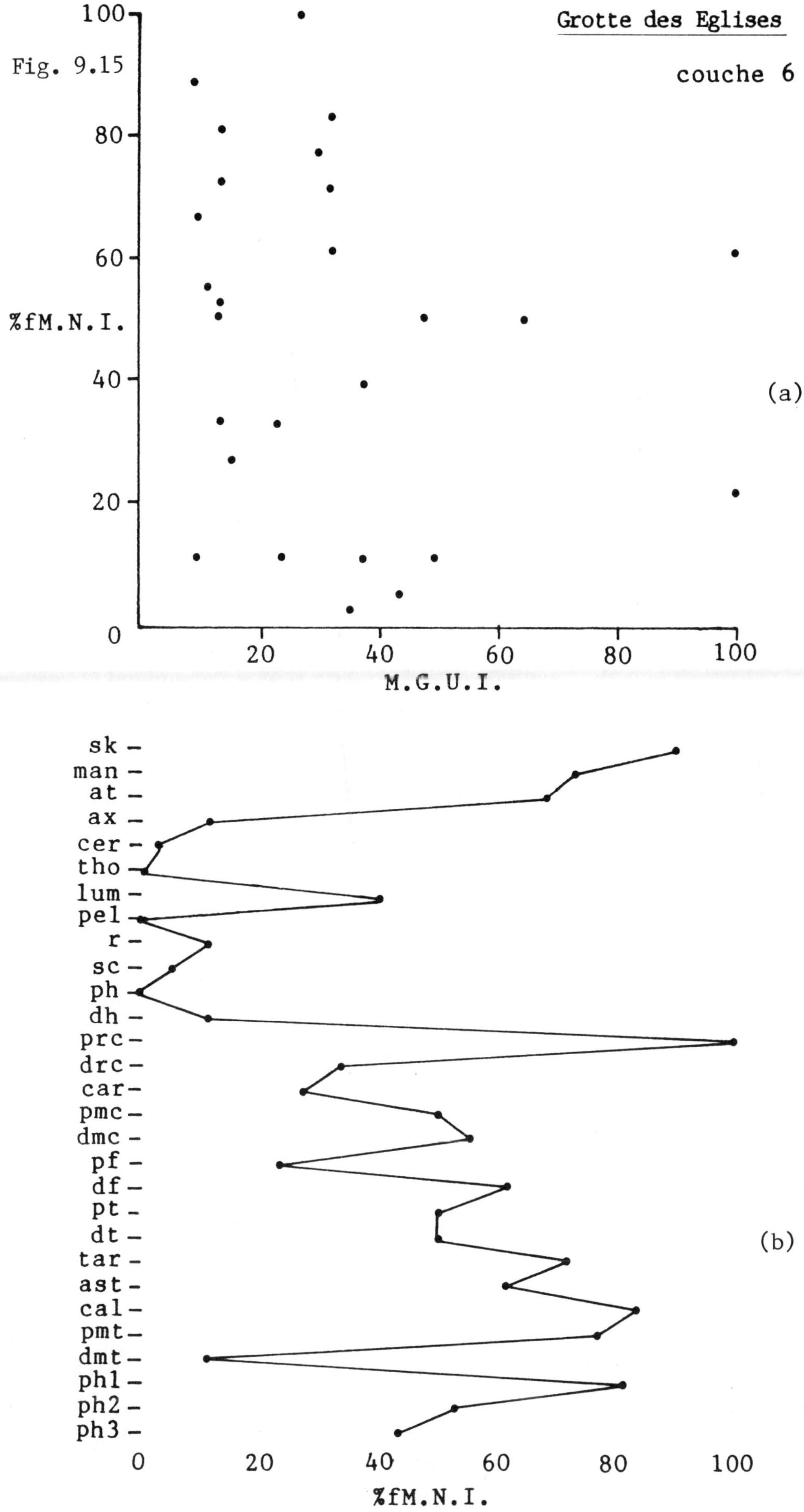

Grotte des Eglises
couche 6
100
80
60
%fM.N.I.
40
20
0
20
40
60
80
100
M.G.U.I.
(a)
sk
man
at
ax
cer
tho
lum
pel
r
sc
ph
dh
prc
drc
car
pmc
dmc
pf
df
pt
dt
tar
ast
cal
pmt
dmt
ph1
ph2
ph3
0
20
40
60
80
100
%fM.N.I.
(b)

In the Upper Perigordian levels at La Gravette we have available both horse and reindeer data, and it is instructive to compare the curves obtained, for it provides an opportunity to consider differences in procurement strategies of primary and secondary resources. The data are shown in fig. 9.16. The major differences between the two species include:

1. The rise of the curve occurs at a lower M.G.U.I. value in the horse (15) than reindeer (c.30).

2. The frequencies of most elements fall within the 13 - 29 % range in the case of reindeer, 40 - 60 % in the horse, implying either that the horse was less intensively butchered or that a smaller proportion of the reindeer carcase was brought to the site than in the case of the horse and is therefore less well represented.

3. However, the horse data completely lacks all high element pieces, whereas in the case of the reindeer even femurs are observed at a relatively high frequency (c. 25 %).

The closer resemblance of the reindeer curve to the gourmet curve than the 'bulk' horse curve implies that the horse was more intensively processed. Thus we may be observing the result of differences in resource kill-rate or availability, for the gourmet curve indicates that the reindeer formed a larger portion of the total kill than did the horse at this winter site.

In addition to a range of utility indices we now need to begin to study a variety of species at each site, for the majority of the rarely published data which are suitable for analysis of the type discussed here, commonly comprise only those elements of the most abundant taxon. An attempt is now made to consider element representation for all the large herbivore species (with an N.I.S.P. count exceeding 50) encountered at sites, so as to enable us to determine the extent to which a range of complementary resources were exploited at a single location. Thus, in order to conclude the present discussion of 'resource use and food management' (Gamble 1986:297) several species will be considered at two sites: Roc de Marcamps (Marcamps, Gironde) and Reignac (Tursac, Dordogne).

Three collections of faunal material are available from the SW-facing Roc de Marcamps, four species being discussed here. In addition these four main species (reindeer, saiga, horse and bovids) are considered as 'combined assemblages', ie: Ferrier, Maziaud and Nicolai collections are amalgamated and considered as 'total' assemblages. The individual collections are discussed only when a sufficiently large sample is available. For horse and bovids we find the Nicolai and Ferrier collections are too small to be considered separately; they are of course included in the 'total' species counts. At the same time it was decided not to add the data collected from the Grotte des Fees - despite the fact that the two sites may in fact represent just one site, albeit two parts thereof. These data are presented in Appendix I. The results obtained for those species from the site of Roc de Marcamps yielding sufficient material in the

Fig. 9.16

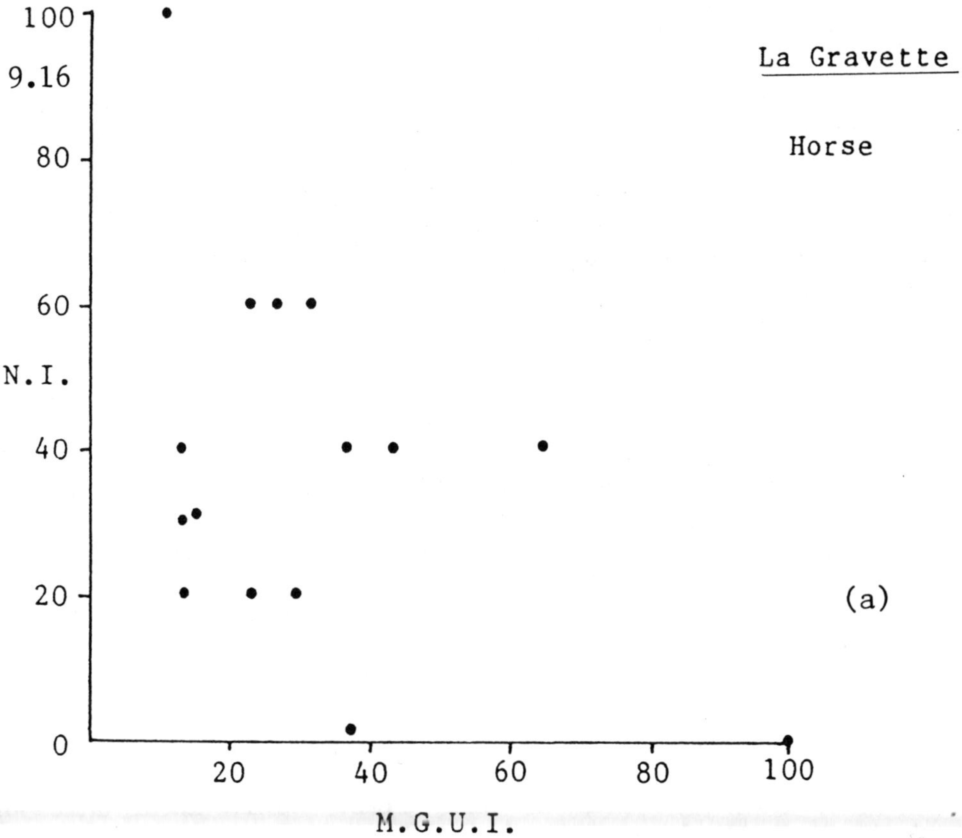

La Gravette
Horse
%fM.N.I.
100
80
60
40
20
0
20
40
60
80
100
M.G.U.I.
(a)

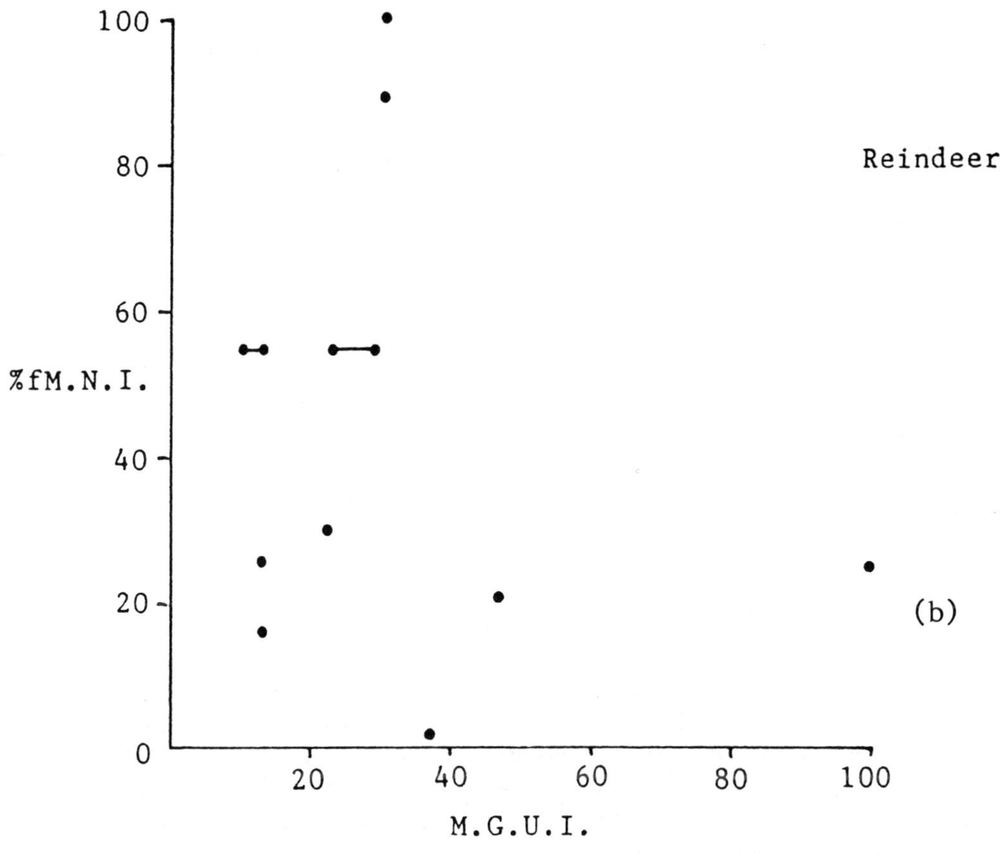

Reindeer
%fM.N.I.
100
80
60
40
20
0
20
40
60
80
100
M.G.U.I.
(b)

separate collections are shown in figures 9.17 to 9.20, as are those for the combined assemblages.

Figure 9.17 shows the patterns observed for reindeer element representation. It is only in the Nicolai collection that we observe elements of M.G.U.I. value exceeding 50.00 (prox. femur M.G.U.I.=100.00). In this and the Maziaud collection, the maximum (100%) frequency is derived from distal metacarpals as it is for the total assemblage (see fig. 9.20). In the Ferrier collection the maximum is yielded by the astragalus, elements of medium utility being relatively well represented - perhaps as a result of the relatively small size of the sample. In the Nicolai collection most parts are represented by %fM.N.I. values of 10-45%, in the Maziaud by values of 10-25%, while in the Ferrier collection half the %fM.N.I. (>0%) values fall between 25% and 40%. In the combined assemblage however, the range of values is larger, from 3% to 34%, a situation closer to Binford's bulk curve being observed in this and the Maziaud collection as may be expected in the case of a second-ranking species.

Saiga (fig. 9.18) is the major species at Roc de Marcamps, and in all cases is represented by a gourmet curve. Both femur and proximal tibia are represented, although frequencies are low. The astragalus yields the maximum %fM.N.I. value in the Ferrier collection with relatively high calcaneus, tarsal, metatarsal and distal tibia frequencies, indicating that perhaps we are dealing with an identifiable unit - the lower hind limb. This is a unit which is clearly seen in both saiga and reindeer in the combined collections (fig. 9.20). Examination of saiga anatomy shows, however, that little meat is to be gained from the lower limb and lower part of the upper limb bones. Perhaps we have evidence of the discard of such bones at the site and removal of much of the rest of the skeleton to elsewhere. Also of note is the lower forelimb unit seen here in both taxa.

Horse is also characterised by abundant lower limb bones, distal metatarsal and metacarpal representing 100% fM.N.I. recorded. In addition the phalanges frequencies are unusually high, representing 22-41% in the combined sample. In contrast to reindeer and saiga, the mandible is poorly represented, although overall the profiles shown for these three species are remarkably similar (see fig. 9.20). Once again the elements of the axial skeleton are rare or absent, a phenomenon which Binford (1978) ascribes to the removal for drying of parts, complete with bone, of relatively high meat utility. As already stated, the similarity between the profiles obtained for reindeer, saiga and horse is striking - a fact which may be seen to suggest that these three taxa were processed in broadly the same way. Further evidence of this is seen in fig. 9.19 in which, with the exception of the outlying distal metacarpal the pattern observed is far from random.

Bovids however are different, as the profile seen in figure 9.20 shows. Like the saiga, the maximum %fM.N.I. is represented by the mandible but apart from this, no element exceeds 25% and only one (tarsal) 15%. The representation profile is very similar to that described by Binford (1978:405) for the Tulukana and Disturbed fall sites, at which dog yard, dumps, cooking areas and household debris are represented - all of which may form part of the Upper Palaeolithic archaeological record. Bone grease manufacturing is suggested by Binford, based not only on element representation but also on the fact

Roc de Marcamps Reindeer

Fig. 9.17

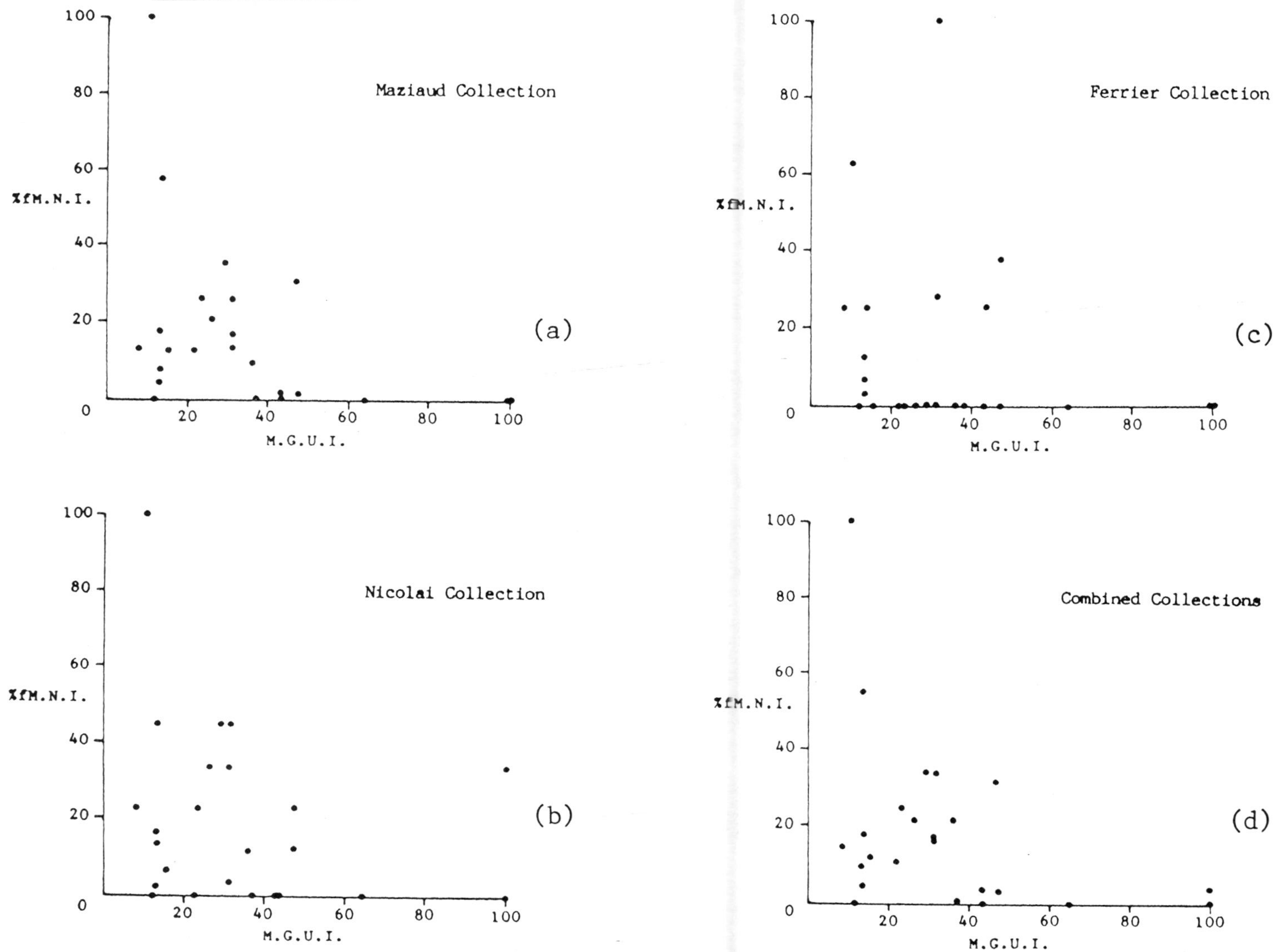

Roc de Marcamps Saiga Fig. 9.18

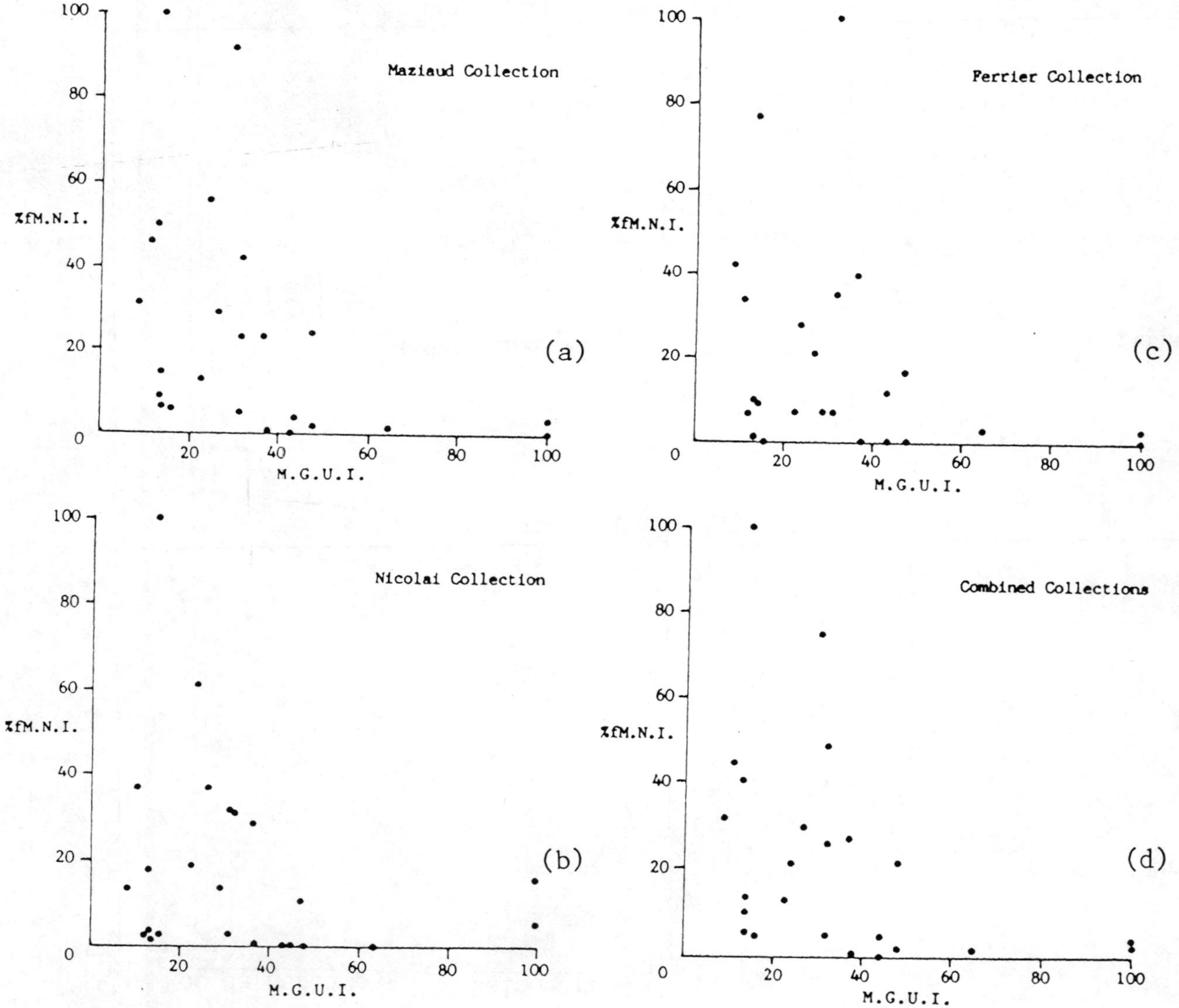

Fig. 9.19

Roc de Marcamps Bovids and Horse

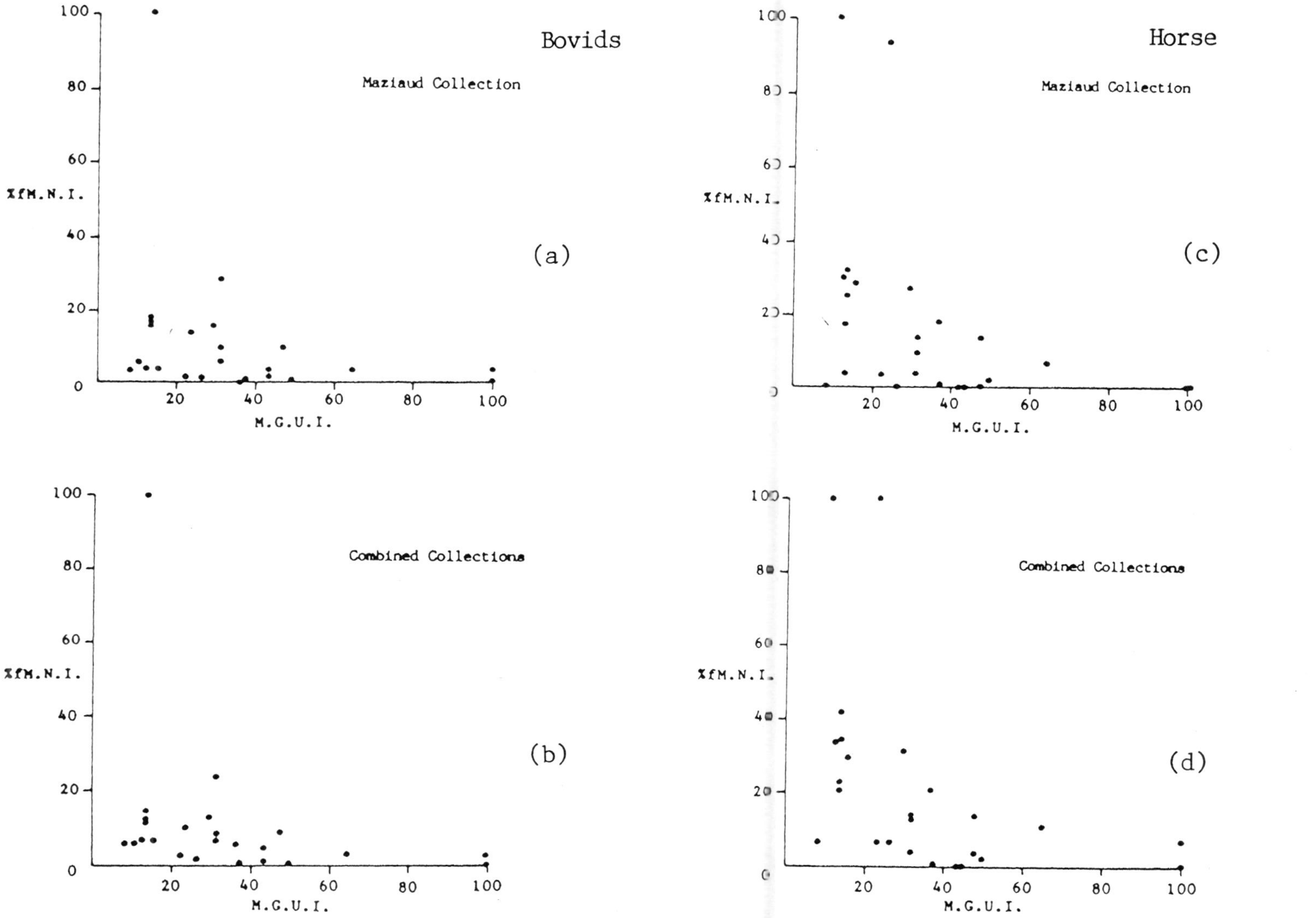

Fig. 9.20

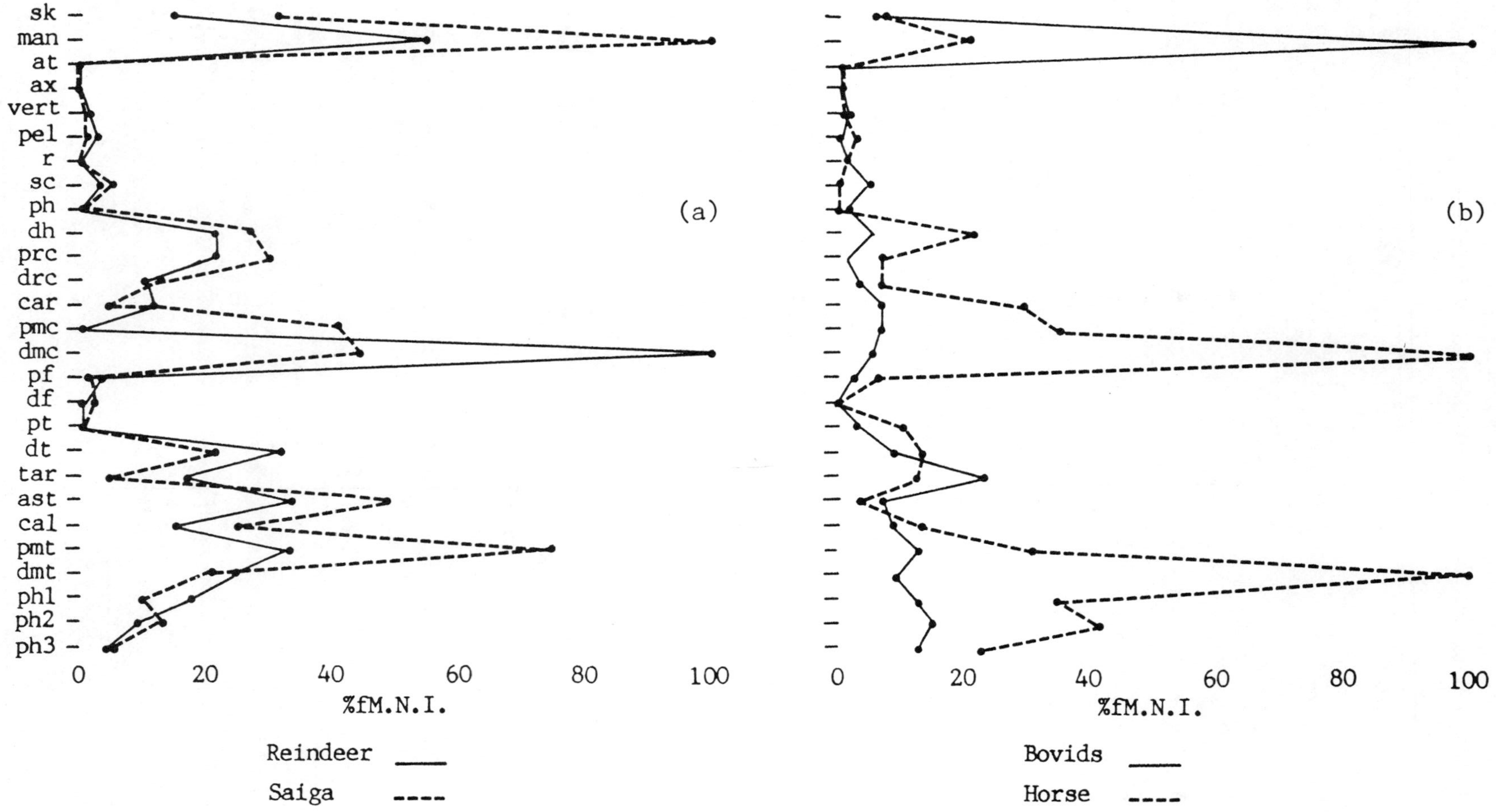

Element Representation Profiles from Le Roc de Marcamps, based on Combined Collections that large amounts of 'pounded-up' bone fragments were recovered, an observation repeated at the Roc de Marcamps, where the mean length of bone fragment is 2.35cm in the Ferrier Collection, 4.67cm in the Maziaud Collection and 5.18cm in the Nicolai Collection, the material being drawn from the boxes of "unidentifiable fragments". Binford also remarks that the bone splinter/articular end ratios at Tulukana are markedly higher than usual, reaching 61.63 per articular end. Although no such quantitative indications are available from Roc de Marcamps, the large number of long bone fragments in the Maziaud collection alone makes it possible to suggest that a high ratio would be seen here, and so perhaps bone grease and marrow processing were an important activity at the site.

The bulk curves seen in fig. 9.19 would further suggest that bovid post-cranial carcases were intensively processed, a feature which is not unexpected for a species which ranks third or fourth in terms of relative abundance in the separate collections.

In conclusion, at Roc de Marcamps, we see saiga, and to a lesser extent reindeer, associated with a gourmet curve, while horse and bovids display bulk curves. The first three species' element representation profiles show similarities to Binford's (1978) (early) spring hunting stands, while the Tulukana site - broadly similar to the bovids assemblage - is autumn occupied. Both pictures agree broadly with Gordon's seasonality conclusions, namely that the Roc de Marcamps represents a Gironde winter camp, occupied until the spring reindeer migration (1988:200).

Figures 9.21 and 9.22 show element representation profiles for six large herbivore species recovered at the Vezere Magdalenian site of Reignac. The material was analysed courtesy of A. Roussot (Musee d'Aquitaine, Bordeaux). In all but two of the profiles (the boar and saiga), lower limb bones predominate. In both species, the profiles for which are broadly similar, half mandibles predominate, the atlas being relatively well represented among boar and red deer remains. The parts of the carcase which are most often lacking in all species are the rib cage and vertebral column. Given the quantity of 'unidentifiable fragments' (fragments falling in the range 1-5 cm, bearing no distinctive features, but often the result of pre-burial processing) one is lead to wonder whether the lack of medium/high utility (meat yielding limb bones such as the ribs and vertebrae) parts reflects transport to the site or the degree of intensive processing.

Although no immediate parallels between Binford's (1978) comparative element representation profiles and those from Reignac are evident, some broad similarities are apparent. The Reignac profiles may be likened to late fall, winter and late spring hunting stands and consumption and residential sites suggesting that the site may have served as a 'base-camp' from which prey was hunted and meat processed, consumed and perhaps stored. Given the differences in the intensity of seasonal environmental conditions between Late Pleistocene South West France and modern-day Alaska and possible associated differences in animal utility, the wider range of seasons indicated by element analyses than by tooth cementum analysis is not surprising. Given these differences in environment, (ie: the prevalence of less extreme conditions in South West France), it seems likely that conditions,

Reignac

Fig. 9.21

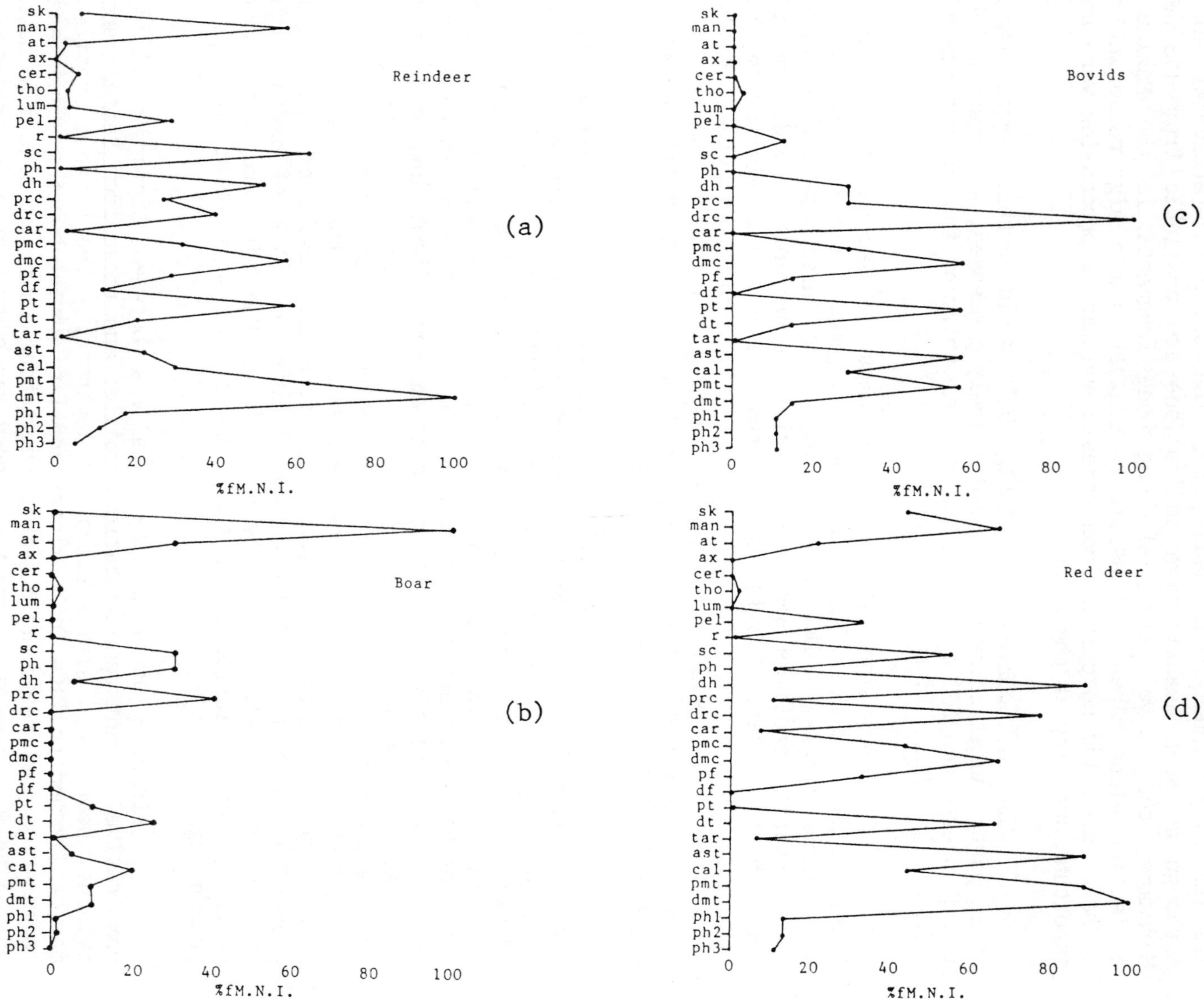

Fig. 9.22

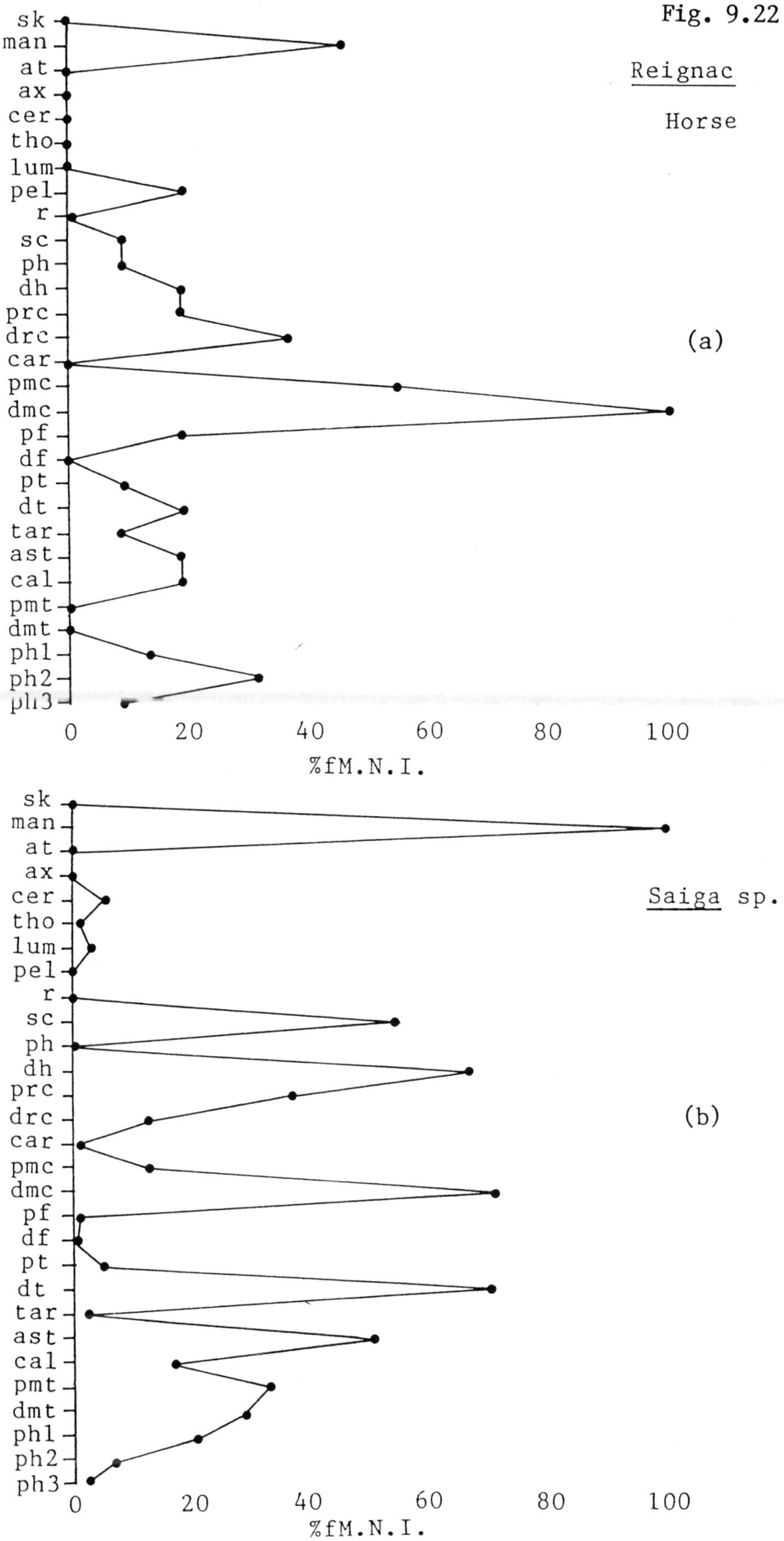
Reignac
Horse
(a)
sk
man
at
ax
cer
tho
lum
pel
r
sc
ph
dh
prc
drc
car
pmc
dmc
pf
df
pt
dt
tar
ast
cal
pmt
dmt
ph1
ph2
ph3
0
20
40
60
80
100
%fM.N.I.
Saiga sp.
(b)
sk
man
at
ax
cer
tho
lum
pel
r
sc
ph
dh
prc
drc
car
pmc
dmc
pf
df
pt
dt
tar
ast
cal
pmt
dmt
ph1
ph2
ph3
0
20
40
60
80
100
%fM.N.I.

suitable for a wide range of species occurred for a longer period of time. In general, results of analyses agree: late autumn, winter and spring occupation of sites and exploitation of all the considered species at this time. No summer site-occupation is suggested. The table below summarises these results:

Table 9.2.

Species	Site / Assemblage Characteristics
Reindeer	(Late) fall hunting 'stand'/residential site
Red deer	Spring hunting stand (& consumption site?)
Bovids	Winter processing / butchery site (bone grease cache ?)
Horse	Late fall hunting 'camp'
Boar	Late fall / winter kill
Saiga	Winter residential & late fall hunt

The reindeer (fig. 9.23(a)) presents a distinctly gourmet curve, once again being the dominant species. fM.N.I. values occur throughout the range 0-100 while only the axis is absent, and thus we see a near complete range of M.G.U.I. values.

The red deer curve (fig. 9.23(c)) is essentially similar to that of the reindeer, while the bovid curve (fig. 9.23(b)) is somewhat bulkier in nature in comparison. The bulk curve seen for the saiga presents a more complex picture in which the frequencies of distal humerus and tibia, the astragalus and scapula exceed their expected values. These elements however represent the upper forelimb and middle/lower hindlimbs, which may be considered as distinct butchery units. The majority of occurrences fall within a utility range of 10-47 and within an fM.N.I. range of 0.36 - 37.50.

When we turn to the boar and horse, we find two additional distinctly bulk curves, the latter perhaps to an even greater extent than the former (see fig. 9.24). Most horse occurrences fall within a wider fM.N.I. range than do those of the boar (0 - 55 % v. 0 - 40 %), and likewise over a wider range of utility values.

9.4 Conclusions.

Several things may be discerned from the study of anatomical element distribution with regard to utility values - given sufficient data of the right type. However, in the absence of detailed age and sexing indications the analyses presented here are of limited value. Nevertheless, some interesting conclusions may be reached regarding assemblage structure and conditions, both ecological and economic, under which they formed.

The "natural" deposit at Jaurens provided an opportunity to consider element representation in a situation where human behaviour is believed to have played no part. The pattern which emerges when element frequencies are compared with M.G.U.I. and meat index values is

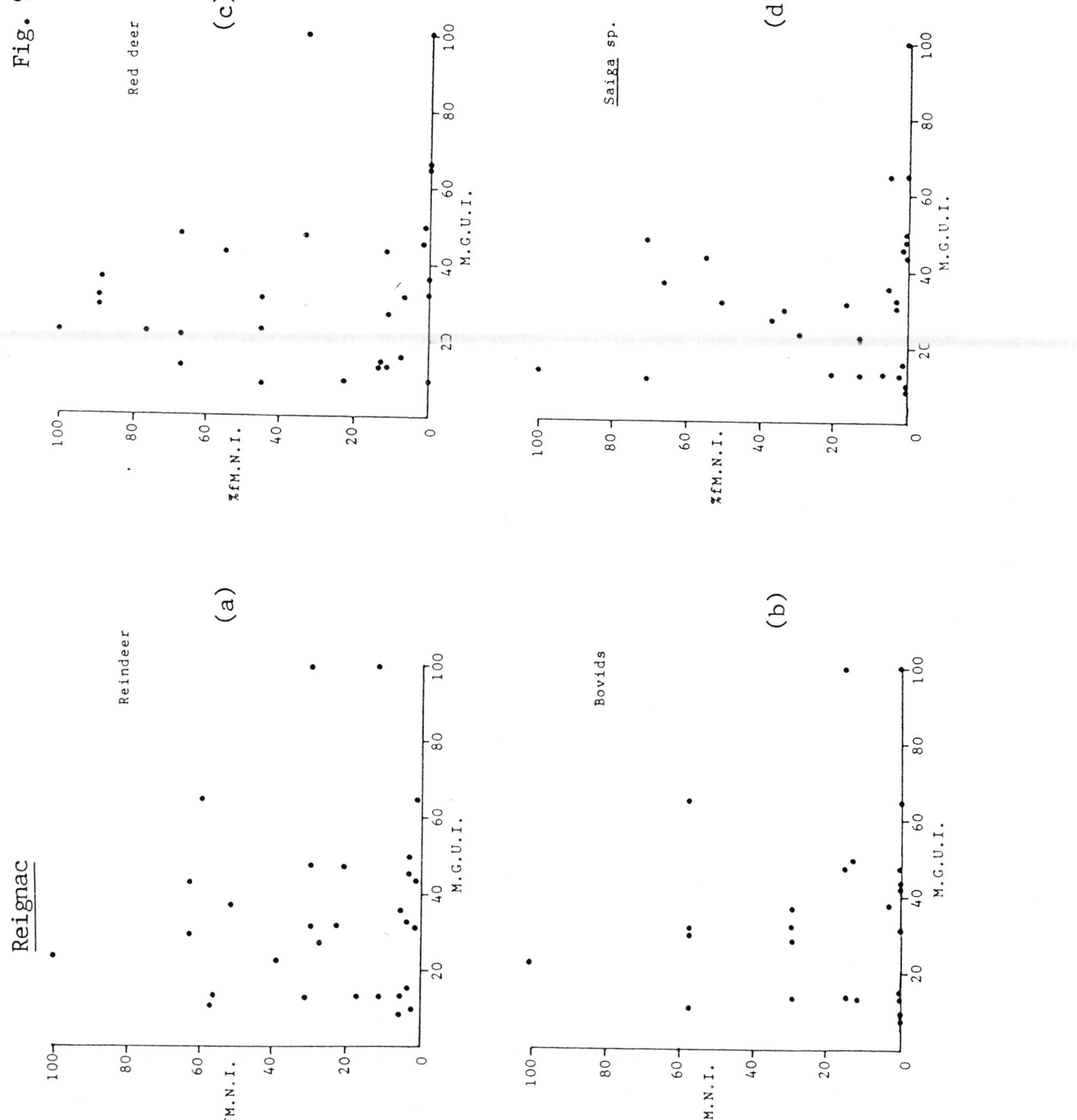
Reignac
(a)
Reindeer
(b)
Bovids
(c)
Red deer
(d)
Saiga sp.
%fM.N.I.
M.G.U.I.
0
20
40
60
80
100

Fig. 9.23

Fig. 9.24

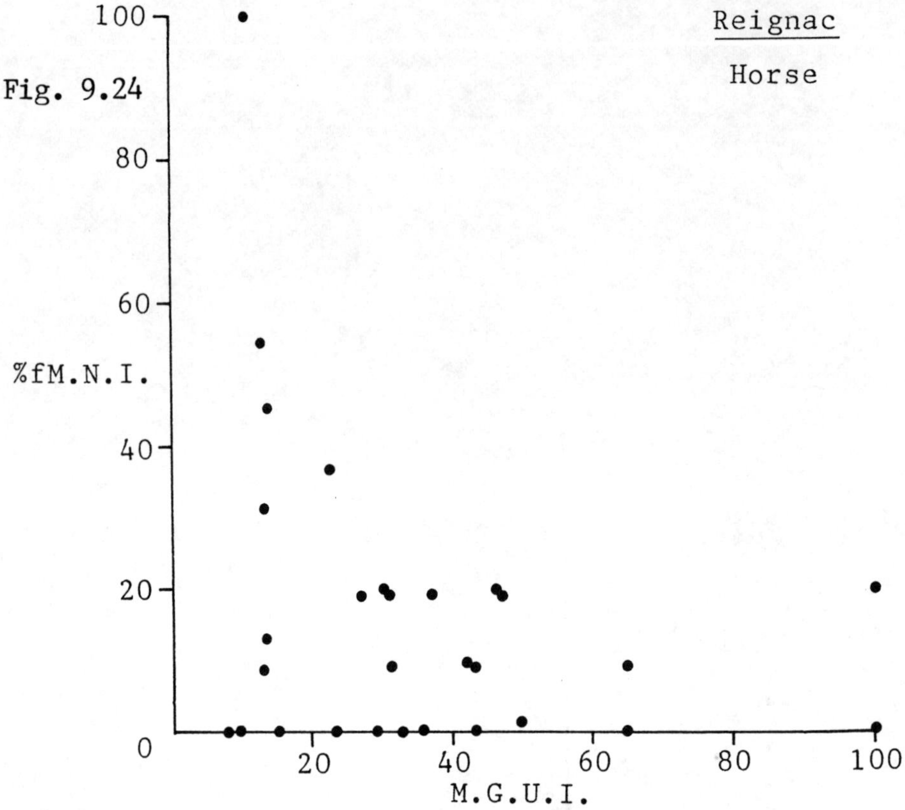

Reignac
Horse
100
80
60
%fM.N.I.
40
20
0
20
40
60
80
100
M.G.U.I.

(a)

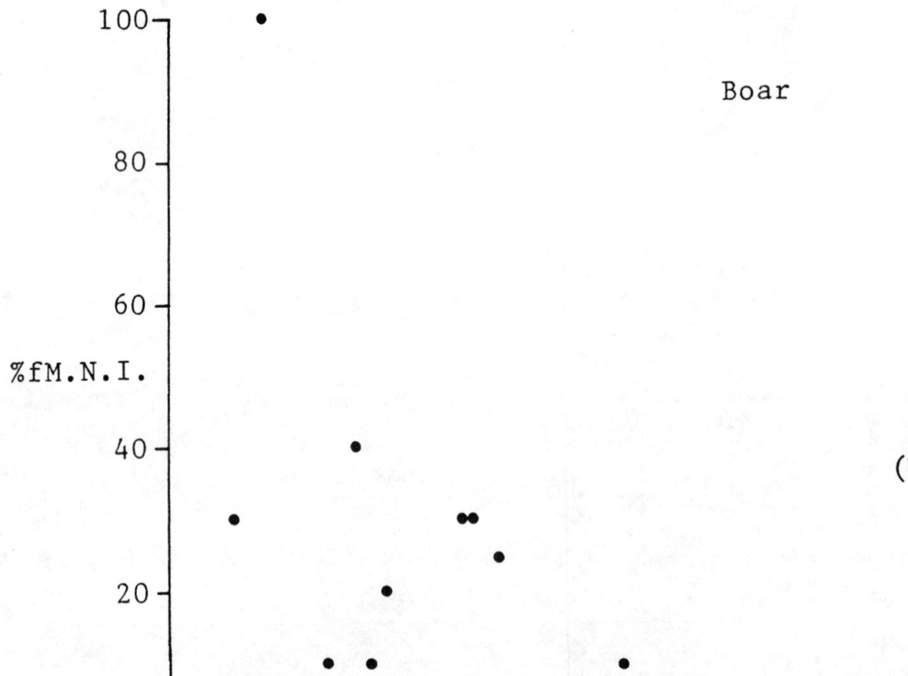

Boar
100
80
60
%fM.N.I.
40
20
0
20
40
60
80
100
M.G.U.I.

(b)

one not seen at other sites where human behaviour is invoked to explain the patterning. Similarly, the relationship between %fM.N.I. and density is not the same as at other sites (see fig. 9.25), lending support to the belief that the degree of element representation observed is not simply a function of preservation.

In general, the more valuable parts of carcases are found at base camps where consumption takes place (eg: Abri Pataud). Primary dismemberment may however also occur, for example at Les Eyzies and Reignac. These camps are generally characterised in South West France by parts representing all, or almost all, of the carcase. Complete animals may have been transferred from the kill, or alternatively (when kill and consumption were some distance apart) the carcase may have been dismembered prior to removal. Thus, at sites such as the Abri Pataud, where only a few elements are missing, carcases were probably on occasion brought back to the site in one piece; more often they were brought back in parts. Similarly at Le Flageolet II primary dismemberment occurred away from the site, the animals being brought to the site already dismembered. Subsequent butchery and processing (for marrow, in the present case) account for the bone fragmentation observed. At Les Eyzies carcases were probably brought back to the site complete, the position of the site being more favourable than that of Le Flageolet II.

Kill-sites are difficult to identify in South West France, perhaps due to the inherent nature of the sites. Kills are likely to have occurred away from the caves and rockshelters from which so much of our data come. Instead, they will have taken place in the open, in valley bottoms or on open plains. Unfortunately it is this type of site which we are lacking (relative to the cave/abri category sites) and those which we do have rarely provide suitable faunal material. It may be argued that we are potentially lacking an important component of the hunting and subsistence system, the major exception to this being the possible location of river crossing kills which White (1985) attributes to such sites as La Madeleine. Perhaps the abundant faunal material buried in front of the Laugerie deposits (White 1985) is just such a case - the area now buried under the road and car park. Here carcases in various stages of disarticulation were discovered by Peyrony, a fact which White (1985:150) invokes to support the notion of a water-crossing kill-strategy. At other sites, such as La Mege, hunting stands may be identified where, once again, high value parts are unlikely to be found. Such sites were probably occupied only on a short term basis, their associated archaeological assemblages possibly reflecting the small number of people occupying the site and their 'limited needs'.

At sites of primary dismemberment and initial butchery (more easily recovered from the archaeological record) it is to be expected that the predominant element type will be of lower value, the higher value parts being rare or absent. Such is the case at Sainte Eulalie (Magdalenian VI) where only two parts of M.G.U.I. value greater than 38.00 occur.

Finally, as the analysis progressed one result emerged repeatedly - that is, the tendency for the major species to be associated with the gourmet curve. This may be explained by the fact that a species (in this case usually reindeer) which is very abundant

Fig. 9.25

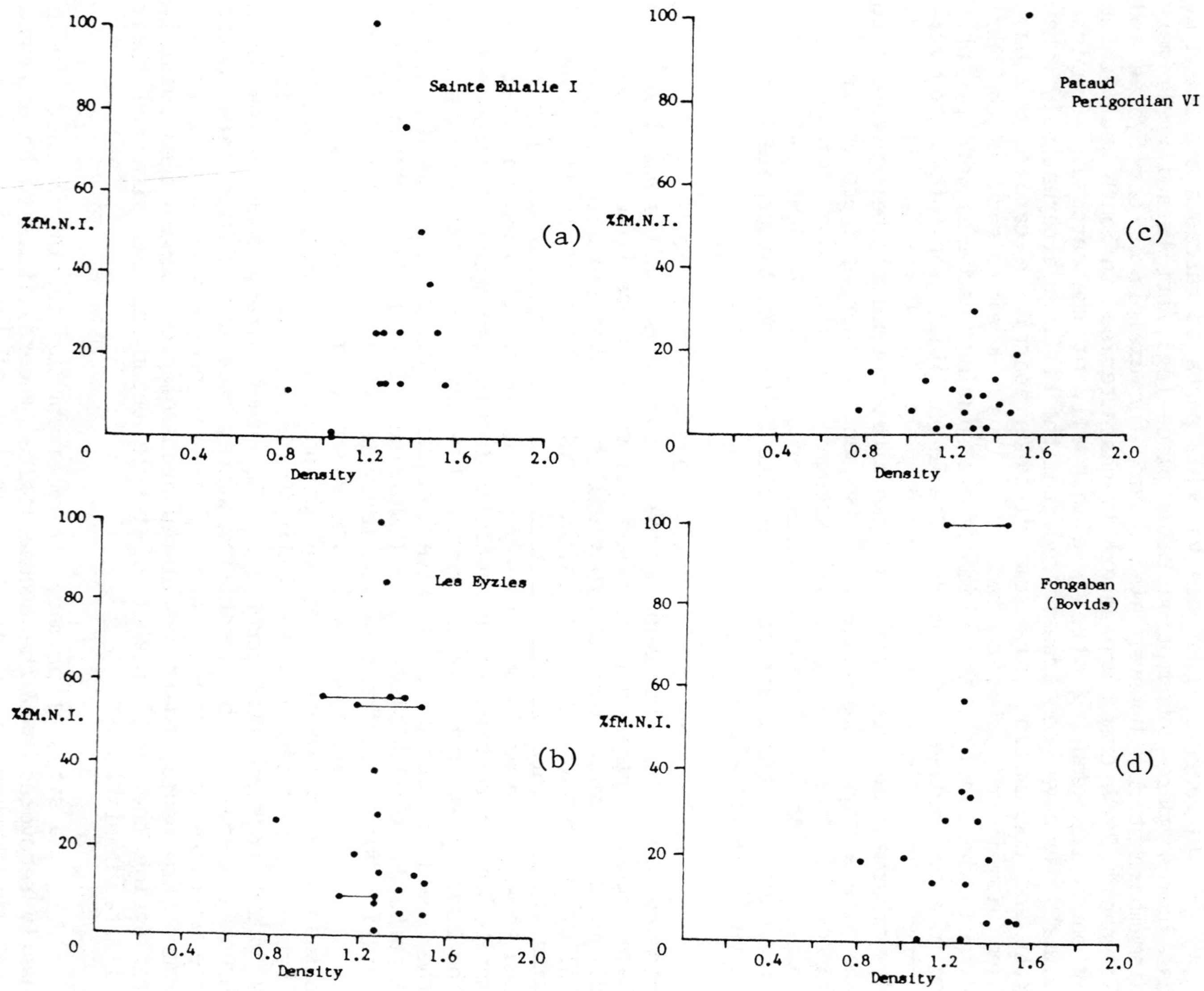

Relationship between Density and %fM.N.I. Values at Selected Sites in the environment and is easily exploited, with minimal risk involvement, does not require intensive processing of all body-parts if a sufficient quantity is to be obtained. Thus medium and low utility parts (and occasionally some high) are discarded. Meanwhile, secondary species are characterised by bulk curves, resulting from more intensive processing of less common taxa. This explains the preponderance of gourmet curves at the above sites where, in the majority of cases, only data for the major species are available (see table 9.3).

Table 9.3. Resource Exploitation at Selected Upper Palaeolithic Sites.

Site/Period	Species	'Curve'	Season	Comments
La Mege Magdalenian	Reindeer	Bulk	Winter/spring	Hunting stand
Flageolet II Magdalenian	Reindeer	Gourmet	Late winter	Marrow extraction
Gare de Couze Magdalenian	Reindeer	Gourmet	Winter/spring	Residential (?)
Pataud Perigordian IV	Reindeer	Gourmet	Autumn/	Major species
Perigordian V	Reindeer	Gourmet	Winter/	
Perigordian VI	Reindeer	Bulk(?)	Spring	
Protomagd.	Reindeer	Bulk		
Le Piage Protomagd.	Reindeer	Gourmet		Major species
Sainte Eulalie Magdalenian	Reindeer	Gourmet	Winter	Occupation site
La Gravette U. Perigordian	Reindeer	Gourmet		Major species
	Horse	Bulk		Secondary species
Fongaban Magdalenian	Bovids	Gourmet	Spring	Major species
Enlene Magdalenian	Reindeer	Gourmet	Winter/spring	Major species
Les Eglises Magdalenian	Ibex	Gourmet	Late autumn	Mountain field-camp
Les Eyzies Magdalenian	Reindeer	Gourmet		Consumption site

Chapter Ten

SUMMARY AND CONCLUSIONS.

In this study the significance of Upper Palaeolithic herbivore faunal assemblages from South West France has been investigated through the use of various statistical and display techniques. The major implications of the methods employed in the analysis, and the results thereof, are considered here.

10.1 Chronological Patterning:

Earliest Upper Palaeolithic assemblages (Chatelperronian) are characterised by a generalised nature associated with the relatively mild conditions of the Hengelo Interstadial. In the Dordogne/Vezere we see reindeer averaging approximately 50% with bovids coming second at a mean value of 17%. To the east and north horse represents approximately 40% of all large herbivore remains, with the reindeer in second place in the east. However, the assemblages in the latter region are made up of markedly fewer taxa than are those of the Central zone where there are seven species. In the east this figure falls to five. The qualitative data available generally confirm these patterns, although the number of taxa in Zone 2 is substantially lower than that in Zone 1 at this time. In the north of the region bovids replace reindeer as the second species, although by a small mean-value margin only. All other species at this time, in all regions, are relatively unimportant.

During the Early Aurignacian reindeer frequencies rise sharply, accompanied, in general, by a marked decline in alternative species. Only horse frequencies remain relatively constant while ibex mean values rise somewhat. In both the north and centre of our region we see a definite increase in bovids during the Later Aurignacian (corresponding with the milder 'Arcy' Interstadial phase), although in the Central zone, where a 'patchy' distribution is suggested, reindeer retains its average, predominant position; during the latter period, however, there is greater diversity in assemblages. Once again, non-numerical material patterning correlates well with the quantitative data.

Upper Perigordian-Protomagdalenian and Solutrean assemblages are available only from the centre of our area where the preponderance of reindeer continues to be associated with alternative, cold (arctic tundra and cold steppe) species.

During the Early Magdalenian the nature of regional assemblages differs somewhat. In the north we see reindeer-dominated assemblages in which few other species are of importance; in the south of the region bovids and horse dominate the weaker reindeer. In the centre of the region however, reindeer continues to predominate, although to a lesser extent, the mean value being lower still only during the markedly milder phases of the Chatelperronian, Later Aurignacian and Azilian. During the greater part of the Magdalenian (II - VI) reindeer is by far the dominant species in both the north and south. To the west we see a change in the structure of assemblages in which saiga appears and in places predominates. Here the regional assemblages are more diverse, no one taxon predominating to the exclusion of others. Horse, bovids and saiga dominate various assemblages - but none of them compares with reindeer in the Perigord, although the non-numerical data do point out the importance of regional saiga assemblages. In the east, assemblages remain of a much more generalised nature.

It is only during the early post-last-glacial that we find another taxon consistently gaining ascendancy over the reindeer. In the northern zone and Vezere/Dordogne red deer replaces reindeer in terms of frequency, associated with alternative temperate 'woodland' species. In the east, however, there is a rise in ibex frequencies, possibly accounted for by the fact that we may here be dealing with specialised ibex hunting-camps, for other assemblages in this region do comprise relatively high red deer totals and witness the re-emergence of roe deer and boar as consistent, if not important, elements of the regional faunas.

To conclude, we have a chronological sequence in which the cold climatic phases (glacials/stadials) of the Early Aurignacian, Solutrean and Middle Magdalenian are closely associated with reindeer, ibex, horse and saiga, and warmer, interstadial conditions (eg: Chatelperronian, Later Aurignacian etc.) are accompanied by reduced reindeer totals and higher red deer, boar, roe deer and bovid frequencies. During the later and shorter milder periods (post-Early Aurignacian), reindeer continues to dominate assemblages however; vegetation is slower to change than is climate, and herbivores are known to react to vegetational rather than temperature fluctuations. Communities, plant or animal, which are adapted to 'cold' conditions are more likely to survive in milder environments than are 'temperate' species during cold, extreme phases. Early or prolonged frost or snow cover alone may seriously affect local vegetation and the herbivores dependent thereon.

10.2 Ecological Implications.

Employing both bivariate and multivariate statistical techniques in our consideration of Upper Palaeolithic faunas, certain regularities occur which may be explicable in terms of natural, broadly ecological factors.

Throughout the latter part of the Upper Palaeolithic, significant positive correlations are observed between red deer, Bos sp., roe deer and boar, suggesting the presence of a relatively temperate, woodland component in the fauna. This component is negatively correlated with the reindeer, a traditionally arctic-tundra species which is also often negatively correlated with those steppic species which characterise a drier, more open, cold group. This steppe component emerges mainly during the Magdalenian, based upon observed frequencies of horse, bovids and saiga. Finally, throughout the Upper Palaeolithic a montane group of species recurs, generally defined by the co-occurrence of the ibex and chamois. Seasonal variations in these may be apparent during isolated episodes at specified sites.

In order to reduce the large mass of numerical data derived from Upper Magdalenian and Pre-Upper Magdalenian faunas to a simpler, more manageable and more easily 'explained' data set, principal components analysis was employed.

Five principal components were obtained in the consideration of Pre-Upper Magdalenian data, and four attributed to the Upper Magdalenian.:

1 - A (Temperate) Woodland Element: boar, red deer, roe deer, ass. (Reindeer is negatively correlated)

2 - A Montane/Skeletal Element: chamois, ibex.

3 - An Open, herbaceous Vegetation Element of low grazing grass.

4 - A Cold, Dry, Flat Element. Saiga is typical of this form of steppe community.

5 - Mammoth: Open, cold grassland, some shrub or light tree-cover. Scavenged (?)

The Upper Magdalenian components comprised:

1 - (Temperate) Woodland: boar, red deer, roe deer, horse.

2 -(Eastern) Cold tundra, reindeer-dominated communities.

3 - Montane: ibex, chamois

4 - A fourth factor characterised by chamois and ass - possibly representing a 'forest edge' environment, perhaps exploited on a seasonal basis.

These groups compare with Chase's (1987) 'arbitrarily chosen' groups - tundra, steppe, prairie/wooded prairie, forest and alpine - as they do, to some extent, with those of Delpech (1984) and Bouchud (1975).

The local nature of open environments, as opposed to their distribution (which is determined largely by topography), varied throughout the Upper Palaeolithic to a greater degree than did those of woodland and montane communities. Woodland and montane components occur in both data sets, their constituent species varying only very slightly. It is among the more open components that we see greater variation in associated faunal species, no doubt reflecting a variety of more specific, localised habitat types during the different stages of the Upper Palaeolithic. Small variations in climatic conditions result in marked changes in open vegetation where the rates of evapotranspiration and run-off are more quickly affected. In woodland conditions the occurrence of dense undergrowth etc. regularises waterflow and may prevent flooding. Furthermore, the degree of soil freezing is less under woodland cover due to protection offered.

10.3 Zoogeographical Patterning

By mapping the frequencies of various species during the Aurignacian and Upper Magdalenian several distribution patterns emerge.

Throughout the Upper Palaeolithic reindeer appears to be a ubiquitous, perhaps 'cosmopolitan', species, occurring throughout the region in quantities which outnumber all other taxa. However, this 'tundra' species is of greatest importance in the centre of the region where assemblages are more specialised than they are in the north and west. Here horse and other steppic species establish a position of importance, perhaps in a north/west 'range' which, on occasion, is associated with pockets of tree and shrub cover and may represent a 'forest-steppe' ecosystem or what Chase (1987) terms "wooded prairie".

Montane species are more commonly encountered in the Dordogne and Vezere area (where Taxonomic Richness is highest during the Later Aurignacian and Upper Magdalenian) than elsewhere, both chamois and ibex being best suited to the rockfaces associated with the Vezere, Dordogne and Lot valleys. It is only the distribution of the more temperate woodland species which varies noticeably during the Upper Palaeolithic - reflecting, in part, changing climate and associated vegetation conditions. In the Aurignacian these species occur largely in the south of the region while in the Upper Magdalenian the west, centre and Dronne see more common woodland species. In the case of the other species (tundra, steppe and montane), it is perhaps the topography of the region rather than the climate which governs distribution. This is particularly so in the case of saiga, horse, bovids, ibex and chamois, the first three preferring flat terrain, the last two choosing rugged, rocky or alpine territory.

The division of South West France into changing landscape types however does reflect spatial variability in a number of ecological factors. These include snow cover, frozen-soil depth (as indicated by fossil periglacial features in chapter 3), direction of wind, extremes in temperatures, relative humidity and, inevitably, resulting vegetation. However, the fact remains that within the heterogeneous mosaic landscape of South West France, topography and associated quality of land surfaces (eg: firm, frozen ground v. dry, sandy soils, v. wet, river-bank and marshland) remain of primary importance in the determination of the distribution of animal species. Such factors form a 'constant' against which fluctuating climatic conditions interact to generate changing patterns of faunal distribution. The fact that these distribution patterns also reflect the behaviour of prehistoric hunter-gatherers (ie. exploitation of prey species) must, however, be borne in mind.

10.4 Butchery and Carcase Management.

By comparing 'fractional Minimum Number of Individuals' and expected frequencies of elements, based upon values of various utility indices for each element type, it has been possible to isolate 'gourmet' and 'bulk' curves, and associated butchery/processing strategies. In general, it has been found that the 'gourmet' curve (the result of the removal of high value elements) is encountered when the species concerned dominates an assemblage, ie. perhaps forms the major prey. Subsidiary resources are more commonly (but not invariably) associated with the 'bulk' curve (the result of removing and intensively processing elements of high and medium value). These results might be expected on the basis of elementary Optimal Foraging Theory. We may here be seeing evidence of intensive processing required to gain a sufficiently high yield from a species taken in smaller quantities, or on a less regular basis, due either to limited availability or problems in exploitation.

By adopting the above approach to the study of element representation we have been able to form some hypotheses concerning the possible function of various sites. For example, element profile examination suggests that La Mege may represent a hunting stand, Le Flageolet II a butchery/processing camp, Reignac a 'base camp', and Fongaban a (seasonal) hunting camp. It is also possible to suggest the season during which assemblages formed, although such must be considered in conjunction with the results of Gordon (1988) and Le Gall (1984), concerning the seasonality of occupation sites in France.

10.5 Seasonal Subsistence Specialisation.

There is one question which repeatedly raises its head when the significance of archaeofaunas, associated with presumed hunter-gatherers, is considered - namely the degree to which the faunal assemblages considered, and the patterns therein, reflect environmental or cultural 'processes'.

In an attempt to account for the possible ecological patterning and associated resource availability, the decision was made to consider recurrent, statistically identifiable patterns. Davis (1987:62) maintains that 'assemblages which accumulate through human agency reflect whatever man was able to collect from the environment.' While the picture may be distorted by such occurrences as a single special-purpose kill-site, faunal material which accumulated over several millennia should, to some extent, reflect the resource base available. Although one would be mistaken in reaching conclusions regarding resource availability based simply on faunal assemblage structure, the latter does provide an essential starting-point. However, the relationship between the number of individuals represented by an assemblage and the number of bones encountered is little understood; likewise, the degree to which archaeozoological assemblage composition reflects the structure of assemblages at the time of burial (Grayson 1981). The composition of the original living animal community is totally unknown.

A major problem encountered when considering large herbivore faunas as palaeoenvironmental (usually palaeoclimatic) indicators in South West France, lies in the fact that there is a great deal of local climatic variation and therefore variety in biotope, each of which may have been exploited on a year-round or seasonal basis. The phenomenon of palaeo-microclimates is one which has received little attention and is a difficult problem which requires consideration. Both temperate woodland and open arctic tundra species may be encountered in substantial numbers in a single level at a site, due to the fact that, given the valley-side position of the site, several biotopes may be exploited during the course of a year or single season. The simultaneous exploitation of several environmental groups will surely result in a relatively diverse, heterogeneous faunal assemblage. Thus at sites such as Caminade Est, Le Flageolet I and Faustin we find mixed faunal elements (tundra, boreal forest, steppe etc.), including relatively high frequencies of reindeer, red deer, horse, bovids and other woodland and montane species in a single deposit, giving us no reason to believe that extensive or specialised (reindeer) hunting strategies were adopted.

The environmental interpretation of diversity depends upon the biogeographic principle stating that that more species are found in areas of mild climatic conditions and potentially higher primary biomass than in areas subjected to more extreme conditions. Tundra environments are generally poor with regard to the number of species observed and total biomass is low. Mean tundra biomass amounts to only approximately 100kg. ha^{-1} (Kistschinski 1976). In taiga and boreal forest zones more species are encountered, accompanied by an observable rise in primary productivity. According to Kistschinski (1976) figures vary from 100 to 400kg ha^{-1}. In steppe environments the number of species remains low. However, here large populations of specific species (usually medium or large-sized mammals, such as saiga) are seen. Such high resource population levels may be permanent or seasonal in nature, due to the influx of such species as the migrating reindeer. It may be suggested that South West France was a region of highly localised areas of high productivity, especially close to running water (eg: in river valleys, such as close to La Marche) where sedges, bryophytes, lichens, herbs and 'abundant' shrubs formed the main vegetation cover. In the area seasonal bird and fish populations were encountered, in addition to the periodic influx of large mammals and the permanent water-edge/wetland communities, forming a diverse resource base available for exploitation. Such an influx increases the available prey density, thereby raising the predictability of the resources and associated security, as the species is more predictable in both its location and number. In general assemblages show a marked degree of specialisation (as indicated by B) despite a broad spectrum of resources (N) throughout the region as a whole. Some chronological changes may be seen in the value of B/N (measuring selectivity) as shown in figure 10.1.

The low diversity shown by assemblages may be inferred to reflect some degree of specialisation, or selectivity, practised by hunter-gatherers. Christenson (1980) discusses such a situation, in which the response to declining resource availability (measured in terms of relative prey density) is overcome by the narrowing of the diet (niche) in order to specialise. Here, diversity (B) is used to measure assemblage rather than environmental specialisation. Until we know how faunal assemblages reflect prehistoric environments and faunal resource availability, this is the most appropriate use of the technique. Meanwhile, Jochim (1981) discusses the inverse relationship between subsistence security and niche width, for as the availability of preferred resources declines so, he claims, the inherent risk may be 'spread' by the exploitation of several other, secondary resources.

The emergence of 'specialised' hunting is frequently equated with the transition from the Middle to Upper Palaeolithic, although this is not universally accepted. Freeman, for example, attributes

Chronological Distribution of Assemblage Diversity (B/N)

Fig. 10.1

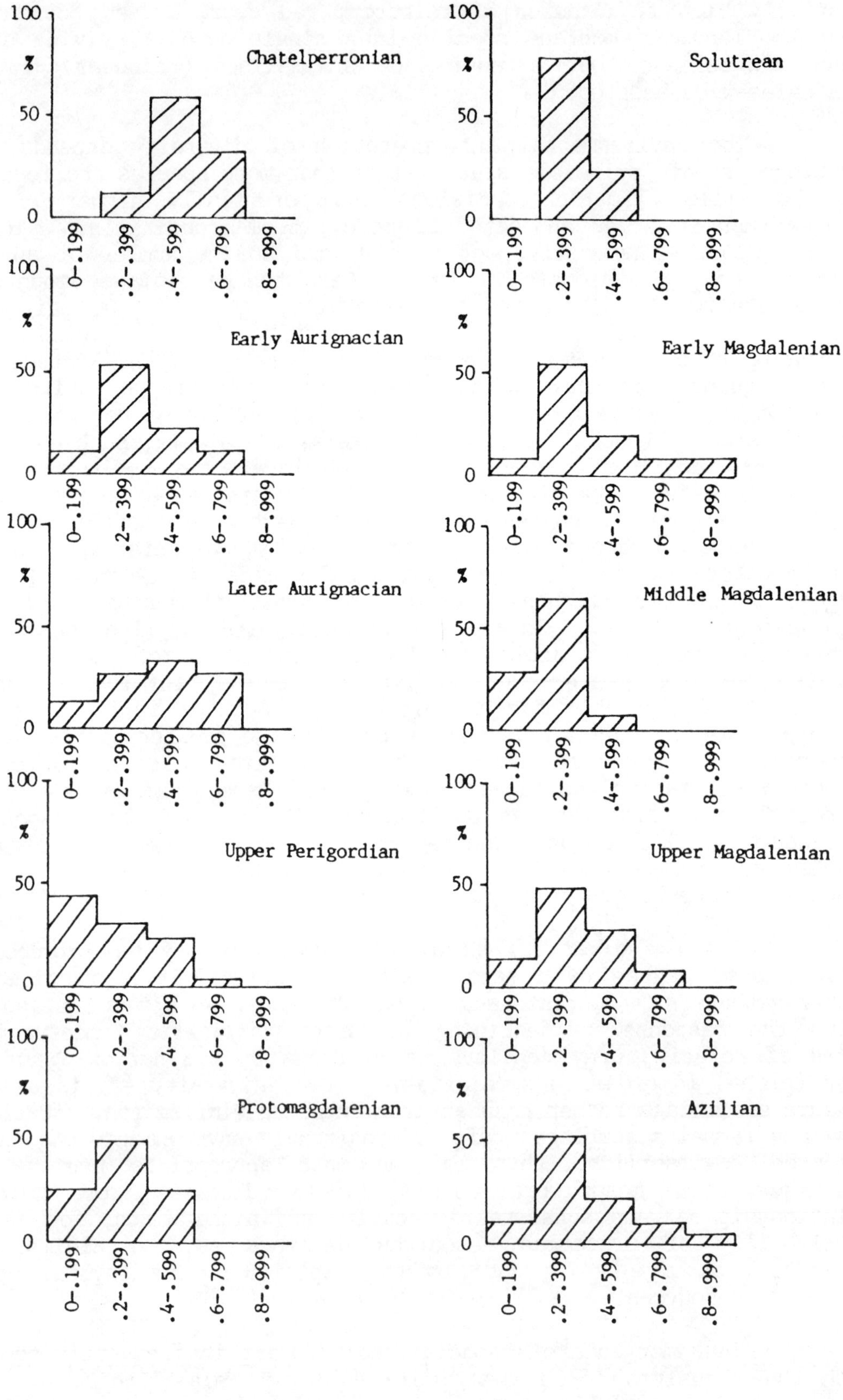

"continued opportunism in food-gathering strategies" (1981:148) to the Early Upper Palaeolithic of Spain although organised game drives are postulated throughout much of the Palaeolithic as a whole. On the other hand, Girard and David (1982) ascribe "la chasse specialisee" to the Middle Palaeolithic at the open-air site of Mauran (Haute Garonne) where bovids make up almost 98% of the recovered material. Chase (1986) also attributes "specialised big-game" hunting to the Middle Palaeolithic - "in the sense that the great majority of their [the hunters'] meat came from ungulates and in some cases from large carnivores, rhinoceri or mammoths" (1986:142).

The 'specialist' is one who selects one of several resources and concentrates attention upon it, albeit on a seasonal or year-round basis. The more specialised the strategy, the less important are resources other than that upon which most attention is focused. The target may be a single species (usually medium- or large-bodied (Orquera 1984)), an ecological component or ethological group (eg: migratory herd animals). In South West France we have evidence to support the hypothesis that several resources were available for exploitation, and thus the high frequencies in which the reindeer often occurs must be seen, to at least some extent, as a result of a specialised rather than random hunting strategy, for it is unlikely that, during the Magdalenian, the reindeer formed at least 90% (and occasionally 100%) of the then-extant fauna. However, it is only when the food supply is scarce that an attempt must be made to exploit every possible resource (Pianka 1978). In a food-rich environment one can bypass a large proportion of what is found, in the expectation of locating preferred resources. Thus it is suggested that 'selective foraging' (defined as the purposeful selection of a single or small number [<4] of resources) formed the norm during much of the Upper Palaeolithic, enabling human groups to make the most of a fluctuating resource base.

Chase (1986) identifies a similar strategy which he terms 'purposeful eclectic hunting'; various species are taken, not at random but deliberately seeking out different resources at different times in different places. This contrasts with 'opportunistic hunting' where prey is taken when the opportunity permits. Furthermore, as Geist (1978:251) points out in his discussion of opportunistic hunting, no elaborate preparations for the kill are undertaken. The result is a mosaic of specialised and generalised faunal assemblages made up of single (special) purpose, temporary sites (eg: kills) and occupational base camps from which several biotopes may have been exploited.

'Specialised hunting', as defined by Chase (1986:142), in which hunter-gatherers depend upon a single resource throughout all or most of the year, is not, in general, supported by the available data, for non-reindeer species are important despite the high reindeer percentages observed. The relative frequency with which alternative species occur suggests that secondary resources provided at least an element of variety in the Upper Palaeolithic diet.

A diet based on a single, high protein/low fat species such as the reindeer (see table 10.1), is unlikely to support a healthy population for long. A diet composed totally of such species results in malnutrition and potentially even death (Speth 1983:150-51) for fat is the most important element of a tundra diet. When the diet

Table 10.1 Typical Nutritional Values of Selected Resources (per 100g)

	Energy (kcal)	Fat (g)	Protein (g)	
Horse	121	3.7	20.5	
	110	2.0	22.0	1
	120			4
Antelope	150	2.2	30.4	
Reindeer	217	14.4	20.5	
	117	1.2	26.6	2
	120	1.2	27.0	
	100			4
Red deer		2.5		
	124			3
Bison	300	7.5	21.8	
	120			4
	300	5.5		
	204			3
Mammoth	150			4
	200			3
Woolly Rhinoceros	175			3
Salmon	182	12.0	18.4	
Eel	168	11.3	16.6	
Trout	179	10.8	18.8	
Hare	162	9.0	28.0	
Rabbit	129	5.0	21.0	
Ass	100			4

(Sources: Tyler pers. comm.
1 Ducommum (1982)
2 Stern et.al. (1980)
3 Soffer 1985
4 Saffirio 1983

comprises a high protein intake, substantially higher food consumption is required in order to generate sufficient energy (Wing and Brown 1979:22, Anderson et. al. 1982:205). In brief, additional energy for metabolism is required. Kilocalorie intake needs to rise by as much as 20% (Lieberman 1987:234) in order for stored body fat and skeletal muscle protein not to be used for energy. Almost one third of calories derived from protein are required to metabolise consumed protein and generate body heat, once again increasing consumption requirements. We might, therefore, expect to find large quantities of reindeer material - especially if taken during the season when the nutritional quality of the meat is declining. During the rutting season up to 80% of adult male saiga are known to starve themselves to death (Time Life 1985). In so doing the nutritional quality of meat will decline. Thus, as in the case of the reindeer, the hunter may be forced to take large numbers of saiga - a phenomenon seen in the west of our region. We may also witness the selective killing of animals and processing of elements which are rich in fat (eg. female saiga). Late winter and spring are likely to be the most stressful (highest risk) time of the year - for both predator (man) and prey. If the hunter practises storage, a substantial portion of the reserves may have been used by this time and there will be little plant material available for exploitation; the hunter-gatherer is forced to increase his fresh meat intake, increase hunting and, in the Dordogne/Vezere region, exploit a seasonally abundant, but perhaps nutritionally poor, resource base.

The inclusion of either adequate fat or carbohydrates in the diet reduces the specific dynamic action of food (see Speth 1983:148-159) by a factor of 2-3 or 5 respectively (Lieberman 1987; Speth & Spielman 1983) and thus the role of protein in the diet decreases. In the archaeological record evidence of fat-rendering may take the form of abundant bone fragments. A high degree of fragmentation is seen when bone grease or marrow is obtained through intensive post-kill carcase processing, as at Le Flageolet II. As a result of the nutritional constraints placed upon hunters, and their attempt to counteract them (eg: reducing complete bones to fragments and thereby increasing N.I.S.P. totals), the degree of apparent specialisation increases markedly, although the fact that animals were killed need not prove that they were in fact consumed at all. Reindeer may have been taken for their hide, the thick winter coat of this species being warmer and more flexible than that of many mammals (Orquera 1984). The abundance of reindeer bone used in the production of both tools and art objects implies that the animal was not sought simply as a source of food - but also as a source of raw material. The equating of an abundant species with a specialised diet may be over simplistic, for a specialised hunting strategy and a specialised (or narrow) diet are not synonymous. Until results of chemical analysis of human bone, such as that reported by Hayden et. al. (1987), and detailed pathological and dental studies are available in quantity, the role of meat (as opposed to fish etc.) and the taxa providing it remains relatively unknown.

Finally, the specialised seasonal occupation of the Dordogne and Vezere valleys during the Upper Magdalenian may, in addition to the seasonal resource availability or predictability, be accounted for in terms of the advantages offered by the topographic character of the region. The area west and northwest of the Dordogne/Vezere confluence (essentially the modern Gironde and lower reaches of the Isle and

Dronne) was associated with an open, steppe and light woodland cover, offering scant winter protection from the elements (low temperatures, snow and harsh winds from the Bay of Biscay). The existence of fossil periglacial phenomena in the Angouleme area, and elsewhere, points to low winter temperatures which may have been avoided by congregating in the cave-offered protection of the Dordogne, Vezere and, to a lesser extent, Charente areas. Such a predictable resource base offering subsistence security during the hardest time of the year was not scorned, as the degree of 'selective foraging' discerned shows. Furthermore, as ethnographic research has shown, hunter-gatherers with a secure food-source frequently indulge in non-subsistence related activities, due to their apparent abundance of spare time (Lee 1968, Riches 1982), which may be reflected by the overwhelming abundance of parietal and mobiliary art which was clearly (or at least historically) the cause of the development of interest in the Upper Palaeolithic of South West France.

10.6 Suggestions for Further Research.

It is a great nuisance that knowledge can only be acquired by hard work.

W. Somerset Maughan.
10 Novels and Their Authors. Ch. 1. sect. i.

As the research reported here progressed, several subjects emerged which still require study and consideration.

10.6.1. Changes in the quantitative structure of faunal assemblages should be related to established patterns of climatic change, rather than be employed to reconstruct the climatic conditions themselves, as has often been the case. In order to consider the relationship between climate and faunas we have to be able to establish, at least to some extent, the degree to which recovered faunas reflect the animal resource base available at the time of deposition. Can we assume, for example, that 96-100% reindeer in levels 3 - 13 at La Madeleine reflects an environment in which reindeer made up at least 95% of the 'large' herbivores present? Is this a reasonable assumption given the variety of taxa which are uncovered in the area or are we, instead, witnessing the result of intensive hunting strategies? We must begin by relating non-faunal environmental data (eg: pollen) and archaeozoological material to each other.

10.6.2. Seasonality indications based upon a wide range of prey species should be obtained. It is tempting to base conclusions concerning the occupation of South West France upon the data already derived from reindeer material. However, this assumes reindeer to be the only prey species consistently taken, a fact which we know to be untrue. It is quite possible that various prey species were taken at different times of year, depending upon the availability and (nutritional) quality of all species. The impact of this upon settlement and subsistence studies would be immense. It is essential to consider all species which were exploited, including the saiga and horse in addition to other cervids and bovids, all of which were taken by man, especially in the western half of our region.

10.6.3. The study of butchery practices and carcase management undertaken here must continue, encompassing all available prey species, regardless of their relative importance. In order to do this a detailed examination of the 'economic anatomy' of non-cervids should be made. Speth (1983, 1987) has applied Binford's (1978) reindeer (caribou) - derived utility indices to the bison, a lead followed here. However, if such indices could be constructed for the horse, saiga, boar etc, taking account of seasonal, age and sexual variations in utility, some degree of the uncertainty associated with the methodology proposed might be overcome.

In order to complete this study we also need to obtain detailed quantitative lists of anatomical element frequencies of each species recovered from sites, accompanied by details of butchery marks and other surface modifications. It should be possible to compare the butchery and processing strategy adopted for both major and minor taxa. Only in this way will we obtain a 'full' range of post-kill subsistence data and avoid the notion of single-species (reindeer) hunting during the Upper Palaeolithic.

10.6.4. Traditionally, faunal analysis in much of Western Europe, and especially in France, has followed a site-by-site, or single species (once again usually reindeer) approach. This is no doubt the result of the abundance of sites in such areas as the Perigord. However, of greater value now is the production of works of synthesis, be they chronological or geographic in nature. Research will need to focus on specific regional studies, concentrating on areas such as the Vezere catchment, the Agenais and gradually more extensive and widespread regions. Attempts at regional synthesis of this kind have been made with regard to lithic data and severely limited faunal material (Le Tensorer 1979, Rigaud 1982), and should now be extended to, and integrated with, the local and regional patterns of faunal exploitation.

10.6.5. Finally, some formal attempt should now be made to integrate faunal and artefactual material in a coherent study of the French Palaeolithic. Traditional emphasis on stone-tool assemblages has resulted in the lagging behind of faunal studies - studies which are now gaining ground. Both are the result of a single subsistence system and the combined study of the two data-types should shed light upon either component as well as the larger, more complex whole.

Bibliography

L'A.	L'Anthropologie
B.A.F.E.Q.	Bulletin de l'Association Française pour l'Etude Quaternaire
B.A.R.	British Archaeological Report
Bull. Inst. Géol. Bassin d'Aquitaine	Bulletin de l'Institut Géologique du Bassin d'Aquitaine
Bull. Mus. Natn. Hist. Nat. Paris	Bulletin du Musée National d'Histoire Naturelle de Paris
Bull. Soc. Bot. Fr.	Bulletin de la Société Botanique Française
Bull. Soc, Linn. Bordeaux	Bulletin de la Société Linnéene de Bordeaux
B.S.P.A.	Bulletin de la Société Préhistorique de l'Ariège
B.S.P.F.	Bulletin de la Société Préhistorique Française
C.A.	Current Anthropology
Congr. Prêh. Fr.	Congrès Préhistorique de France
C. R. somm. séances Soc. Géol. Fr.	Compte Rendu sommaire des séances de la Société de la Géologie française.
Doc. Lab. Géol. Lyon	Document de la Laboratoire Géologique de la Faculté des Sciences, Lyon
G.P.	Gallia Préhistoire
Geol. Surv. Czech.	Sbornik of the Geological Survey of Czechoslowakia
Israel Jnl. Zool.	Israeli Journal of Zoology
Kroeber Anthr. Soc. Papers	Kroeber Anthropological Society Papers
Mem. Soc. Pré. Fr.	Mémoire de la Société Préhistorique Française

Nouv. Arch. Mus. Hist. nat. Lyon
Nouvelles Archives du Muséum d'Histoire naturelle de Lyon

Nutr. Abst. Rev.
Nutritional Abstracts Review

P.-V. Soc. Linn. Bordeaux
Procès Verbaux de la Société Linnéene de Bordeaux

Quat. Res. Quaternary Research

Rev. Géo. Phys. et Géol. Dyn.
Revue de la Géographie Physique et de la Géologie Dynamique

Soc. Arch. Champenoise, Reims
Société Archéologique Champenoise, Reims

Ahlen, I. (1965) Studies on the red deer, Cervus elaphus L. in Scandinavia. Viltrevy 3:1-376

Alados, C. L. (1985) Group size and composition of the Spanish Ibex (Capra pyrenaica Schinz) in the sierras of Cazorla and Segura. In S. Lovari (ed) The Biology and Management of Mountain Ungulates. Croom Helm, London p.134-147

Allee, W. C. & K. P. Schmidt (1951) Ecological Animal Geography. 2nd edition. John Wiley & Sons, Inc. New York

Anderson, A. C. (1877) Remarks on the reindeer. In E. Lartet & H. Christy (eds) Reliquiae Aquitanicae: Being Contributions to the Archaeology and Palaeontology of Perigord and Adjoining Provinces of S. France. H. Bailliere, London p. 142-6

Anderson, L., Dibble, M. V., Turkki, P. R., Mitchell, H. S., & H. J. Rynbergen (1982) Nutrition in Health and Disease. Lippincott Company, Philadelphia.

Arthur, G.W. (1975) An Introduction to the Ecology of Early Historic Communal Bison Hunting among the Northern Plains Indians. Mercury Series, Archaeological Survey of Canada Paper 37. National Museum of Man, Ottawa.

Austen, N. L. (1877) Notes on the Scandinavian reindeer. In E. Lartet & H. Christy (eds) Reliquiae Aquitanicae: Being Contributions to the Archaeology and Palaeontology of Perigord and Adjoining Provinces of S. France. H. Bailliere, London p.213-8

Bahn, P. G. (1977) Seasonal migration in south-west France during the late glacial period. Journal of Archaeological Science 4:245-57

(1979) La paléoéconomie magdalénienne du bassin de Tarascon (Ariège). B.S.P.A. 34:37-46

(1982) Inter-site and inter-regional links during the Upper Palaeolithic: the Pyrenean evidence. Oxford Journal of Archaeology 1: 247-68

Bailey, J.A. (1984) Principles of Wildlife Management John Wiley & Sons, New York

Baker, R.R. (1978) The Evolutionary Ecology of Animal Migrations. Hodder & Stoughton, London

Banfield, A.W.F. (1974) The Mammals of Canada. University of Toronto Press, Toronto

Banque Nationale de Paris (1979) Guide de l'Art et de la Nature - Dordogne. Ed. Berger-Levralt Paris

Bay-Petersen, J. (1975) Pre-Neolithic Faunal Exploitation in S. France and Denmark. Unpublished Ph.D. Thesis, University of Cambridge.

Beaufort, L. F. de (1951) Zoogeography of the Land and Inland Waters. Sidgwick & Jackson Ltd., London

Beaulieu, J. L., Pons, A. & M. Reille (1982) Recherches pollen-analytiques sur l'histoire de la végétation de la bordure nord du Massif de Cantal (Massif Central). Pollen et Spores 24:251-300

Bertouille, H., Bertouille, M. & J. Bouchud (1969) La prospection géologique et paléontologique des avens. Etudes préliminaires faites dans la grotte Bernard (Ariège). B.S.P.F. 65: 139-43

Binford, L. R. (1973) Inter-assemblage variability - the Mousterian and 'functional' argument. In A. C. Renfrew (ed) The Explanation of Culture Change: Models in Prehistory. Duckworth, London p.227-54

(ed)(1977) For Theory Building in Archaeology: Essays on Faunal Remains, Aquatic Resources, Spatial Analysis and Systemic Modeling. Studies in Archaeology. Academic Press, London

(1978) Nunamiut Ethnoarchaeology. Studies in Archaeology. Academic Press, London

(1980) Willow smoke and dogs' tails: hunter-gatherer settlement systems and archeological site formation. American Antiquity 45(1):4-20

(1981) Bones: Ancient Men and Modern Myths. Studies in Archaeology, Academic Press, London.

(1982) Comments on White. C.A. 23:177-182

(1983) In Pursuit of the Past: Decoding the Archaeological Record. Thames & Hudson, London.

Binford, L. R. & J. B. Bertram (1977) Bone frequencies and attritional processes. In L. R. Binford (ed) For Theory Building in Archaeology: Essays on Faunal Remains, Aquatic Resources, Spatial Analysis and Systemic Modeling. Studies in Archaeology, Academic Press, London p.77-153

Birkeland, P. W. (1984) Soils and Geomorphology. Oxford University Press, Oxford.

Birks, H. J. B. & H. H. Birks (1980) Quaternary Palaeoecology. Edward Arnold, London.

Bjärval, A. & S. Ulström (1986) The Mammals of Britain and Europe. Croom Helm, Beckenham, Kent.

Bokönyi, S. (1974) The Przewalski Horse. Souvenir Press, London.

Bonifay, M.-F. (1986) Intérêt des études taphonomiques au Pléistocène ancien: Soleilhac et Ceyssaguet (Blansac, Haute Loire). Bull. Mus. Natn. Hist. nat., Paris 6(C):269-281

Bordes, F. (1984) Leçons sur le Paléolithique, Tome II. Le Paléolithique en Europe. Cahiers du Quaternaire no. 7. C.N.R.S. Paris

Borowski, S., Krasinski, Z. & L. Mitkowski (1967) Food and role of the European bison in forest ecosystems. Acta Theriologica (Warsaw) 12:367-376

Bouchud, J. (1952a) Etude des rongeurs et des oiseaux de l'Abri Castanet. B.S.P.F. 49:267-71.

(1952b) Etude des rongeurs et des oiseaux de Lachaud. B.S.P.F. 49:262-7

(1953) Signification climatique des faunes paléolithiques. B.S.P.F. 50:431-5

(1954a) Le renne et le problème des migrations. L'A. 58:79-85

1954b) Dents de renne, bois de rennes et migrations. B.S.P.F. 51:340-45

(1956) Faune de l'Abri Pagès (a Rocamadour). L'A. 60:444-6

(1957) La faune de l'abri de la grotte à Melon. L'A. 61:433-4

(1959) Les paléolithiques, ont-ils domestiqué le renne? L'A. 63:93-100

(1961) Etude de la faune du gisement des Cottés (Haute Vienne) L'A. 65:258-70

(1965) Le Cervus mégacéros dans le Sud et le Sud-Ouest de la France. Israel Jnl. Zool. 14:24-37

(1966) Essai sur le Renne et la Climatologie du Paléolithique Moyen et Supérieur. Imprimerie Magne, Perigueux.

(1967) Etude d'un crane de renne de fossile (R. tarandus guettardi Desmarest) découvert dans le Sud de la France. Problèmes Actuels de Paléontologie 163 : 557-67. C.N.R.S. Coll. internat. Paris.

(1968a) L'abri de Facteur à Tursac (Dordogne). La faune et sa signification climatique. G.P. 11:113-120

(1968b) Les apports de la Paléontologie du Quaternaire à la connaissance de l'homme fossile. L'Homme de Cro-Magnon. Anthropologie et Archéologie. Art et Métiers graphiques, p.213-218

(1970) Détermination de la faune du Roc d'Abeilles. G.P. 13:23

(1975) Etude de la faune de l'Abri Pataud. In H. L. Movius (ed) Excavation of the Abri Pataud, Les Eyzies (Dordogne). American School of Prehistoric Research Bulletin 30:69-153

Bouchud, P. & J. Bouchud (1955) La faune de Blassac. B.S.P.F. 52:364-70

Bourque, B. J., Morris, K. & A. Spiess (1978) Determining the season of death of mammal teeth from archaeological sites: a new sectioning technique. Science 199:530-1.

Bouyssonie, J. & H. Delsol. (1930) L'Abri préhistorique de Jolivet, près Terrasson, (Dordogne). Revue Anthropologique 40:367-77.

Bradley, R. S. (1985) Quaternary Paleoclimatology: Methods of Paleoclimatic Reconstruction. Allen & Unwin. London

Brain, C. K. (1980) Some criteria for the recognition of bone collecting agencies in African caves. In A. K. Behrensmeyer & A. P. Hill (eds) Fossils in the Making. University of Chicago Press, Chicago p.107-30

(1981) The Hunters or the Hunted? University of Chicago Press, Chicago

Breuil, H. (1906) Les Cottés, une grotte du vieil age du renne à St.- Pierre de Maille (Vienne). Revue de l' Ecole

d'Anthropologie de Paris
16:47-62

(1912) Les subdivisions du Paléolithique supérieur et leur significantion. Comptes rendus de 14e Congrès International d'Anthropologie et d'Archéologie Préhistorique. Genève. p.165-238

Bricker, H. M. & P. A Mellars (1987) Datations ^{14}C de l'Abri Pataud (Les Eyzies, Dordogne) par le procédé "accélerateur - spectrometre de masse". L'A. 91:227-234

Brown, J. H. (1988) Species Diversity. In A. A. Myers & P. S. Giller (eds) Analytical Biogeography: An integrated approach to the study of animal and plant distributions. Chapman & Hall, London p.57-89

Buckland, W. (1823) Reliquiae Diluvianae: or, Observations on the Organic Remains Contained in Caves, Fissures, and Diluvial Gravel, and on other Geological Phenomena Attesting the Actions of a Universal Deluge. John Murray, London.

Burch, E.S. (1972) The caribou - wild reindeer as a human resource. American Antiquity 37:339-68

Buzas, M.A., Koch, C.F., Culver, S.J. & N.F. Sohl (1982) On the distribution of species occurrence. Paleobiology 8:142-50

Candolle, A. de (1855) Géographie botanique raisonnée ou exposition des faits principaux et des lois concernant la distribution géographique des plantes de l'époque actuelle. 2 vols. Masson, Paris

Capitan, L. & J. Bouyssonie (1924) Limeuil: son gisement à gravures sur pierre de l'Age de Renne. Nourry, Paris

Capitan, L., Breuil, H., Bourrinet, P. & D. Peyrony (1906) L'Abri Mège: une station magdalénienne à Teyjat (Dordogne). Revue de l'Ecole d'Anthropologie de Paris 16:196-212

(1908) La Grotte de la Mairie à Teyjat (Dordogne): fouilles d'un gisement magdalénien. Revue de l'Ecole d'Anthropologie de Paris. 18:153-73

Capitan, L. & D. Peyrony (1912) Trois nouveaux squelettes humains fossiles II. Le Cap Blanc. Revue Anthropologique Paris 11:439-42.

Castaing, P., Feral, A. & A. Klingebiel (1971) Paléogeographie de l'Holocène sur le plateau continental au large de l'embouchure de la Gironde. C. R. somm. seances Soc. Géol. Fr. 6:325-7

Cauvin, M.-C. (1971) Les Industries Post-Glaciaires du Périgord. Publications du Centre de Recherches d'Ecologie et de Préhistoire. C.N.R.S. Paris

Chadwick, D.H. (1979) Our wildest wilderness: the Arctic wildlife range. National Geographic 56:737-69

Chaline, J. (1972) Les Rongeurs du Pléistocène Moyen et Supérieur de France. Cahiers de Paléontologie, C.N.R.S. Paris.

Champagne, F. & R. Espitalié (eds) (1981) Le Piage: site préhistorique du Lot. Mem. Soc. Pré. Fr. 15. Paris

Chaplin, R.E. (1971) The Study of Animal Bones from Archaeological Sites. Seminar Press London

(1975) The ecology and behaviour of deer in relation to their impact on the environment of prehistoric Britain. In J.D. Evans, S. Limbrey & H. Cleere (eds) The Effect of Man on the Landscape: the Highland Zone C.B.A. Research Report 11, London p. 40-42.

Chase, P. G. (1985) On the use of Binford's utility indices in the analysis of archaeological sites. P.A.C.T. 11:287-302

(1986) The Hunters of Combe Grenal: Approaches to Miidle Paleolithic Subsistence in Europe. B.A.R. International Series 286.

(1987) Spécialisation de la chasse et transition vers le Paléolithique supérieur. L'A. 91:175-88.

Chauvet,G. (1872) Station de l'époque du renne au Menieux. Commune d'Edon, Charente. Afas, Bordeaux.

(1897) Stations humaines quaternaires de la Charente. Bull. Soc. Arch. Hist. Charente 1896, 221-336.

Chernov Y.I. (1985) The Living Tundra. Studies in Polar Research Cambridge University Press, Cambridge

Christenson, A. L. (1980) Change in the Human Niche in Response to Population Growth. In T. K. Earle & A. L. Christenson (eds) Modeling Change in Prehistoric Subsistence Economies. p.31-71. Studies in Archaeology, Academic Press, London

Christy, H. (1877) The prehistoric cave-dwellers of S. France - I. Stone implements. II. The Dordogne caves III. The reindeer Period. In E. Lartet .& H. Christy Reliquiae Aquitanicae: Being Contributions to the Archaeology and Palaeontology of Perigord and the Adjoining Provinces of S. France. H. Balliere, London p. 11-18

CLIMAP (1976) The surface of the ice-age earth. Science 191:1131-44

(1981) Seasonal Reconstructions of the Earth's Surface at the Last Glacial Maximum. Geological Society of America. New York

Clutton-Brock, J. (1981) Domesticated Animals from Early Times. Heinemann. B.M.N.H.. London

Clutton-Brock, T.H., Guinness, F.E. & S.D. Albon (1982) Red Deer: Behaviour and Ecology of Two Sexes Edinburgh University Press, Edinburgh

Collinson, A. S. (1977) Introduction to World Vegetation. Allen & Unwin London

Corbet, G. B. & D. Ovenden (1980) The Mammals of Britain and Europe. Collins, London

Corbet, G. B. & H. N. Southern (1977) The Handbook of British Mammals 2nd ed., Mammal Society, Blackwell Scientific Publications, Oxford

Crawley, M.J. (1983) Herbivory: The Dynamics of Animal - Plant Interactions. Studies in Ecology Vol. 10, Blackwell Scientific Publications, Oxford.

Daniel,R. (1969) Quelques mots sur la grotte de Malbarrat, Les Eyzies (Dordogne). B.S.P.F. 66:111-15

Daryll-Forde, C. (1934) Habitat, Economy and Society: a Geographical Introduction to Ethnology. Methuen & Co. Ltd. London

David, N. (1973) On Upper Palaeolithic society, ecology and technological change: the Noaillian case. In A. C. Renfrew (ed) The Explanation of Culture Change: Models in Prehistory Duckworth, London p.277-303

David, N & H. M. Bricker (1987) Perigordian and Noaillian in the Greater Perigord. In O. Soffer (ed) The Pleistocene Old World: Regional Perspectives. Interdisciplinary Contributions to Archaeology. Plenum Press, New York

Davis, S. J. M. (1987) The Archaeology of Animals. Batsford, London

Debard, E. (1979) Le gisement Pléistocène supérieur de la grotte de Jaurens à Nespouls, Corrèze, France: étude sédimentologique du remplissage. Nouv. Arch. Mus. Hist. nat. Lyon 17:17-24

Debard, E. &. F. Moser (1976) Evolution climatique de la fin du Wurm récent et du post-glaciaire d'après les remplissages des grottes et abris sous basaltiques du Vélay. B.A.F.E.Q. 12/13:87-93

Delibrias, G. & Evin, J. (1980) Sommaire des datations ^{14}C concernant la préhistoire en France. B.S.P.F. 77:216-24

Delibrias, G & Guillier, M.-T. & Evin, J. & J. Chevallier (1987) Sommaire des datations ^{14}C concernant la préhistoire en France. III: dates éffectuées de 1979 à fin 1984. B.S.P.F. 84:207-223

Delpech, F. (1967) Recherches Paléontologiques Concernant Quelques Gisements du Magdalénien VI: Station de la Gare de Couze (Dordogne), du Morin (Gironde) et de Duruthy (Landes). Thèse de doctorat en Géologie approfondie option paléontologie, Bordeaux I.

(1968) Faunes du Magdalénien et de l'Azilien du gisement de Duruthy, Commune de Sorde-l'Abbaye, Landes. Actes Soc. Linn. Bordeaux 105:2-17

(1970) L'abri magdalénien du Flageolet II - paléontologie (Bézenac, Dordogne). B.S.P.F. 67:494-9

(1972) Fouilles de sauvetage dans le gisement magdalénien de Fongaban, commune de St. Emilion (Gironde). La faune. L'A. 78:615-29

(1973) Signification paléoclimatique des associations d'herbivores reconnues dans un gisement archéologique. B.S.P.F. 70:187-9

(1975) Les Faunes du Paléolithique Supérieur dans le Sud-Ouest de la France. Thèse de Doctorat d'Etat ès Sciences Naturelles. Université de Bordeaux I.

(1983) Les Faunes du Paléolithique Supérieur dans le Sud-Ouest de la France. Cahiers du Quaternaire No. 6. C.N.R.S. Paris

(1984) Les ongules en Périgord et Nord-Ouest du Quercy durant le Wurm III. Chronoclimatologie, paléobiogeographie, palethnologie. Géobios 17 (5):531-48

Delpech, F. & O. LeGall (1983) La faune magdalénienne de la grotte des Eglises. B.S.P.A. 38:91-118

Delpech, F. & J.-P. Rigaud (1974) Etude de la fragmentation et de la répartition des restes osseux dans un niveau d'habitat paléolithique. In H. Camps-Faber (ed) L'Industrie de l'os dans la Préhistoire Editions de l'Université de Provence, Aix-en-Provence p. 47-55

Delpech, F.& C. Suire (1974) La faune mésolithique et post-mésolithique du gisement de Rouffignac. In C. Barriere (ed) 'Rouffignac - l'Archéologie'. Publ. Inst. Art Préhist. Univ. Toulouse-le-Mirail. Mem. II, fasc. 2:49-94.

Delpech, F., Donnard, E., Gilbert, A., Guadelli, J.-L., Le Gall, O., Martini-Jacquin, A., Paquereau, M.-M., Prat, F. & J.-F. Tournepiche (1983) Contribution à la lecture des paléoclimats quaternaires d'après les données de la paléontologie en milieu continental. Quelques exemples de flores et de faunes d'ongulés pris dans le Pléistocène supérieur d'Aquitaine. Bulletin de l'Institut de la Géologie du Bassin d'Aquitaine Bordeaux 34:165-77

Delporte, H. (1956) La grotte des Fées de Châtelperron (Allier). Congr. Préh. Fr. 15:452-77

(1976) L'Organisation du Périgordien supérieur en France et ses rapports avec le Périgordien d'Europe Occidentale. In B. Klima (ed) Périgordien et Gravettien en Europe. IXe Congres U.I.S.P.P., Nice p.7-50

(ed) (1984) Le Grand Abri de la Ferrassie: Fouilles 1968 - 1973. Etudes Quaternaires Géologie, Paléontologie, Préhistoire. Mémoire 7. Paris

Delporte, H. & Truffeau, A. (1973) Les industries du Périgordien supérieur de la Ferrassie. Quartar 24:93-123

Demars, P.-Y. (1982) L'Utilisation du Silex au Paléolithique Supérieur: Choix, Approvisionnement, Circulation. Cahiers du Quaternaire, No. 5. C.N.R.S. Paris

(1985) La signification de l'Aurignacien V dans l'évolution des cultures lithiques au Paléolithique supérieur en France. In M. Otte (ed) La Signification Culturelle des Industries Lithiques. Actes du Colloque de Liège. B.A.R. International Series 239. Oxford p. 328-336

Dennell, R. W. (1983) European Economic Prehistory: A New Approach. Academic Press, London.

Desbrosse, R. & C. Mourer-Chauviré (1973) Les oiseaux magdaléniens de Pierre-Chatel (Ain). Quatar 24:149-64

Donner, J. J. (1975) Pollen composition of the Abri Pataud sediments. In H. L. Movius Jr. (ed) Excavation of the Abri Pataud. American School of Prehistoric Research Bulletin 30 p.160-73

Doran, J. E. & F. R. Hodson (1975) Mathematics and Computers in Archaeology. Edinburgh University Press, Edinburgh

Ducommun, J.-F. (1982) Gastronomie de la Viande de Cheval. Fédération de la Boucherie Hippophagique de France. Paris

Dunn, G. & B. S. Everitt (1982) An Introduction to Mathematical Taxonomy. Cambridge Studies in Mathematical Biology 1. C.U.P., Cambridge

Duplessy, J. C., Delibrias, G., Turon, J. L. & J. C. Duprat (1981) Deglacial warming of the northeastern Atlantic Ocean: correlation with the paleoclimatic evolution of the European continent. Palaeogeography, Palaeoclimatology & Palaeoecology 35:121-144

Durand-Delga, M. (ed) (1980) La chaine des Pyrénées et son avant-pays aquitain-languedocien. France: Introduction à la géologie du Sud-Ouest. 26e Congrès Géologique International. Imprimerie Louis-Jean, Gap

Eicher, V., Siegenthaler, V. & S. Wegmuller (1981) Pollen and oxygen isotope analyses on late- and post-glacial sediments of the Tourbiere de Chirens (Dauphine, France). Quat. Res.15 (2):160-70

Elsner-Schack, I. von (1985) What is good chamois habitat? In S. Lovari (ed) The Biology and Management of Mountain Ungulates. Croom Helm, Beckenham Kent p. 71-6

Eltringham, S.,K. (1984) Wildlife Resources and Economic Development. John Wiley & Sons, Chichester

Fabre, F. (1964) Les paléoclimats en Basse-Provence du Magdalénien V ancien à l'Azilien ancien. Bulletin du Muséum d'Histoire Naturelle de Marseille 24:165-176

Fénélon, P. (1951) Le Périgord. Etude Morphologique. Lahure, Paris.

Flerov, K.K. (1952) Fauna of the U.S.S.R. Mammals Vol. 1. Musk deer and deer. Israel Program for Scientific Translation, Jerusalem 1960

Francisci, F., Focardi, S. & L. Boitani (1985) Male and female alpine ibex: phenology of space use and herd size. In S. Lovari (ed) The Biology and Management of Mountain Ungulates. Croom Helm, Beckenham p.124-33

Freeman, L. G. (1981) The fat of the land: notes on palaeolithic diet in Iberia. In R. S. O. Harding & G. Teleki (eds) Omnivorous Primates: Gathering and Hunting in Human Evolution. Columbia University Press, New York

Freethy, R. (1983) Man and Beast: the Natural and Unnatural History of British Mammals. Blandford Press, Poole

Frenshen, P. & F. Salomonsen (1955) The Arctic Year. J. Cape, London

Gamble, C.S. (1982) Interaction and alliance in palaeolithic society. Man 17:92-107

(1986) The Palaeolithic Settlement of Europe. Cambridge World Archaeology. Cambridge University Press, Cambridge

Gauch, H. G. (1982) Multivariate Analysis in Community Ecology. Cambridge Studies in Ecology. Cambridge University Press Cambridge

Geist, V. (1978) Life Strategies, Human Evolution, Environmental Design: Toward a Biological Theory of Health. Springer-Verlag New York

Giller, P. S. (1984) Community Structure and the Niche. Chapman & Hall, London

Girard, C. & F. David (1982) A propos de la chasse specialisée au paléolithique moyen: l'exemple de Mauran (Haute-Garonne). B.S.P.F. 79:11-12

Gordon, B. C. (1980) Reviews: Reindeer and Caribou Hunters: An Archaeological Study by A. E. Spiess, Academic Press, New York 1979. Arctic 33 (2):370-4

(1982a) Of men and reindeer herds in French Magdalenian prehistory. Paper presented at the Fourth I.C.A.Z. Conference, London.

(1982b) Summary of National Museum of Man Involvement in French Prehistoric Research. National Museum of Man, Ottawa.

(1982c) Tooth sectioning as an archaeological tool. Canadian Studies Report No. 14e. Archaeological Survey of Canada. National Museum of Man, Ottawa.

(1986a) Of men and reindeer in French magdalenian prehistory: conclusions. Paper presented at Fifth I.C.A.Z. Conference, Bordeaux

(1986b) Of men and reindeer in French magdalenian prehistory. Paper presented at World Archaeology Conference Southampton

(1988) Of Men and Reindeer Herds in French Magdalenian Prehistory. B.A.R. International Series 390, Oxford

Goss, F. (1931) Memoirs of a Stag Harbourer: a Record of Twenty-eight Years with the Devon and Somerset Stag Hounds, 1894-1921 H. F. & J. Witherby, London

Grayson, D. K. (1981) A critical view of the use of archaeological vertebrates in paleoenvironmental reconstruction. Journal of Ethnobiology 1:28-38

(1984) Quantitative Zooarchaeology: Topics in the Analysis of Archaeological Faunas. Studies in Archaeological Science. Academic Press, London

Groves, C. P. (1974) Horses, Asses and Zebras in the Wild. David and Charles, Newton Abbot

Guérin, C. (1980) Les Rhinocéros (Mammalia, Périssodactyla) du Miocène Terminal au Pléistocène Supérieur en Europe Occidentale. Comparison avec les Espèces Actuelles. Doc. Lab. Geol. Lyon, No.79.

Guérin, C. & M. Faure (1983) Les hommes du Paléolithique Européen ont-ils chassé le rhinocéros ? In F. Poplin (ed) 'La Faune et l'Homme Préhistoriques'. Mem. Soc. Préh. Fr. 16:29-36

Guérin, C., Philippe, M. & R. Vilain (1979) Le gisement Pléistocène supérieur de la grotte de Jaurens à Nespouls, Corrèze, France: historique et généralites. Nouv. Arch. Mus. Hist. nat. Lyon 17:11-16

Guillien, Y. & G. Henri-Martin (1968) Dentures de rennes et saison de chasse: l'abri aurignacien de La Quina. L'A2:337-48

Guillien, Y. & S. de Saint-Mathurin (1976) Le gisement du Roc aux Sorciers et la séquence climatique du Magdalénien. B.S.P.F. 73:15-21

Guthrie, R. D. (1982) Mammals of the mammoth steppe as paleoenvironmental indicators. In D. M. Hopkins, J. V. Matthews, Jr., C. S. Schweger & S. B. Young (eds) Paleoecology of Beringia. Academic Press, New York p.307-328

Hamr, J. (1985) Seasonal home range size and utilisation by female chamois (Rupicapra rupicapra L.) in Northern Tyrol. In S. Lovari (ed) The Biology and Management of Mountain Ungulates. Croom Helm, Beckenham p.106-116

Hannon, G. E., Turner, C. & W. A. Watts. (1987) Vegetational evidence for quaternary climatic changes in S. W. Europe. Paper given at The Royal Society Discussion Meeting. The Past Three Million Years: Evolution of Climatic Variability in the North Atlantic Region, 25th - 26th February 1987

Harris, R.A. & K.R. Duff (1970) Wild Deer in Britain. David & Charles, Newton Abbot

Harrold, F. B. (1988) The Chatelperronian and the Early Aurignacian in France. In J. F. Hoffecker & C. A. Wolf (eds) The Early Upper Palaeolithic: Evidence from Europe and the Near East. B.A.R. International Series 437, Oxford p.157-191

Hayden, B., Chisholm, B. & H. B. Schwarcz (1987) Fishing and foraging: marine resources in the Upper Palaeolithic of France. In O. Soffer (ed) The Pleistocene Old World: Regional Perspectives. Interdisciplinary Contributions to Archaeology. Plenum Press, New York. p.279-291

Hemingway, M. F. (1980) The Initial Magdalenian in France. B.A.R. International Series 90. Oxford

Hill, A. (1979) Disarticulation and scattering of mammal skeletons. Paleobiology 5:261-274

Hodder, I. & C. Orton (1976) Spatial Analysis in Archaeology. New Studies in Archaeology. C.U.P. Cambridge

Hokr, Z. (1951) A method of the quantitative determination of the climate in the Quaternary period by means of mammal associations. Geol. Surv. Czech. 18:209-218

Holmes, D. L. (1969) Elements of Physical Geology: the Study of Landforms. Thomas Nelson & Sons. London.

Hue, E. (1907) Musée Ostéologique. Mus. Ost. Et. Faune Quat. 2. Schleicher Frères Ed. Paris.

Humboldt, F. H. A. von & Bonpland, A. J. A. (1805) Essai sur la Géographie des Plantes: accompagné d'un Tableau Physique des Régions Equinoxiales. Levrault, Schoell & Compagnie, Paris

Huntley, B. & H. J. B. Birks (1983) An Atlas of Past and Present Pollen Maps for Europe: 0 - 13 000 years ago. C.U.P. Cambridge

Illies, J. (1974) Introduction to Zoogeography. Macmillan London.

Ingold, T. (1980) Hunters, Pastoralists and Ranchers: Reindeer Economies and their Transformations. Cambridge University Press, Cambridge.

Jacobi, A. (1931) Das Rentier. Zool. Anz. Erg. Bd. 96

Janis, C.M. (1984) The use of fossil ungulate communities as indicators of climate and environment. In P. J. Brenchley (ed) Fossils and Climate. John Wiley & Sons Ltd. Chichester. p.85-104

Jochim, M. A. (1981) Strategies for Survival: Cultural Behavior in an Ecological Context. Academic Press, London

(1983) Palaeolithic cave art in ecological perspective. In G. N. Bailey (ed) Hunter-Gatherer Economy in Prehistory: a European Perspective. New Directions in Archaeology. C.U.P. Cambridge p.212-8

Johnston, R. J. (1978) Multivariate Statistical Analysis in Geography. A Primer on the General Linear Model. Longman London

Kay, N. R. B. (1985) Seasonal variation of appetite in ruminants. In W. Haresign (ed) Recent Advances in Animal Nutrition. Butterworth, London. p.205-208

Kay, N. R. B. & B.W. Staines (1981) The nutrition of red deer. Nutr. Abst. Rev. B, 51:601-22

Kay, N. R. B., Milne, J.A. & W.J. Hamilton (1984) Nutrition of red deer for meat production. Proceedings of the Royal Society of Scotland Edinburgh, B, 828:231-42

Kelsall, J.P. (1968) The Migratory Barren-ground Caribou of Canada. Department of Indian Affairs and Northern Development. Canadian Wildlife Service, Queen's Printer, Ottawa

Kistschinski, A. A. (1976) Cold latitudes and the high mountains: cold zones. In B. Grzimek, J. Illies & W. Klaysewitz (eds) Grzimek's Encyclopedia of Ecology. Van Nostrand Reinhold Co. New York, p. 176-186

Klein, D.R. (1968) The introduction, increase and crash of reindeer on St. Matthew Island. Journal of Wildlife Management 32: 350-67

Klein, R. G. & K. Cruz-Uribe (1984) The Analysis of Animal Bones from Archaeological Sites. Prehistoric Archaeology and Ecology Series. Chicago University Press, Chicago.

Koch, C.F. (1987) Prediction of sample size effects on the measured temporal and geographic distribution patterns of species. Paleobiology 13:100-7

Krebs, C. J. (1985) Ecology: The Experimental Analysis of Distribution and Abundance. 3rd ed Harper & Row, New York

(1989) Ecological Methodology. Harper & Row New York

Kurten, B. (1968) Pleistocene Mammals of Europe. Weidenfeld & Nicolson, London

Kutzbach, J. F. & H. E. Wright (1985) Simulation of the climate of 18000 years B.P.: results for the North American/North Atlantic / European sector and comparison with the geologic record of North America. Quaternary Science Reviews 4 (3):147-88

Laborde, E. D. (1961) Western Europe. London University Press. London

Lacorre, F. (1939) La grotte des Fées, à Marcamps (Gironde) ou Roc de Marcamps. P.-V. Soc. Linn. Bordeaux 90:35-51

(1953) Sur les bois et les dents de renne de Badegoule. B.S.P.F. 50:547-55

(1956) Les migrations des rennes dans la province préhistorique des Eyzies. B.S.P.F. 53:302-10

Lamb, H. H. (1977) Climate: Present, Past and Future. 2 vols. Methuen & Co. Ltd. London

Larsen, J.A. (1980) The Boreal Ecosystem. Academic Press, London

Lartet, E. & H. Christy (1877) Reliquiae Aquitanicae: Being contributions to the archaeology and palaeontology of Perigord and the adjoining provinces of Southern France. H. Bailliere, London

Laville, H. (1973) Climatologie et Chronologie du Paléolithique en Périgord: Etude Sédimentologique de Dépôts en Grottes et sous Abris. Thèse de doctorat ès Sciences. Université de Bordeaux I.

(1975) Climatologie et Chronologie du Paléolithique en Périgord: Etude sédimentologique de dépôts en grottes et sous abris. Etudes Quaternaires Memoire No. 4 Université de Provence, Marseille

(1976) Deposits in calcareous rockshelters: analytical methods and climatic interpretation. In D. A. Davidson & M. L. Shackley (eds). Geoarchaeology: Earth Science and the Past. Duckworth, London.

(1988) Recent developments on the chronostratigraphy of the Paleolithic in the Perigord. In H. L. Dibble & A. Montet-White (eds) Upper Pleistocene Prehistory of Western Eurasia. University Museum Symposium Series volume 1. University Museum Monograph 54. University of Pennsylvania, Philadelphia p.147-157

Laville, H & Rigaud, J.-P. (1973) The Perigordian V industries in Perigord. Typological variations, stratigraphy and relative chronology. W.A. 4:330-38

Laville, H., Rigaud, J. & J. Sackett (1980) Rockshelters of the Perigord: Geological Stratigraphy and Archaeogical Succession. Studies in Archaeology. Academic Press, London

Laville, H., Turon, J.-L., Texier, J.-P., Raynal, J.-P., Delpech, F., Paquereau, M.-M., Prat, F. & A. Débénath (1983) Histoire paléoclimatique de L'Aquitaine et du Golfe de Gascoigne au Pléistocène supérieur depuis le dernier interglaciaire. Bull. Inst. Géol. Bassin d'Aquitaine Bordeaux 34:219-41

Lawrence, R.D. (1974) Wildlife in North America: Mammals. Michael Joseph, London

Lee, R. B. (1968) What hunters do for a living, or, How to make out on scarce resources. In R. B. Lee & I deVore (eds) Man the Hunter. p.30-48. Aldine Publishing Company. New York.

Le Gall, O. (1982) Les Poissons de Quelques Gisements Préhistoriques du Sud-Ouest de la France. Etude Ostéologique. Intérêt Paléoécologique et Palethnologique. Thèse de 3e cycle en Géologie du Quaternaire et Préhistoire, Université de Bordeaux I

(1984) L'Ichtyofaune d'Eau Douce dans les Sites Préhistoriques. Ostéologie - Paléoécologie - Palethnologie. Cahiers du Quaternaire No. 8. C.N.R.S. Paris

Legge, A. J. & P. A. Rowley-Conwy (1988) Star Carr Revisited: a re-analysis of the large mammals. The Archaeological Laboratory, Centre for Extra-Mural Studies, Birkbeck College, University of London. London

Lemee, G. (1956) Sur l'évolution de la végétation dans le massif du Cantal au tardiglaciaire et au post-glaciaire. Bull. Soc. Bot. Fr. 103:83-94

Lenoir, M. (1983) Le Paléolithique des Basses Vallées de la Dordogne et de la Garonne. Thèse de Doctorat d'Etat ès Sciences Université de Bordeaux I.

Leonard, R. D. & G. T. Jones (eds) (1989) Quantifying Diversity in Archaeology. New Directions in Archaeology. C.U.P. Cambridge

LeResche, R. & S. Linderman (1975) Caribou trail systems in Northeastern Alaska. Arctic 28:54-61

Leroi-Gourhan, A. (1953) L'Interpretation des vestiges osseux. Cong. Préh. Fr. 14:377-94

Leroi-Gourhan, Arl. (1973) Le paysage au temps des gravures de la grotte de La Marche (avec une analyse zoologique par J. Chaline) In J. Maluquer de Motes (ed) Estudios Dedicados a Luis Pericot. Instituto de Arqueologia y Prehistoria, Universidad de Barcelona.

(1984) Paysages et climats il y a 35 000 ans. Histoire et Archéologie (Les Dossiers) 87:20-23

Leroi-Gourhan, Arl. & J. Renault-Miskovsky (1977) La palynologie appliquée à l'archéologie. Approche écologique de l'Homme fossile. B.A.F.E.Q. Supp. 47:35-49

Leroyer, C. & Arl. Leroi-Gourhan (1983) Problèmes de chronologie: le castelperronien et l'aurignacien. B.S.P.F. 80:41-4

Léson, A. (1978) Je découvre les Animaux Sauvages, II, Grands Mammifères Européens Zoo de Haye - G.E.C.Na.L. Lesot, Paris.

Le Tensorer, J.-M. (1979) Recherches sur le Quaternaire en Lot-et-Garonne: Stratigraphie, Paléoclimatologie et

Préhistoire Paléolithique. Thèse de Doctorat d'Etat ès Sciences Université de Bordeaux.

(1981) Le Paléolithique de l'Agenais. Cahiers du Quaternaire No.3 C.N.R.S. Paris.

Leveque, F. & J.-Cl. Miskovsky (1983) Le Castelperronien dans son environnement géologique. L'A. 87:369-391

Levine, M. A. (1977) Interpretation of macro-fauna from archaeological deposits. The ecosystem approach. B.A.F.E.Q. Sup. 47:57-62

(1979) Archaeozoological Analysis of Some Upper Pleistocene Horse Bone Assemblages in W. Europe. Unpublished Ph.D. thesis, University of Cambridge

(1983) Mortality models and the interpretation of horse population structure. In G. N. Bailey (ed) 'Hunter-Gatherer Economy in Prehistory' p.23-46. New Directions in Archaeology. C.U.P. Cambridge

Lieberman, L. S. (1987) Biocultural consequences of animals versus plants as sources of fats, proteins and other nutrients. In M. Harris & E. B. Ross (eds) Food and Evolution: Toward a Theory of Human Food Habits. Temple University Press, Philadelphia. p.225-258

Lowe, J. J. & M. J. C. Walker (1984) Reconstructing Quaternary Environments. Longman, London

Lumley, H. de (ed)(1976) La Préhistoire Française. C.N.R.S. Paris

Lyman, R. L. (1985) Bone frequencies: differential transport, in situ destruction, and the M.G.U.I. J.A.S. 12:221-236

MacArthur, R.H. (1972) Geographical Ecology: Patterns in the Distribution of Species. Princeton University Press Princeton, New Jersey.

Malvésin-Fabre, G. (1948)Essai sur la Faune Pléistocène de la Gironde: Paléobiologie et Paléoclimatologie. Imprimerie E. Drouillard, Bordeaux

M^c^Cowan, T. D. (1961) Animals, climate and palaeolithic man. Kroeber Anthr. Soc. Papers 25:221-30

McDonald, J. N. (1981) North American Bison: Their classification and evolution. University of California Press, Berkeley

M^c^Intyre, A., Kipp, N. G., Bé, A. W. H., Crawley, T., Kellogg, T., Gardner, J. V., Prell, W. & W. F. Ruddiman (1976) Glacial North Atlantic 18 000 years ago: A CLIMAP reconstruction. In R. M. Cline & J. D. Hays (eds) Investigation of Late Quaternary

Paleooceanography and Paleoclimatology. Geological Society of America, Mem. 145, p.43-76

Magurran, A. E. (1988) Ecological Diversity and Its Measurement. Croom Helm, London

Masini, F. (1985) Wurmian and Holocene chamois of Italy. In S. Lovari (ed) The Biology and Management of Mountain Ungulates. Croom Helm, Beckenham p.31-44

Meddis, R. (1984) Statistics Using Ranks: A Unified Approach. Blackwell, Oxford.

Mellars, P. A. (1985) The ecological basis of social complexity in the Upper Paleolithic in Southwestern France. In T. D. Price & J. A. Brown (eds) Prehistoric Hunter-Gatherers: The Emergence of Cultural Complexity. p. 271-98. Studies in Archaeology. Academic Press, London

Mellars, P. A., Bricker, H. M., Gowlett, J. A. J. & R. E. M. Hedges (1987) Radiocarbon Accelerator Dating of French Upper Palaeolithic Sites. C.A. 28:128-133

Memoire, N. (1986) Quels Etaient les Animaux de la Préhistoire à Pair-non-Pair en Gironde? Muséum d'Histoire Naturelle de Bordeaux. Presses de l'Atelier Municipal de la Ville de Bordeaux, Bordeaux

Metcalfe, D. & K. T. Jones (1988) A reconsideration of animal body-part utility indices. Am. Ant. 53:486-504

Miller, F. L., Broughton, E. & A. Gunn (1988) Mortality of Migratory Barren-ground Caribou on the Calving Grounds of the Beverley Herd, Northwest Territories. Canadian Wildlife Service Occasional Paper 66. Canadian Wildlife Service, Ottawa

Mohr, E. (1971) The Asiatic Wild Horse. J.J. Allen & Co. London

Moore, P. D. & J. A. Webb (1978) An Illustrated Guide to Pollen Analysis. Hodder & Stoughton, London

Morala, A. (1985) Périgordien et Aurignacien en Haut-Agenais: Etude d'Ensembles Lithiques. Archives d'Ecologie Préhist. 7. Toulouse.

Mortillet, G. de & A. de Mortillet (1881) Musée Préhistorique. Paris

(1900) Le Préhistorique: Origine et antiquité de l'homme. 3rd ed. Librairie C. Reinwald, Schleicher Freres, Editeurs. Paris

Moss, R., Watson, A. & J. Ollason (1982) Animal Population Dynamics Outline Studies in Ecology. Chapman & Hall, London.

Mourer-Chauviré, C. 1975) Les Oiseaux du Pléistocène moyen et supérieur de France. Doc. Lab. Géol. Lyon. 64

Mourer-Chauviré, C. (1980) Le gisement Pléistocène supérieur de la grotte de Jaurens, à Nespouls, Corrèze, France: les équidés (Mammalia, Périssodactyla). Nouv. Arch. Mus. Hist. nat. Lyon 18:17-60

Mourer-Chauviré, C. & G. Carbonnet (1976) La recherche des facteurs dans l'analyse de groupe. Un exemple: les oiseaux du Quaternaire français. B.A.F.E.Q. 12/13: 59-69.

Movius, H. L. (1975) Excavation of the Abri Pataud, Les Eyzies (Dordogne). American School of Prehistoric Research, Bulletin 30.

Myers, A. A. & P. S. Giller (eds) Analytical Biogeography: an integrated approach to the study of animal and plant distributions. Chapman & Hall, London

Nelson, R.K. (1973) Hunters of the Northern Forest. Chicago University Press, Chicago

Newhouse, J. (1952) Reindeer are Wild Too. The Travel Book Club London

Obermaier, H. (1939) Streiflichter in das Leben der Spateizeitlichen Rentierjager der Urschweiz. Jahrb. d. Schwiez. f. Rogesc., p. 123-32

Olsen, S. R. (1987) Magdalenian reindeer exploitation at the Grotte des Eyzies, S.W.France. Archaeozoologia 1:171-182

Orquera, L. A. (1984) Specialization and the Middle/Upper Paleolithic Transition. C.A. 25:73-98

Paquereau, M.-M. (1970a) Flores et climats paléolithiques dans le Sud-Ouest de la France. Rev. Géo. Phys. et Geol. Dyn. 12:109-16

(1970b) Etude palynologique du gisement du Flageolet II B.S.P.F. 67:487-93

(1978) Flores et climats du Wurm III dans le Sud-Ouest de la France. Quaternaria 20:124-64

(1980) Chronologie palynologique du Pléistocène dans le Sud-Ouest de la France. In J. Chaline (ed) Problèmes de Stratigraphie Quaternaire en France et dans les Pays Limitrophes. B.A.F.E.Q. suppl. 7:298-306

Patte, E. (1941) Le Paléolithique dans le Centre-Ouest de la France. Libraires de l'Académie de Médecine. Masson & Cie. Eds. Paris

Perle, A. & J. Hamr (1985) Food habits of chamois (Rupicapra rupicapra L.) in Northern Tyrol. In S, Lovari (ed) The Biology and Management of Mountain Ungulates Croom Helm, Beckenham Kent p. 77 - 84

Petersen, R.L. (1966) The Mammals of Eastern Canada. Oxford University Press, Toronto

Peyrony, D. (1932a) Les abris Lartet et du Poisson à Gorge d'Enfer (Dordogne). L'A. 36:341-68

(1932b) Les gisements préhistoriques de Bourdeilles (Dordogne). Archives de l'Institut de la Paléontologie Humaine 10. Paris

(1934a) La Ferrassie. Moustérien, Périgordien, Aurignacian. Préhistoire 3. Libr. E. Leroux. Paris

(1934b) Station préhistorique de Longueroche (Magdalénian et Azilien). Revue Anthropologique 44:226-47

(1935) Le gisement Castanet, Vallon de Castelmerle, Cne. de Sergeac. B.S.P.F. 32:418-43

(1936) L'abri de Villepin (Commune de Tursac, Dordogne). Magdalénien superieur et Azilien. B.S.P.F. 33:253-71

Pianka, E. R. (1978) Evolutionary Ecology. Harper & Row. New York.

Pielou, E. C. (1977) Mathematical Ecology. Wiley, New York

Polunin, O. & M. Walters (1985) A Guide to the Vegetation of Britain and Europe. O.U.P. Oxford

Poplin, F. (1977) Problèmes d'ostéologie quantitative rélatifs à l'étude de l'écologie des hommes fossiles. B.A.F.E.Q. Sup. 47:63-8

(1983) Incisives de renne sciées de Magdalénien d'Europe Occidentale. In F. Poplin (ed) La Faune et l'Homme Préhistoriques. Mem. Soc. Pré. Fr. 16:55-67

Prat, F. (1962) La faune du gisement de la Gare de Couze. L'A. 66:248-254

(1968) Sur les Vestiges de Mégacéros découverts dans Quelques Gisements Paléolithiques du Sud-Ouest de la France. Thèse de Doctorat d'Etat ès Sciences Naturelles. Université de Bordeaux I.

Prat, F. & D. de Sonneville-Bordes (1969) Découvertes récentes de Paléolithique supérieur de la grotte de Font de Gaume (Dordogne). Quaternaria 11:115-40

Pujol, C. & J.-L. Turon (1974) Paléoclimtologie et stratigraphie du Quaternaire terminal du Golfe de Gascogne déduites de l'analyse foraminifères planctoniques et des ensembles sporopolleniques des sédiments marins. Boreas 3:99-104

Putman, R. J. (1984) The geography of animal communities. In J.A. Taylor (ed) Themes in Biogeography Croom Helm Beckenham. Kent. p.163-90

(1986) Grazing in Temperate Ecosystems: Large Herbivores and the Ecology of the New Forest. Croom Helm, Beckenham, Kent

Putman, R. J. & S. D. Wratten (1984) Principles of Ecology. Crrom Helm London

Reiger, J.F. (1980) The history of hunting. In R. Elman (ed) Hunting A.B. Nordbok, Gothenburg p.121-44

Remmert, H. (1980) Arctic Animal Ecology. Springer-Verlag. New York

Renault-Miskovsky, J. (1986) L'Environnement au Temps de la Préhistoire: Méthodes et Modèles. Masson, Paris.

Reynolds, H. W., Glaholt, R. D. & A. W. L. Hawley (1982) Bison. In J. A. Chapman & J. A. Feldhamer (eds) Wild Mammals of North America: Biology, Management and Economics. John Hopkins University Press, Baltimore p.972-1007

Rigaud, J.-P. (1982) Le Paléolithique en Périgord: les données du Sud-Ouest sarladais et leurs implications. Thèse de Doctorat d'Etat, Université de Bordeaux I.

(1988) The Gravettian Peopling of Southwestern France: Taxonomic problems. In H. L. Dibble & A. Montet-White (eds) Upper Pleistocene Prehistory of Western Eurasia. University Museum Symposium Series volume 1. University Museum Monograph 54. University of Pennsylvania, Philadelphia

Riches, D. (1982) Northern Nomadic Hunter-Gatherers: A Humanistic Approach. Studies in Anthropology, Academic Press London.

Roe, F.G. (1970) The North American Buffalo: A Critical Study of the Species in its Wild State. David & Charles, Newton Abbot.

Rozoy, J. G. (ed)(1978) Les Derniers Chasseurs. L'Epipaléolithique en France et en Belgique. Essai de Synthese. Soc. Arch. Champenoise, Reims.

Ruddiman, W. F. & A. McIntyre (1981) The North Atlantic Ocean during the last deglaciation. Palaeogeography, Palaeoclimatology & Palaeoecology. 35:145-214

Ryder, M.L. (1977) Seasonal coat change in grazing red deer (Cervus elaphus). Journal of Zoology London 181:137-41

Saffirio, L. (1983) Problema alimentare durante l'ultima glaciazione, in Europa. Quartar 33/34:141-161

Saint-Perier, R. de (1920) Les migrations des tribus magdaléniennes des Pyrénées. Revue Anthropologique 30:136-41

Seronie-Vivien, R. (1971) Note Préliminaire sur la faune des niveaux aziliens de la grotte de Pégourié (Caniac, Lot). Bull. Soc. Linn. Bordeaux. N.S. 1(2)

Shaw, G. & D. Wheeler (1985) Statistical Techniques in Geographic Analysis. Wiley & Sons, Chichester

Shennan, S. (1988) Quantifying Archaeology. E.U.P. Edinburgh

Shipman, P. (1981) Life History of a Fossil: An Introduction to Taphonomy and Paleoecology. Harvard University Press, Cambridge, Massachusetts.

Silk, J. (1979) Statistical Concepts in Geography. Allen & Unwin, London

Skogland, T. (1983) The effects of density dependent resource limitation on size of wild reindeer. Oecologica (Berlin) 60:156-68

(1985) The effects of density dependent resource limitations on the demography of wild reindeer. Journal of Animal Ecology 54: 359-74

Slade, L. M. & E. B. Godfrey (1982) Wild Horses. In J. A. Chapman & G. A. Feldhamer (eds) Wild Mammals of North America: Biology, Management and Economics. John Hopkins University Press, Baltimore. p.1089-1098

Smith, B. D. (1975) Middle Mississippi Exploitation of Animal Populations. Museum of Anthropology , University of Michigan, Anthropological Paper 57.

Smith, P. E. L. (1966) Le Solutréen en France. Publication de l'Institut de Préhistoire Université de Bordeaux. No. 5.

(1969) The Solutrean culture. Scientific American 211:85-94

Soffer, O. (1985) The Upper Paleolithic of the Central Russian Plain. Studies in Archaeology. Academic Press London

Sonneville-Bordes, D. de (ed)(1979) La Fin des Temps Glaciaires. C.N.R.S. Paris

Sonneville-Bordes, D. de & P. Laurent (1983) Le phoque à la fin des temps glaciaires. In F. Poplin (ed) 'La Faune et l'Homme Préhistoriques'. Mem. Soc. Pré. Fr. 16:69-80

Southwood, T.R.E. (1978) Ecological Methods: with Particular Reference to the Study of Insect Populations. Chapman & Hall, London.

Sparks, B. W. (1972) Geomorphology. Longman, London

Speth, J. D. (1983) Bison Kills and Bone Counts: Decision Making by Ancient Hunters. Prehistoric Archaeology and Ecology Series, Chicago University Press, Chicago.

Speth, J. D. & K. Spielmann (1983) Energy source, protein metabolism and hunter-gatherer subsistence strategies. Journal of Anthropological Archaeology 2:1-31

Spiess, A.E. (1979) Reindeer and Caribou Hunters: An Archaeological Study. Studies in Archaeology. Academic Press, London

Stern, R. O., Arobio, E. L., Naylor, L. L., & W. C. Thomas (1980) Eskimos, Reindeer and Land. School of Agriculture & Land Resource Management, University of Alaska, Fairbanks

Straus, L. G. (1981) On the habitat and diet of Cervus elaphus. Munibe 33:175-182

(1983) Terminal Pleistocene faunal exploitation in Cantabria and Gascony. In J. Clutton-Brock & C. Grigson (eds) 'Animals and Archaeology:1. Hunters and their Prey'. p.209-25. BAR Int. Series 163

(1986) Cantabrian Spain circa 18 000 B.P.: the 'Solutrean'. Paper presented at the World Archaeology Conference, Southampton.

(1987) Upper Paleolithic ibex hunting in Southwest Europe. J.A.S. 14:163-178

Straus, L. G. & C. W. Heller (1988) Explorations of the Twilight Zone: the Early Upper Paleolithic of Vasco-Cantabrian Spain and Gascony. In J. F. Hoffecker & C. A. Wolf (eds) The Early Upper Paleolithic: evidence from Europe and the Near East. B.A.R. International Series 437 Oxford p.97-133

Stuart, A.J. (1982) Pleistocene Vertebrates in the British Isles. Longman, London.

Sturdy, D.A. (1975) Some reindeer economies in prehistoric Europe. In E.S. Higgs (ed) Palaeoeconomy, Cambridge University Press, Cambridge. p.55-95

Styles, B.W. (1981) Faunal Exploitation and Resource Selection: Early Woodland Subsistence in the Lower Illinois Valley.

Northwestern University Archeological Program. Scientific Paper No. 3. Evanston.

Sutcliffe, A.J. (1985) On the Track of Ice Age Mammals. British Museum (Natural History), London

Taylor-Page, F.J. (1982) Field Guide to British Deer. British Deer Society. Blackwell Scientific Publications, Blackwell, Oxford 3rd. Ed.

Thévenin, R. (1943) La Faune disparue de France depuis les Origines jusqu'à nos jours. Payot, Paris

Thibault, Cl. (1979) L'évolution géologique de l'Aquitaine méridionale à la fin des temps glaciaires. In D. de Sonneville-Bordes (ed). La Fin des Temps Glaciaires en Europe. C.N.R.S. Paris.

Thomas, D. H. (ed)(1983) The Archaeology of Monitor Valley, 2. Gatecliff Shelter. Anthropological Papers of the American Museum of Natural History, New York. 59 (1).

Thomas, D. H. & D, Mayer (1983) Behavioural faunal analysis of selected horizons. In D.H. Thomas (ed) The Archaeology of Monitor Valley, 2. Gatecliff Shelter. Anthropological Papers of the American Museum of Natural History, New York. 59 (1):353-91

Time Life (eds) (1985) Grasslands and Tundra. Time-Life Books Inc. Amsterdam

Turner, C. & G. E. Hannon (1988) Vegetational evidence for late Quaternary climatic changes in southwest Europe in relation to the influence of the North Atlantic Ocean. Philosophical Transactions of the Royal Society of London B 318:451-485

Veyret, Y. (1986) Quaternary glaciations in the French Massif Central. Quaternary Science Reviews 5:395-6

Vézian,J. & J. Vézian (1970) Les gisements de la grotte de St. Jean-de-Verges (Ariège). B.S.P.A. 25:29-77

Vigneaux, M. (1975) Aquitaine Occidentale. Masson, Paris.

Vire, A (1905) Grotte préhistorique de Lacave (Lot). L'A. 16:4111-29

(1908) La Crozo de Gentillo ou Grotte de 'Combe-Cullier', Commune de Lacave (Lot). L'A 19:409-24

Vita-Finzi, C. (1978) Archaeological Sites in their Setting. Thames & Hudson, London

Walker, E.P. (1968) Mammals of the World. John Hopkins Press Baltimore

(1975) Mammals of the World. 3rd ed. John Hopkins Press, Baltimore

Wallace, A. R. (1860) On the zoological geography of the Malay Archipelago. Journal of the Linnean Society of London 14: 172-184

(1876) The Geographical Distribution of Animals. 2 vols. Macmillan, London

Washburn, A. L. (1979) Geochryology: a survey of periglacial processes and environments. Edward Arnold, London

White, R. K. (1980) The Upper Palaeolithic Occupation of the Perigord: A Topographic Approach to Subsistence and Settlement. Doctoral Thesis, Department of Anthropology, University of Toronto

(1985) Upper Paleolithic Land Use in the Perigord: A Topographic Approach to Subsistence and Settlement. B.A.R. (Int. ser.) 253. Oxford

Williamson, M. (1988) Relationship of species number to area, distance and other variables. In A. A. Myers & P. S. Giller (eds) Analytical Biogeography: An integrated approach to the study of animal and plant distributions Chapman & Hall, London p.91-116

Wilson, J. (1975) The last glacial environment at the Abri Pataud: a possible comparison. In H. Movius (ed) Excavation of the Abri Pataud, Les Eyzies (Dordogne) American School of Prehistoric Research 30, p.175-86

Wing, E. S. & A. B. Brown (1979) Paleonutrition: Method and Theory in Prehistoric Foodways. Studies in Archaeology, Academic Press, London.

Winterhalder, B. & E. A. Smith (1982) Hunter-Gatherer Foraging Strategies: Ethnographic and Archaeological Analyses. Prehistoric Archaeology and Ecology Series, Chicago University Press. Chicago

Woodward, F. I. (1987) Climate and Plant Distribution. Cambridge Studies in Ecology. C.U.P. Cambridge

Zlotin, A. I. & K. S. Khodashova (1980) The Role of Animals in Biological Cycling of Fores-Steppe Ecosystems. Dowden, Hutchinson & Ross, Inc. Stroudsburg

Appendix I

Data Tables

(Species and Element % Frequencies)

Chatelperronian.
(large herbivores)

	R.t.	C.e.	Bov.	E.c.	C.c.	C.i.	S.s.	R.r.	C.a.	M.p.
Chatelperron B1	10.93			78.13						10.93
B2	34.04			55.32	10.63					
B3	53.26	9.67	9.67	17.74	9.67					
B3a	16.67		16.67	50.00					16.67	
B4	22.81	14.03	7.02	42.10					7.02	7.02
B4a	12.50		12.50	75.00						
B5	27.27		32.73	36.36					3.64	
Bmoyen	27.67	3.95	15.81	45.45	1.98				3.95	1.19
C	19.46	1.49	22.46	47.90	0.75		2.99		4.49	0.45
La Ferrassie										
L2bj-L5 fr M2e sg	42.12	5.88	39.71	7.35			1.47		1.47	
L2b2 fr L3b-M1b sg	33.93	10.70	46.43	5.36				1.79		
L2a-L2b1 fr - L1b et L2a sg	57.53		36.99	4.11			1.37			
Les Cottes	22.72	9.10	13.60	45.50		4.50			4.50	
Pair non Pair E-D'	28.19		21.43	40.41	0.19	0.19	1.12		8.45	0.93
Le Piage F-1	73.30	3.30	16.70	5.00				1.60		
La Plematrie Sj/Ej	44.34	3.77	17.92	33.02					0.95	
Sm/Em	33.33	4.17	16.67	39.58					4.17	
Roc de Combe 8	40.96	2.40	39.15	15.66		1.20		0.60		

a

	R.t.	C.e.	Bov.	E.c.	C.c.	C.i.	S.s.	R.r.	C.a.	M.p.
Saint Cesaire Ejop	40.73	8.06	22.58	16.13			0.81	1.21	5.65	2.42

a = *Equus hydruntinus* present (N.I.S.P. = 2)

Early Aurignacian.
(large herbivores)

	R.t.	C.e.	Bov.	E.c.	C.c.	C.i.	S.ś.	R.r.	C.a.	M.p.	
Battuts 3	69.20	4.17	6.67	0.83		15.00		3.33			
Bourgeois-Delauny	19.66	0.97	12.24	63.59		0.54	0.54		1.18	0.54	a
La Chevre	78.90		4.10	15.60		0.40			0.40		
La Ferrassie L1 sag. et K6 a K4	77.38	1.36	21.37	3.17		0.90	0.45				
Fontaury C2	62.29		1.64	32.78							
C3	68.69		2.61	28.26							
La Gravette	95.00	*	*	3.00	*		*	*			
Pair-non-Pair KD	42.38	1.51	16.96	28.44			0.86		9.81		
Pataud 11	62.83	0.19	0.69	35.86			0.09			0.09	
12	72.88		4.52	22.03		0.50					
13	98.60			1.40							
13/14	100.00										
14	99.00	0.60	0.13	0.06		0.06					
Le Piage F	66.70	0.27	19.60	13.50							
G-I	82.21	0.41	11.48	5.04		0.81					

	R.t.	C.e.	Bov.	E.c.	C.c.	C.i.	S.s.	R.r.	C.a.	M.p.
J	94.50		2.60	2.13		0.70				
K	86.91	0.19	4.36	6.45		1.33				
Roc de Combe 7a	93.90	0.40	0.80	3.70		0.80	0.40			
7b	89.40		2.35	7.05				1.18		
indef	97.40		1.22	0.87		0.17		0.35		
7 tot.	95.09	0.32	1.41	2.39			0.10	0.32		

[a] = Megaceros sp. (0.54%), Equus hydruntinus (0.21%)

Later Aurignacian.
(large herbivores)

	R.t.	C.e.	Bov.	E.c.	C.c.	C.i.	S.s.	R.r.	E.h.	M.p.	
Caminade Est D	7.14	10.70	39.30	17.80			3.60	10.70	10.70		
La Chevre	31.70	5.90	48.90	11.80			1.40	0.45			
La Ferrassie											
K3-K1 & J	6.45	18.71	51.61	3.23	2.68		13.55	1.29	1.94		*
L1-H1	16.86	18.60	34.88	18.02	2.33	3.49	1.16	2.91	1.74		
G1 sag. G1-G3 fr.	13.68	38.46	21.37	4.27	12.82			7.69			*
G0 fr. G0 & F sag.	72.73	9.09	11.04	3.25	1.95		0.65	1.30			
F4-F1 fr. Ga-e sag.	8.03	65.69	2.92	4.38	8.03	0.73	1.46	5.84	1.46		*
Font de Gaume	69.46		2.74	13.89		13.89					
Maldidier											
5 + 6	38.80	17.91	10.44	4.47		25.37		2.98			
5	38.71	16.13	16.13	6.50		22.58					
6	41.20	20.60	5.90	2.94		29.40		2.06			
Pair-non-Pair K	40.76	1.23	27.63	25.99		0.27				4.10	
Pataud 7	69.60	14.10	1.80	10.60	1.20		0.60	0.80	1.20	0.40	

	R.t.	C.e.	Bov.	E.c.	C.c.	C.i.	S.s.	R.r.	E.h.	M.p.	
Roc de Combe 6	83.60	1.09	5.46	2.73	0.54	0.54	2.73			2.73	$
5	92.49	0.33	0.91	0.33		0.75		0.16		5.00	

* = Megaloceros giganteus, K3-K1 & J = 0.65%
G1 sag. G1-G3 fr. = 1.71%
F4-F1 fr. Ga-e sag. = 1.46%

= Megaceros sp. 0.05½

$ = Coelodonta antiquitatis, 0.54% Roc de Combe 6

@ = Le Flageolet VII som. = Perigordien a Noailles (Delpech 1983:57)

Upper Perigordian.
(large herbivores)

	R.t.	C.e.	Bov.	E.c.	C.c.	C.i.	S.s.	R.r.	E.h.	M.p.
Battuts 5	0.60	47.25	9.75	8.84		29.87	0.30	3.35		
7	56.79	2.46	4.93	3.70		14.81		17.28		
9	7.27	10.90	1.81	30.90		47.27		1.81		
La Ferrassie D4 &										
E sag	30.00	50.00	5.56	3.33		4.44		6.67		
E4-D2fr D3 & D2sag	2.12	80.10	5.25	1.62	5.86	0.20	0.71	1.52	2.63	
D1-B6	5.19	77.92	3.90	5.19	2.60			1.30	5.19	
Le Flageolet I										
VII	35.70	50.25	7.62	2.07	1.21	0.69		2.07	0.34	
TOT VI + VI/VII	40.68	28.13	5.70	13.30	1.90	9.88		0.38		
VI	35.61	31.22	4.39	15.61	0.98	11.71		0.49		
V	94.44	1.85	0.69	2.08		0.23	0.23	0.46		
IV	95.78			3.15		1.05				
I-III	90.47	7.61	0.95	0.95						
Gavaudun	46.80	2.50	16.40	32.30		0.60				
Laraux C3	61.54	1.77	3.55	28.99				0.59		
Laugerie Haute Est										
Per VI - VII	84.19	8.57	0.45	0.45		4.06		2.25		
Pataud 3	91.50	0.84	3.12	2.90	0.04	1.06	0.04	0.44	0.04	
eb. 3-4	94.99	1.63	1.37	1.32	0.26	0.37			0.05	
4 sup.	96.97	1.60	0.40	0.30	0.37	0.20	0.01	0.14	0.01	
4 moy.	97.23	1.17	0.56	0.68	0.09	0.05		0.22		

		R.t.	C.e.	Bov.	E.c.	C.c.	C.i.	S.s.	R.r.	E.h.	M.p.
Pataud	4 inf.	93.49	1.56	1.87	1.87	0.21	0.24		0.76		
	eb. 4-5	85.40	10.89	1.31	0.43		0.43				
	5 sup.	98.60	0.77	0.26	0.11	0.06	0.10		0.09		0.01
	5 moy.	98.72	0.70	0.12	0.29		0.08		0.04		0.04
	5 inf.	99.02	0.45	0.04	0.19	0.08	0.16		0.04		0.01
	extension 2	98.08		0.90	0.36		0.06		0.12		0.06 c
Roc de la Belle			13.04	30.43	39.13				13.04	4.35	
Roc de Combe	4	94.44	1.38		0.69	0.69	2.08		0.69		
	3	86.66		10.66	1.33		1.33				
	2	85.56		4.12	6.18		3.09		1.03		
	1	87.53	3.00	2.12	3.46	0.15	0.72		2.94		0.05

c = Ovibos moschatus 0.42% (N.I.S.P. = 7)

Proto-Magdalenian.
(large herbivores)

	R.t.	C.e.	Bov.	E.c.	C.c.	C.i.	S.s.	R.r.	E.h.	M.p.
Lachaud Ib	93.47			6.50						
Ic	93.75		3.13					3.13		
Pataud 2	87.18	2.68	1.98	2.82	0.07	2.54		2.61		0.07
Le Piage C-E	83.97	2.40	5.70	5.14	1.71			0.60		

Solutrean
(large herbivores)

	R.t.	C.e.	Bov.	E.c.	C.c.	C.i.	S.s.	R.r.	S.t.	M.p.
Badegoule Sol IV	79.68		1.19	9.56		4.70		4.70		
Grotte de l'Eglise	57.27	1.36	11.81	28.63	0.45		0.45			
Laugerie-Haute Est										
+ Mag 0	64.60	0.99	1.73	25.99				5.44	1.23	
Sol sup. 22&23	82.29			17.22		0.47				
Sol moy. 29	92.75			5.07		2.17				
Sol moy. 30	96.26		0.74	1.49		1.49				
Sol inf. 31	90.11	0.56	0.56	6.49		1.69		0.56		
Laugerie-Haute Ouest										
Sol fin. & sup. 2	89.73	1.14	0.38	0.38						7.60
3	99.34			0.66						
4	98.48			1.52						
5	99.31			0.46						0.23
6	98.96			1.04						
7	99.49			0.50						
(2-7	97.40	0.21	0.07	0.48						1.47)
Sol moy. 8	98.07			1.95		0.49				0.49
9	79.82					0.92				19.27
10	78.07			8.56				0.53		12.83
(8-10	86.22			3.99		0.39		0.19		9.18)
Sol inf. 11A	83.44			8.27		0.68				7.58
12a	97.13			2.02						0.84
12b	96.40	0.19		2.84		0.38				0.19
(12a & 12b	97.78	0.09		2.40		0.17				0.53)
12c	97.69		0.25	1.53						0.51

	R.t.	C.e.	Bov.	E.c.	C.c.	C.i.	S.s.	R.r.	S.t.	M.p.
12d	94.51			2.44						3.05
(12c & 12d	96.75		0.18	1.80						1.25)
Sainte Eulalie IV	86.44	1.69	3.38	1.69		5.08		1.69		

Early Magdalenian (Badegoulian)
(large herbivores)

	R.t.	C.e.	Bov.	E.c.	C.c.	C.i.	S.s.	R.r.	S.t.	M.p.
Cottier n II-III	17.33			8.00		69.33		5.33		
Laugerie-Haute Est										
Magd. 0	64.50	1.00	1.75	26.00				5.50	1.25	
Magd. I	95.26	0.08	0.48	2.72		0.96		0.48		
Fritch 6	64.60		0.86	30.92		2.93	0.23			
5d	86.49	0.06	0.06	8.45		3.98	0.95			
5c	85.79			7.10		7.10				
5b	80.36		2.98	11.74		3.97				
5a	77.79		0.52	18.76		2.93				
4	23.02		0.42	76.27		0.29				
3b	26.92		3.85	69.23						
3a	44.45			55.55						

Middle Magdalenian
(large herbivores)

	R.t.	C.e.	Bov.	E.c.	C.c.	C.i.	S.s.	R.r.	S.t.	M.p.
Combe Cullier 4'1	90.50	1.50	3.00			1.50		3.00		
4'2	92.60		3.70					3.70		
5	78.04	3.65	17.07	?	1.21					
8	90.76		1.53	3.07		1.53		3.07		
9	91.26			1.98		2.77		3.97		
11	92.15		0.80	0.20		0.60		2.41	3.82	
12	88.00	2.00		1.33		2.00		6.66		
13a	89.85	2.25	0.56	1.12		3.09		2.81	0.28	
13c	85.71	3.17		1.19		4.76		5.15		
14, 15 & 16	89.19					6.76	2.70	1.35		
Lachaud	71.43			12.25				8.16	8.16	
Laugerie Haute Est										
Mag II & III	91.66	0.72	0.62	4.39		0.41		0.51	1.62	
Mag III	91.89	0.56	0.73	4.46		0.38		0.51	1.46	
Ste-Eulalie	85.86	1.08	3.26	1.08		5.43		3.26		

Upper Magdalenian.
(large herbivores)

	R.t.	C.e.	Bov.	E.c.	C.c.	C.i.	S.s.	S.t.	R.r..	E.h.
Bergerie	76.56	15.62	3.13	1.56		3.13				
Bois Ragot 6	91.35			8.65						
5b	58.33	6.06	5.30	24.24	3.03		2.27			0.76
Bruniquel	73.73	0.99	4.95						20.32	
Cap Blanc	93.85	0.84	0.91	4.20			0.06	0.12		
Faustin	11.57	15.80	18.92	48.40	4.20		1.05			
Grotte des Fees	6.23	4.89	20.49	26.50			1.78	39.64	0.40	
Le Flageolet IX	94.60	2.45	0.54	0.68		0.13		1.22	0.40	
Fongaban 2	2.63		87.96	4.38				7.01		
3	0.09		98.60	0.63				0.72		
4-6			76.70	10.27				13.01		
Fontarnaud	11.72	1.30	18.90	60.30				7.50		
Gabillou	48.94		5.30	24.50			2.13	19.15		
Gare de Couze B	85.90	7.15	2.24	0.81	0.40		3.47			
C	88.50	4.34	1.73	1.44	2.90		1.15			
D	91.60	3.48	1.16	1.45	1.74		0.50			
E	90.90	5.45	1.51	0.60	1.21		0.30			
F	94.20	4.19	0.26	0.26	0.52		0.26		0.26	

		R.t.	C.e.	Bov.	E.c.	C.c.	C.i.	S.s.	S.t.	R.r.	E.h.
Gare de Couze	G	91.02	5.80	1.03	1.31	0.79		0.40			
	G_0	89.70	6.25	1.49	0.81	1.35					
	G_1	98.30	0.76	0.41	0.23	0.23		0.15			
	H	97.80		1.11	0.31		0.15	0.15		0.31	
La Madeleine	2	96.19		2.18	0.91					0.60	
	4	97.40		2.59							
	5	100.00									
	6	98.56		0.85	0.28					0.28	
	7	97.16		2.69						0.14	
	8	96.90		2.74	0.34						
	9	96.41		3.05	0.52						
	10	97.59		2.00	0.40						
	11	96.87		2.27	0.85						
	12	97.21	0.10	0.51	1.85					0.30	
	13	93.70		0.27	5.67					0.37	
	14	63.00		1.27	35.87						
	15	87.50		0.17	12.11		0.17				
	16	89.07			9.83		0.54			0.54	
La Mairie		53.30	6.70	26.70	6.70			6.70			
Marcamps (Total)		14.21	1.20	11.78	10.89	0.05		0.02	60.99	0.10	0.64
Coll. Maziaud		14.21	1.25	11.06	10.35	0.06			62.44	0.08	0.51
Coll. Ferrier		12.67	1.09	18.77	11.89			0.15	54.14	0.31	0.93
Coll. Nicolai		16.06	0.80	9.43	15.86				56.22		1.40
La Mege		72.70		9.10	9.10			9.10			
Montmorillon		51.39	0.46	2.80	38.81			1.55	4.81		
Le Morin AI-II		39.02	11.72	27.39	16.51	0.75		4.59			

	R.t.	C.e.	Bov.	E.c.	C.c.	C.i.	S.s.	S.t.	R.r.	E.h.	
Le Morin AIII	41.61	10.51	23.43	18.18	0.42		5.82				
AIV	49.74	10.03	21.71	13.75	0.51		4.13			0.10	
BI	55.99	0.97	30.31	12.22			0.24				d
BII & B?	78.46		13.84	7.67							$
Reignac	98.00	*	*	*	*	*	*	*			
Coll. Hulin	89.95	1.61	1.36	2.72	0.14	0.03	1.29	2.89			
Roc d'Abeilles	65.23	13.04	8.70						13.04		
Rond du Barry D	1.68	11.95	9.07	5.87	5.87	24.95	30.39				
E	42.46	4.58	13.93	20.05	0.34	14.53	4.10				
Sainte Eulalie I	64.48	4.35	3.26	7.61		5.43			13.72		
III	85.86	1.08	3.26	1,08		5.43			3.26		

d = <u>Megaceros</u> sp. 0.24% (N.I.S.P. = 1)

$ = couches BII and B indet. (B?) have been amalgamated in this table, attributed to Magdalenian V.

<u>Azilian</u>

	R.t.	C.e.	Bov.	E.c.	C.c.	C.i.	S.s.	R.r.	E.h.	O.c.	
Blassac		42.86	14.28	28.57		14.28					
Bois Ragot 4b	6.70	77.80	5.56	1.11	3.33		4.44	1.11			
3		85.30	2.94	2.94	5.88		2.94				
Campalou	0.11	84.27	0.23	0.93	0.11	11.31	0.34				
La Faurelie II 2	35.00	40.00					25.00			*	
[(2 incl. rabbit)	2.97	3.39					2.11			91.52]	
3	15.22	65.22	2.20	4.35			13.04			*	
[(3 incl. rabbit)	1.75	7.52	0.25	0.50			1.50			88.47]	
Pages		83.33	2.80	11.11			2.80				
Pont d'Ambon 2		37.37	33.33	24.24	2.02		2.02	1.00			
3		79.46	4.46	6.25	0.89		8.93				
3a		82.05	2.05	2.05	6.15		7.69				
3b		84.05			7.24		8.69		2.25		
4	0.09	89.47		0.87	5.26		3.50		0.87		
Roc d'Abeilles	28.60	47.60		19.64				4.76			
Grotte du Tai SN SX	6.03	41.38	0.86			48.28		2.58			e
C'1	10.32	4.89	0.54			79.89		2.71			f
C'1 C"1	14.20	10.05	1.77			69.23		4.73			
C"1	20.93	7.36				69.76		1.93			

e = <u>Alces alces</u> (0.86%) f = <u>Alces alces</u> (0.54%)

Abri Pataud.

Reindeer

	Protomag.	Perig VI	Perig. V	Perig. IV
Antlers	126	203	344	521
Skull	52	17	78	106
Maxilla	13	24	46	22
Mandible	78	132	613	561
Scapula	10	17	107	62
Vertebrae (ctl)	1	30	116	105
Sacrum			2	1
Vertebrae (caud.)	1	2	5	4
Sternum		2	15	9
Inominate	2	4	27	26
Humerus (p)	1		16	109
(d)	2	9	36	64
Radius (p)	7	7	19	30
(d)	3	2	44	44
Cubitus (p)	2	12	7	10
(d)	4	1	6	16
Metacarpals(p)	6	12	13	18
(d)	12	38	117	116
Femur (p)	4	6	61	69
(d)	2	1	25	75
Tibia (p)			25	16
(d)	1	6	106	63
Metatarsals(p)	35	2	240	10
(d)	8	14	227	193
Fibula		4	63	60
Patella	2	5	50	172
Astragalus	3	12	39	85
Calcaneus	2	4	18	61
Phalanx I	7	117	165	814
II	5	41	76	446
III	4	27	11	167
Teeth (perm.)	200	168	881	1074
(dec.)	20	27	98	164

(Source: After Poplin 1977)

Element	Flageolet I reindeer	Flageolet II reindeer	La Mege reindeer	Fongaban bovids
Antlers			5	
Skull		11		
Maxilla		199		13
Mandible	21	22	17	5
Scapula		4	2	4
Vertebrae		1	11	12
Sacrum			3	2
Humerus	10	13	6	18
Radio-cubitus	26	50	7	31
Metacarpals	21	33	11	32
Femur	17	15	2	13
Tibia	24	27	6	90
Metatarsals	61	117	7	26
Astragalus			4	51
Calcaneus			1	41
Phalanges	78	54	6	
Phalanx I				105
II				73
III				35
Ribs		15		2
Carpals		7		
Os coxis		4		
Tarsals		7		
Sesamoid		9		
Metapodia (unid.)		28		
Mc.+ access.ph.		9		

Element	Gare de Couze inf. reindeer	Gare de Couze sup. reindeer	Le Piage reindeer	Enlene reindeer
Antlers	70	19		44
Skull				19
Maxilla			1	
Mandible			2	39
Scapula		14	5	19
Vertebrae				136
Sacrum			2	
Sternum				3
Humerus				36
(p)				
(d)	114	21		
Radius				65
(p)	4	4		
(d)	4	3		
Metacarpals			12	22
Femur			2	40
(p)	1			
(d)	15			
Tibia			2	56
(p)	4			
(d)	7	8		
Metatarsals			7	84
Astragalus	119	11		
Calcaneus	15	2		
Phalanges			26	
Phalanx I	148	8		100
II	73	8		89
III				27
Metapodia	63	17		
Cubo-scaphoid		9		
Tarsals			14	62
Carpals			9	62
Ribs				54

Element	Ste.Eulalie I reindeer	Ste.Eulalie III reindeer	La Gravette reindeer	La Gravette horse
Maxilla	2			
Mandible	1	4		
Scapula		3		2
Vertebrae			18	2
Humerus				
(p)				
(d)	4	5		2
Radius				
(p)	2	2		3
(d)	1	3	23	3
Metacarpals				
(p)	2	3		2
(d)	2	5		5
Femur				
(p)			19	
(d)		2		
Tibia				
(p)				
(d)	3		16	2
Metatarsals			1	
(p)	6	5		
(d)	8	13		
Astragalus	1	1	78	3
Calcaneus	1		69	
Phalanges	10	3		
Phalanx I			78	3
II			48	2
III				
Ribs	1			
Carpals				9

Frequencies of Reindeer Elements and M.N.I. counts from Les Eyzies.

	Axial / Side indet.	Left Frag.	Right Frag.	M.N.I.
Antler			1	1
Isolated teeth		16	33	7
Maxilla		2	3	2
Mandible		3	4	2
Atlas	4			4
Axis	3			3
Cerv.vert.	11			
Thor.vert.	5			
Pelvis	3			1
Sternebra	1			1
Scapula		3		3
Humerus		13	18	14
Radius		20	11	9
Ulna		15	4	7
Scaphoid		6	9	9
Lunate		5	11	11
Cunciform		7	8	8
Magnum		6	8	8
Unciform		5	4	5
Pisiform		1	4	4
Metacarpal	3	9	9	7
Inominate		3	2	3
Femur		1	4	3
Patella		8	5	8
Tibia		16	14	14
Lat. malleolus		5	9	9
Calcaneus		30	24	17
Astragalus		11	10	11
Naviculo-cuboid		12	7	12
Lat./Mid. cuneiform		12	7	12
Med. cuneiform			5	5
Metatarsal	5	24	17	14
Metapodial splint	10			2
Metapodial	4			1
Prox.ax.sesamoid	32			4
Prox.peri.sesamoid	39			5
Distal sesamoid	6			2
Prox.phalanx	59			4
Mesial phalanx	48			4
Ungual phalanx	30			4
Dew claw - p. phal.	14			3
m. phal.	15			2
u. phal.	15			4

(After Olsen 1987:178)

Ibex Material from Les Eglises

	c.4	c.6
Skull	5	16
Maxilla	4	1
Mandible	4	13
Scapula	2	1
Vert. cerv.		1
caud.		5
unid.	14	75
Humerus (d)		2
Radio-cub. (p)		18
(d)		6
(shaft)		1
Metacarpal (p)	1	9
(d)		5
(shaft)		1
Femur (p)	4	4
(d)	1	11
(shaft)		5
Tibia (p)		9
(d)		9
(shaft)	1	8
Metatarsal (p)	2	14
(d)		2
Access. metatarsal	1	3
Astragalus	1	11
Calcaneus	1	15
Phalanx I	14	59
II	11	38
III	3	31
Ribs	3	28
Atlas		6
Axis		1
Carpal	12	34
Access. metacarpal	1	2
Os coxis	1	7
Tarsal	26	90
Metapodia	1	3

(Source: after Delpech 1975)

Roc de Marcamps, Reindeer Material

	Maziaud	Nicolai	Ferrier
Maxilla	10	2	2
Mandible	31	4	1
Scapula	1		2
Vertebrae	3		
Pelvis	1	1	
Humerus (p)			
(d)	8	1	
(shaft)			
Radio-cub. (p)	17	3	
(d)	10		
(shaft)			
Metacarpal (p)			
(d)	29	9	5
(shaft)	52		
Femur (p)		3	
(d)			
(shaft)			
Tibia (p)			
(d)	25	2	3
(shaft)			
Metatarsal (p)	27	4	
(d)	20	2	
(shaft)	2		
Astragalus	21	3	8
Calcaneus	11	4	
Phalanges	96	12	11
I	58	6	2
II	24	5	8
III	14	1	1
Carpal	72	4	
Tarsal	96	2	16
Acces. metapodia	12		2
Metapodia			2

Roc de Marcamps, Saiga

	Maziaud	Nicolai	Ferrier
Maxilla	88	5	18
Mandible	275	38	33
Scapula	11		5
Vertebrae	11	4	
Pelvis	4		
Humerus (p)			
(d)	65	11	17
(shaft)			
Radio-cub. (p)	73	11	9
(d)	28	4	3
(shaft)	7	3	
Metacarpal (p)	137	1	3
(d)	126	14	15
(shaft)			
Femur (p)		5	1
(d)	1	1	
(shaft)	8	2	
Tibia (p)	3		1
(d)	64	3	7
(shaft)		1	
Metatarsal (p)	241	5	3
(d)	142	23	12
(shaft)	23		
Astragalus	115	12	23
Calcaneus	62	12	15
Astr / Calc. artic.	1		
Phalanges	333	34	33
I	98	26	16
II	162	6	16
III	73	2	1
Tarsal	93	7	22
Carpal	110	6	
Metapodia	3	2	3

Roc de Marcamps, Bovids

	Maziaud	Nicolai	Ferrier
Skull		2	
Maxilla	2		
Mandible	50	4	16
Scapula	2		1
Vertebrae	4	2	
Humerus (p)	1		
(d)			1
(shaft)		3	
Radio-cub. (p)			
(d)			
(shaft)	1		1
Metacarpal (p)	2	2	1
(d)	2		
(shaft)	1		
Femur (p)	2		
(d)			
(shaft)			
Tibia (p)			
(d)	3		1
(shaft)	3		
Metatarsal (p)	7		1
(d)	6		1
(shaft)	2		
Astragalus	3		2
Calcaneus	5	1	
Phalanges	104	3	8
I	35	2	
II	37		5
III	32	1	3
Ribs	5		
Tarsal	104	4	9
Carpal	14	3	18
Metapodia	6	1	1

Roc de Marcamps, Horse

	Maziaud	Nicolai	Ferrier
Maxilla		2	
Mandible		3	2
Vertebrae	5		
Pelvis			1
Humerus (p)			
(d)	2		1
(shaft)			
Radio-cub. (p)		2	
(d)	1		1
(shaft)			
Metacarpal (p)	3		
(d)	10		
(shaft)	24		
Femur (p)		2	
(d)			
(shaft)			
Tibia (p)			1
(d)	4		
(shaft)			
Metatarsal (p)	3		1
(d)	11		
(shaft)	21		
Astragalus	1		
Calcaneus	4		
Phalanges	44	3	10
I	15	2	3
II	19	1	4
III	10		3
Rib	7		1
Carpal	59	1	
Tarsal	19	1	5
Acces. metapodia	4	8	3
Metapodia		3	

Reignac, Dordogne (Hulin Collection)

	Reindeer	Bovids	Horse	Antelope
Skull	104	1	3	2
Maxilla	33			
Mandible	353		5	24
Scapula	382		1	13
Vertebrae	254	4		8
Pelvis	175		2	
Sternebrae	11			
Humerus (p)	8		1	
(d)	316	2	2	16
Radio-cub. (p)	169	2	2	9
(d)	241	7	4	3
(shaft)	1			
Metacarpal (p)	194	2	6	3
(d)	353	4	11	17
(shaft)	1			
Femur (p)	120	1	2	1
(d)	15			
(shaft)	53			
Tibia (p)	305	4	1	1
(d)	62	1	2	17
(shaft)	61			
Metatarsal (p)	385	2		8
(d)	613	1		7
(shaft)				
Astragalus	135	4	2	12
Calcaneus	183	2	2	4
Phalanges	819	10	12	28
Rib	108	12	1	
Metapodia	18	3		
Atlas	14			
Acces. metapodia	33		3	
Carpal	135			1
Tarsal	82		7	5

Reignac, Dordogne (Hulin Collection)

	Red deer	Boar	Roe deer
Skull		1	
Maxilla	4		
Mandible	6	20	2
Scapula	5	6	1
Vertebrae	1	2	
Pelvis	3		
Humerus (p)	1	1	
(d)	8	6	3
(shaft)			
Radio-cub. (p)	1	8	
(d)	7		
(shaft)			
Metacarpal (p)	4		
(d)	3		
(shaft)			
Femur (p)	3		
(d)			
(shaft)			
Tibia (p)		2	
(d)	6	4	3
(shaft)			
Metatarsal (p)	8		
(d)	4		
(shaft)		4	
Astragalus	8	1	
Calcaneus	4	4	
Phalanges	14	2	1
Rib	1		
Atlas	1	3	
Tarsal	4	1	1
Metapodia	8	1	
Carpals	5		

Data from sites not considered in chapter 9
for reasons of sample size and problems of
chronological attribution.

	Pair-non-Pair (Horse)	Pair-non-Pair (Reindeer)
Antlers		124
Skull	1	10
Maxilla	5	
Mandible	27	8
Scapula	7	10
Vertebrae	12	37
Sacrum	29	2
Humerus (p)	1	2
(d)	10	27
Radio-cub. (p)	5	42
(d)	9	27
Metacarpal (complete)	19	
(p)	30	28
(d)	32	47
Femur (p)	1	1
(d)		1
Tibia (p)		1
(d)	20	4
Metatarsal (p)	30	7
(d)	45	34
Astragalus	45	94
Calcaneus	17	59
Phalanges	82	192
Ribs	4	
Carpals	121	60
Tarsals	134	42
Atlas		4
Axis		2

	Grotte des Fees (Horse)	Grotte des Fees (Reindeer)
Astragalus	2	
Metacarpal (p)		
(d)	3	3
Phalanx	4	3
Tarsal	8	5
Metatarsal (d)	1	1
Fibula	3	
Scapula	1	
Metapodia (d)		1
Mandible		2
Rib		1
Antler		2
(Teeth	103	10)

	Grand Moulin (Horse)	Jolias (Horse)
Metacarpal (p)	1	1
(d)	1	3
(complete)		1
Metatarsal (p)		
(d)	2	
(complete)		
Tarsal	12	3
Astragalus	3	1
Tibia (p)		
(d)	1	
Radio-cub.	1	
Phalanges	15	9
Metapodia		1
(Teeth	98	112)

Appendix II
Site Bibliography.

Miscellaneous: Regional Surveys.
Bay-Petersen, J. (1975) Pre-Neolithic Faunal Exploitation in S. France and Denmark. Unpublished Ph.D. thesis, University of Cambridge.

Delpech, F. (1975) Les Faunes du Paléolithique Supérieur dans le Sud-Ouest de la France. Thèse de Doctorat ès Sciences naturelles. Université de Bordeaux I.

(1983) Les Faunes du Paléolithique Supérieur dans le Sud-Ouest de la France. Cahiers du Quaternaire No.6. C.N.R.S. Paris

Hemingway, M. F. (1980) The Initial Magdalenian in France. B.A.R. (Int.) 90.

Lavaud, F. (1980) Les Faunes Paléolithiques du Wurm II et III dans le Sud-Ouest et Centre-Ouest de la France. Thèse de specialité, Université de Poitiers

Lenoir, M. (1983) Le Paléolithique des Basses Vallées de la Dordogne et de la Garonne. Thèse de Doctorat d'Etat ès Sciences. Université de Bordeaux I.

Le Tensorer, J.-M. (1979) Recherches sur le Quaternaire en Lot-et-Garonne Stratigraphie, Paléoclimatologie et Préhistoire Paléolithique. Thèse de Doctorat d'Etat ès Sciences. Université de Bordeaux III.

(1981) Le Paléolithique de l'Agenais. Cahiers du Quaternaire No. 3. C.N.R.S. Paris.

Levine, M. (1979) Archaeozoological Analysis of Some Upper Pleistocene Horse Bone Assemblages in Western Europe. Unpublished Ph.D. thesis, University of Cambridge.

Patte, E. (1941) Le Paléolithique dans le Centre-Ouest de la France. Libraire de l'Académie de Médecine. Masson et Cie, Eds. Paris

Prat, F. (1968) Sur les Vestiges de Megaceros découverts dans Quelques Gisements Paléolithiques du Sud-Ouest de la France. Thèse de Doctorat d'Etat ès Sciences naturelles. Université de Bordeaux I.

Rigaud, J.-P. (1982) Le Paléolithique en Périgord: les Données du Sud-Ouest Sarladais et leurs Implications. Thèse de Doctorat d'Etat ès Sciences. Université de Bordeaux I.

White, R. (1980) The Upper Palaeolithic Occupation of the Perigord: A Topographic Approach to Subsistence and Settlement. Unpublished Ph.D. thesis. University of Toronto.

(1985) Upper Palaeolithic Land-Use in the Perigord: A Topographic Approach to Subsistence and Settlement. B.A.R. (International Series) 253. Oxford.

Site-Specific Surveys:

Aurignac

Bahn, P. G. (1984) Pyrenean Prehistory: A Palaeoeconomic Survey of the French Sites. Aris & Phillips, Ltd. Warminster

Auzary

Perpere,M. (1981) Le Magdalénien final de la grotte Auzary (Puy-de-Dome). B.S.P.F.78:342-53

Baring

(Lenoir 1983)

Battuts

Alaux, J.-F. (1969) Note préliminaire sur l'abri périgordien des Battuts (Tarn). B.S.P.F. 66 : 10 - 15.

(Also Delpech 1975, 1983)

Bergerie

Séronie-Vivien, M.-R. & C. Bensch (1966) Un habitat en fond de la grotte à l'Age du Bronze et du Magdalénien: la grotte de la Bergerie à Caniac (Lot). Actes Soc. Linn. Bord. serie B 103 (13)

Bernous

Peyrony, D. (1932) Les Gisements Préhistoriques de Bourdeilles (Dordogne). Archives de l'Institut de la Paléontologie Humaine de Paris. (Paris) 10.

Blassac

Bouchud, P. & J. Bouchud (1955) La faune de Blassac. B.S.P.F. 52:364-70

Bout, P. (1955) L'abri sous roche de Blassac (Haute-Loire). B.S.P.F. 52:316-22

Bois Ragot

Sonneville-Bordes, D. de (ed)(1979) La Fin des Temps Glaciaires. C.N.R.S. Paris.

Bonhomme

Sauter, M.-R. (1946) Station Paléolithique du Bonhomme (vallon des Rebières, Dordogne). Cahiers de Préhistoire et d'Archéologie 2. Geneva.

Bourgeois-Delauny

(Lavaud 1980)

Brouillaud

Peyrony, D. & P. Bourrinet (1928) Les Fouilles de Tabaterie (Dordogne): Sandongne et Abri Brouillaud. Imprimerie Magne. Perigueux.

Caminade Est
Delpech, F. (1970) Faune aurignacienne de Caminde Est, Commune de la Canéda (Dordogne). Quaternaria 13:137-140

Laville, H., Rigaud, J.-P. & J. Sackett (1980) Rock Shelters of the Périgord: Geological Stratigraphy and Archaeological Succession. Studies in Archaeology. Academic Press London.

Paquereau, M.-M. (1970) Etude palynologique des niveaux aurgnaciens de Caminade Est (Dordogne). Quaternaria 13:133-5.

Sonneville-Bordes, D. de (1970) Les industries aurignaciennes de l'abri de Caminade Est, Commune de la Canéda (Dordogne). Quaternaria 13:77-131

(Also Delpech 1975, 1983)

Campalou

(Delpech 1975)

Cap Blanc
Roussot, A. (1976) Abri du Cap Blanc, Commune de Marquay, Dordogne. Livret-guide de l'excursion A - 4. p.85-7 IXe Congres de l'UISPP, Nice

Cavart

(Le Tensorer 1979, 1981)

Cellier

(White 1980, 1985)

Chaire à Calvin
Bouvier, J.-M. (1986) La Chaire a Calvin. Livret-guide des excursions du 28 août. p.31-7. I.C.A.Z. Bordeaux.

(Also Patte 1941)

Chaise à Vouthon
Bouchud, J. & P. Bouchud (1953) La petite faune de la grotte de la Chaise (Charente). B.S.P.F. 50:170-7

David, P. & F. Prat (1965) Considérations sur les faunes de la Chaise (commune de Vouthon, Charente). Abris Suard et Bourgeois-Delaunay. B.A.F.E.Q. 2:221-31

Débénath, A. (1986) Les gisements de la Chaise à Vouthon. Livret-guide des excursions du 28 août .p.39-45 I.C.A.Z. Bordeaux.

(Also Patte 1941, Prat 1968)

Chancelade

(Patte 1941, Prat 1968)

Château Bernard

(Patte 1941)

Châtelperron
Delporte, H. (1956) La grotte des Fées de Châtelperron (Allier). Congr. Préh. Fr. 15:452-75

Lacaille, A.-D. (1947) Chatelperron: A New Survey of its Palaeolithic Industry. Soc. Antiq. London

La Chèvre.
Laville, H., Rigaud, J.-P. & J. Sackett (1980) Rock Shelters of the Perigord: Geological Stratigraphy and Archaeological Succession. Studies in Archaeology. Academic Press London

Chez-Leix
(Lenoir 1983)

Combe-à-Rolland
(Patte 1941)

Combe Cullier
Vire, A. (1908) La Crozo de Gentillo ou Grotte de Combe Cullier, Commune de Lacave (Lot). L'A. 19:409-24

& A. Niederlander (1921) Nouvelles fouilles à la Crozo de Gentillo, Commune de Lacave (Lot). B.S.P.F. 18:269-70

(Delpech 1975, 1983)

Les Cottés
Bouchud, J. (1961) Etude de la faune du gisement des Cottés (Haute-Vienne). L'A. 65:258 - 70.

Cottier
Bouchud, J. & P. Bouchud (1953) La faune des grottes des Orciers et de Cottier (Haute-Loire). B.S.P.F. 50:444-57

Bout, P. (1953) La grotte de Cottier près Retournac (Haute-Loire). B.S.P.F. 50:437-43

Crabillat
Peyrony, D. & E. Peyrony (1941) Gisement préhistorique de Crabillat. B.S.P.F. 38:245-62

(Also White 1980, 1985)

Dalignac
(Patte 1941)

Grotte de l'Eglise
Coutier, L. (1952) Notes géologiques et paléontologiques sur la Grotte de l'Eglise à Saint-Martin-d'Exideuil (Dordogne). B.S.P.F. 49:312-4

(Also Delpech 1975, 1983)

Face de Fieux, en
(Patte 1941)

La Faurèlie
Laville, R., Rigaud, J.-P. & J. Sackett (1980) Rock Shelters of the Perigord: Geological Stratigraphy and Archaeological Succession. Studies in Archaeologyy. Academic Press. London.

(Also Delpech 1975, 1983)

Fauroux

(Lenoir 1983)

Le Faustin
Delpech, F. (1971) L'Abri Faustin, Commune de Cessac (Gironde). Etude paléontologique. B.S.P.F. 68:328-32

(Also Delpech 1975, 1983)

Grotte des Fees.
Lacorre, F. (1938) La grotte des Fées, à Marcamps (Gironde) ou Roc de Marcamps. P.-V. Soc. Linn. Bord. 90:35-51

(Also Delpech 1975, 1983)

La Ferrassie.
Delpech, F. (1984) La Ferrassie: Carnivores, Artiodactyles et Périssodactyles. In H. Delporte (ed) 'Le Grand Abri de la Ferrassie, Fouilles 1968 - 1973.' p.61 - 89. Etudes Quaternaires, Géologie, Paléontologie, Préhistoire Mem. 7. Paris.

Delporte, H. (1969) Les fouilles du Musée des Antiquites Nationales à la Ferrassie. Antiquités Nationales 1:15-28.

(ed) (1984) Le Grand Abri de la Ferrassie, Fouilles 1968 - 1973. Etudes Quaternaires: Géologie, Paléontologie, Préhistoire. Mem. 7. Paris.

& A. Truffeau (1972 - 3) Les industries du Perigordien superieur de la Ferrassie. Quatar 23/24:93 - 123

Donard, E. (1984) Les lagomorphes de la Ferrassie. In H. Delporte (ed) 'Le Grand Abri de la Ferrassie, Fouilles 1968 - 1973'. Etudes Quaternaires: Géologie, Paléontologie, Préhistoire Mem 7:91-2. Paris

Peyrony, D. (1934) La Ferrassie. Moustérien, Périgordien, Aurignacien. Préhistoire 3. Libr. E. Leroux. Paris.

(Also Delpech 1975, 1983, Prat 1968)

Festons

(White 1980, 1985)

Le Flageolet I
Rigaud, J.-P. (1969) Note préliminaire sur la stratigraphie du Flageolet I, Commune de Bézenac (Dordogne). B.S.P.F. 3:73-5.

Delpech, F. & J.-P. Rigaud (1974) Etude de la fragmentation et de la répartition des restes osseux dans un niveaux d'habitat paléolithique. In H. Camps-Fabrer (ed) L'Industrie de l'Os dans la Préhistoire, p.47-55. Editions de l'Université de Provence. Aix-en-Provence.

(Also Delpech 1975, 1983, Rigaud 1982)

Le Flageolet II

Delpech, F. (1970) L'Abri magdalénien du Flageolet II (Bézenac, Dordogne). Paléontologie. B.S.P.F. 67:494-9

Laville, H., Rigaud, J.-P. & J. Sackett (1980) Rock Shelters of the Perigord: Geological Stratigraphy and Archaeological Succession. Studies in Archaeology. Academic Press London.

(Also Delpech 1975, 1983)

Fongaban

Delpech, F. (1972) Fouilles de sauvetage dans le gisement magdalénien de Fongaban, Commune de St.-Emilion (Gironde). La faune. L'A. 76:615-29

(Also Delpech 1975, 1983)

Fontalès

Darasse, P. & S. Guffroy (1960) Le magdalénien supérieur de l'abri de Fontalès. L'A. 64:1-35

(Also Delpech 1983)

Fontarnaud

Roussot, A. & J. Ferrier (1971) La grotte de Fontarnaud, commune de Lugasson (Gironde). B.S.P.F. 68:505-20

(Also Delpech 1975, 1983)

Fontaury

(Lavaud 1980)

Font de Gaume.

Prat, F. & D. de Sonneville-Bordes (1969) Découvertes récentes de Paléolithique Supérieur a la Grotte de Font-de-Gaume (Dordogne). Quaternaria 11:115-32

(Prat 1968)

Fontéchevade

(Patte 1941)

Fourneau du Diable

Peyrony, D. (1932) Les Gisements Préhistoriques de Bourdeilles (Dordogne). Archives de l' Institut de Paléontologie Humaine de Paris, Mem. 10.

Fritsch
Trotingen, F. (1984) Etudes sur l'Abri Fritsche (Indre). Gallia Préhistoire suppl. 19. Paris

(Also Hemingway 1980)

Gabillou
Gaussen, J. (1964) La grotte ornée de Gabillou (près Mussidan, Dordogne). Publications de l'Institut de la Préhistoire Université de Bordeaux 3. Bordeaux

Gare de Couze
Delpech F. (1967) Recherches Paléontologiques Concernant Quelques Gisements du Magdalénien VI: Stations de la Gare de Couze (Dordogne), du Morin (Gironde) et de Duruthy (Landes). Thèse de Doctorat en Géologie approfondie Option Paléontologie, Bordeaux I.

Prat, F. (1962) La faune du gisement de la Gare de Couze. L'A. 66:248-54

(Also Delpech 1975, 1983)

Garnet

(Lenoir 1983)

St.Germain-la-Rivière

(Delpech 1983, Lenoir 1983)

La Gravette.
Lacorre, F. (1960) La Gravette: le Gravettien et le Bayacien. Laval - Imprimerie Barneoud S.A. C.N.R.S. Paris

(Also Bay-Petersen 1975)

Guimberteau

(Lenoir 1983)

Houleau

(Delpech 1983)

Jardel
Bouyssonie, J. (1956) L'Abri Jardel II, commune de Peyzac (Dordogne). Congr. Pré. Fr. 15:262-9

Jardel, E. & A. Roussot (1967) L'Abri Jardel II, commune de Peyzac-le-Moustier (Dordogne). Note complémentaire. Bulletin de la Société Historique et Archéologique du Périgord 94:21-35.

Jaurens
Debard, E. (1979) Le gisement Pléistocène supérieur de la grotte de Jaurens à Nespouls, Correze, France: étude sédimentologique du remplissage. Nouv. Arch. Mus. Hist. nat. Lyon 17:17-24

Guérin, C., Philippe, M. & R. Vilain (1979) Le gisement Pléistocène supérieur de la grotte de Jaurens à Nespouls, Corrèze, France: historique et généralites. Nouv. Arch. Mus. Hist. nat. Lyon 17:11-16

Lopez Martinez, N. (1980) Les Lagomorphes (Mammalia) du Pléistocène Supérieur de Jaurens. Nouv. Arch. Mus. Hist. nat. Lyon 18:5-16

Mourer-Chauviré, C. (1980) Le gisement Pléistocène supérieur de la grotte de Jaurens à Nespouls, Corrèze, France: les Equidés (Mammalia, Perissodactyla). Nouv. Arch. Mus. Hist. nat. Lyon 18:17-60

Jaurias

Couste, R. (1951) Gisement magdalénien des grottes de Jaurias. B.S.P.F. 48:381-4

(Also Lenoir 1983)

Jumeau

Gaussen, J. (1959) Un gisement magdalénien dans la vallée de l'Isle. L'Abri Jumeau, à Soursac (Dordogne). B.S.P.F. 56:456-92.

Lacave

Vire, A. (1905) Grotte préhistorique de Lacave (Lot). L'A. 16:411-429

Lachaud

Bouchud, J. (1952) Etude des rongeurs et des oiseaux de Lachaud. B.S.P.F. 49:262-7

Cheynier, A. (1965) L'Abri Lachaud, à Terrasson (Dordogne). Préhistoire 16.

Lartet

Peyrony, D. (1932) Les abris Lartet et du Poisson à Gorge d'Enfer. L'A. 36:241-68

Lascaux

Bouchud, J. (1959) Etude de la faune de la grotte de Lascaux (Dordogne). Bulletin de la Société des Etudes et Recherches Préhistoriques des Eyzies. p.1-3

Leroi-Gourhan, Arl. & J. Allain (1979) Lascaux Inconnu. C.N.R.S. Paris

Laugerie Haute Est.

Bordes, F. (1958) Nouvelles fouilles à Laugerie Haute Est. Premiers résultats. L'A. 62:205-44

(1978) Le protomagdalénien de Laugerie Haute Est (fouilles F. Bordes). B.S.P.F. 75:501-21

Peyrony, D. & E. Peyrony (1938) Laugerie-Haute près des Eyzies (Dordogne). Archives de l'Institut de Paléontologie Humaine de Paris, Mem.19

(Also Delpech 1975, 1983)

Laugerie Haute Ouest
(Delpech 1975, 1983)

Lestruque
Blanc, M. S. (1931) Les fouilles de MM. Louis Peyrille et Delmas, à Lestruque, commune de Tremolat (Dordogne) B.S.P.F. 28:236-8

Limeuil
Capitan, L. & J. Bouyssonie (1924) Limeuil: son gisement à gravures sur pierres de l'Age du Renne. Nourry, Paris.

Liveyre
Daniel, R. (1969) Quelques précisions sur la remplissage archéologique de la grotte de Liveyre, commune de Tursac (Dordogne). B.S.P.F. 66:112-6

Longueroche
Peyrony, D. (1934) Station préhistorique de Longueroche (Magdalénien et Azilien). Revue Anthropologique 44:226-47

La Lustre
(Delpech 1983, Lenoir 1983)

La Madeleine
Bouvier, J.-M. (1977) Un Gisement Préhistorique: La Madeleine. P. Fanlac Perigueux.

(1986) Abri de la Madeleine. Livret-guide des excursions du 28 août. p.89-98. I.C.A.Z. Bordeaux.

Capitan, L. & D. Peyrony (1928) La Madeleine: Son Gisement, Son Industrie, Ses Oeuvres d'Art. Nourry, Paris

(Also Delpech 1975, 1983)

Mairie
Capitan, L., Breuil, H., Bourrinet, P. & D. Peyrony (1908) La grotte de la Mairie à Teyjat (Dordogne): fouilles d'un gisement magdlénien. Revue de l' Ecole de l'Anthropologie de Paris 18:153-73

Maldidier.
Laville, H. (1973) Climatologie et Chronologie du Paléolithique en Périgord. Etude sédimentologique de depôts en Grottes et Sous Abri. Thèse de Doctorat d'Etat ès Sciences Naturelles. Université de Bordeaux I.

(Also Delpech 1975, 1983)

Marcamps
Lacorre, F. (1938) La grotte des Fées, à Marcamps (Gironde) ou Roc de Marcamps. P.-V. Soc. Linn. Bordeaux 90:35-51

(Also Delpech 1975, 1983)

Le Martinet

(Le Tensorer 1979, 1981)

Mège

Capitan, L., Breuil, H., Bourrinet, P. & D. Peyrony (1906) L'Abri Mège: une station magdalénienne à Teyjat (Dordogne). Revue de l'Ecole de l'Anthropologie de Paris 16:196-212

Montastruc

(LeTensorer 1981)

Montgaudier

Duport, L (1986) Le site préhistorique de Montgaudier. Bulletins et Mémoires. Société Archéologique et Historique de la Charente, 1986. p. 23-30.

(Also Prat 1968)

Montmorillon

(Delpech 1983)

Moradies

(Patte 1941)

Le Morin

(Delpech 1975, 1983)

Morts, Grotte des

Bouyssonie, J (1945) Le Préhistoire en Corrèze. Imprimerie Lachaise Brive

Moulin-Neuf

(Delpech 1983, Lenoir 1983)

Orgedeuil

(Patte 1941)

Pagès.

Bouchud, J. (1956) Faune de l'abri Pagès (à Rocamadour). L'A. 60:444-66

Pair-non-Pair.

Daleau, F. (1881) La grotte de Pair-non-Pair à Marcamps. Bull. de d'Association Française pour l'Avancement des Sciences. Alger

Sarradet, M. & P. Vidal (1970) La Grotte de Pair-non-Pair et ses Environs. Ed. P. Fanlac, Perigueux.

(Also Prat 1968)

Pasquet

(White 1980, 1985)

Pataud.

Bouchud, J. (1975) Etude de la faune de l'abri Pataud. In H. Movius (ed) 'Excavation of the Abri Pataud, Les Eyzies

(Dordogne). American School of Prehistoric Research, Bulletin 30:69-153. Cambridge, Mass.

David, N. (1985) Excavation of the Abri Pataud, Les Eyzies (Dordogne): The Noaillian (level 4) Assemblages and the Noaillian Culture in Western Europe. American School of Prehistoric Research Bulletin 37. Cambridge, Mass.

Poplin, F. (1977) Problèmes d'ostéologie quantitative rélatifs a l'écologie des hommes fossiles. B.A.F.E.Q. sup. 47:83-8.

Spiess, A. E. (1979) Reindeer and Caribou Hunters: An Archaeological Study. Studies in Archaeology. Academic Press, London.

Le Piage

Beckouche, S. (1981) Les grands mammifères de paléolithique supérieur du Piage (Lot). In F. Champagne & R. Espitalié (ed) 'Le Piage, Site Préhistorique de Lot'. Mem. S.P.F. 15: 165 - 91.

Champagne, F. & R. Espitalié (ed) (1981) Le Piage, Site Préhistorique de Lot. Mem. S.P.F. 15. Paris.

Piganeau

(Lenoir 1983)

Le Placard

Chauvet, G. (1910) Os, ivoires et bois de renne oeuvres de la Charente (hypothèses paléthnographiques). Bulletin de la Société Archéologique et Historique de la Charente 1910. (Extrait)

(Patte 1941)

La Plématrie

(Lavaud 1980)

Pont d Ambon

Célérier, G, & F. Delpech (1978) Un chien dans l'Azilien de Pont d'Ambon (Dordogne)? B.S.P.F. 75:212-15

& E. H. Moss (1983) L'Abri-sous-Roche de Pont d'Ambon à Bourdeilles (Dordogne) - un gisement magdalénien-Azilien. G.P. 26:81-108

Le Gall, O. (1984) L'Ichtyofaune d'Eau Douce dans les sites Préhistoriques: Ostéologie, Paléoécologie, Palethnologie Cahiers du Quaternaire N° 8. C.N.R.S. Paris.

(Also Delpech 1975, 1983)

St. Quéroy

Lehmen, J.-P., (1980) Découvertes de squelettes de rennes dans la grotte du Quéroy (Charente). C.R.Académie des Sciences. Paris (D)290:301-304

La Quina

(Patte 1941)

Reignac
Roussot, A. (1976) Gisement de Reignac, commune de Tursac, Dordogne. Livret-guide de l'excursion A - 4.p.116-8. IXe Congrès de l'UISPP. Nice.

le Roc, Sers
Favraud, A. (1908) La Grotte du Roc, commune de Sers (Charente) avec superposition du Solutréen sur l'Aurignacien. Bulletin de l'Ecole Anthropologique 18:407-23

(Patte 1941)

Roc Allen

(Le Tensorer 1979, 1981)

Roc d'Abeilles
Bouchud, J. (1970) Détermination de la faune du Roc d'Abeilles. G.P. 13:23

Champagne, F & R. Espitalié. (1970) L'Abri du Roc d'Abeilles à Calviac (Dordogne). G.P. 13:1-23.

Roc de Combe
Delpech, F. (1972) Etude paléoclimatique des faunes aurignaciennes et périgordiennes du gisement du Roc de Combe (Payrignac, Lot). Quaternaria 16:247-54

(1973) Présence de Mustela minuta Pomel (1853) = (= Mustela rixosa Bangs 1896) dans les couches aurignaco-périgordiennes du gisement de Roc de Combe (Commune de Payrignac, Lot). B.A.F.E.Q.11:3-6

(Also Delpech 1975, 1983)

Roc de la Belle.
Roussot, A. & F. Delpech (1969) Le Roc de La Belle, gisement paléolithique inédit, Commune de Cubjac (Dordogne). Bulletin de la Société Historique et Archéologique de Périgord 96:219-30.

(Also Delpech 1075, 1983)

Rochaudry

(Patte 1941)

Roche de Castelmoron
(Le Tensorer 1979, 1983)

Rochereil
Jude, P. E. (1960) La Grotte de Rochereil: Station Magdalénienne et Azilienne. Institut de Paléontologie Humaine de Paris Mem. 30.

Rond du Barry
Bayle des Hermens, R. de (1969) Le magdalénien final de la grotte du Rond du Barry, commune de Polignac (Haute-Loire). Cong. Pre. Fr. 19:37-57

(1983) La grotte du Rond du Barry. Archéologia 182:48-62.

Mourer-Chauviré, C. (1974) Etude préliminaire des oiseaux de la grotte du Rond du Barry . L'A. 78:37-48

Poulain, T. (1969) La grotte du Rond du Barry a Surzelles, commune de Polignac (Haute-Loire). Etude de la faune. Cong. Pré. Fr. 19:58-68.

Roque-Saint-Christophe
(White 1980, 1985)

Sainte-Eulalie
Lorblanchet, M. (1973) La grotte de Sainte-Eulalie à Espagnac (Lot). G.P. 16:3-62

(Also Delpech 1975, 1983)

Sergeac, (Vallon de Castelmerle)
Bouchud, J. (1952) Etude des rongeurs et des oiseaux de l'abri Castanet B.S.P.F. 49:267-71

Delage, F. (1935) Les Roches de Sergeac (Dordogne). L'A. 45:281-317.

Peyrony, D. (1935) Le gisement Castanet, vallon de Castelmerle, commune de Sergeac. B.S.P.F. 32:418-43

Tai
(Delpech 1975)

Trou de Cluzeau
(Patte 1941)

Vachons
(Patte 1941)

Vidon
(Lenoir 1983)

Villepin
Peyrony, D. (1936) L'abri de Villepin (commune de Tursac). Magdalénien supérieur et Azilien. B.S.P.F. 33:253-71

Appendix III

Chronological Plots. The occurrences of major taxa are recorded. For the sake of presentation absences are omitted.

Key:

W - West (Zone III)
E - East (Zone IV)
C - Central (Zone I)
N - North (Zone II)

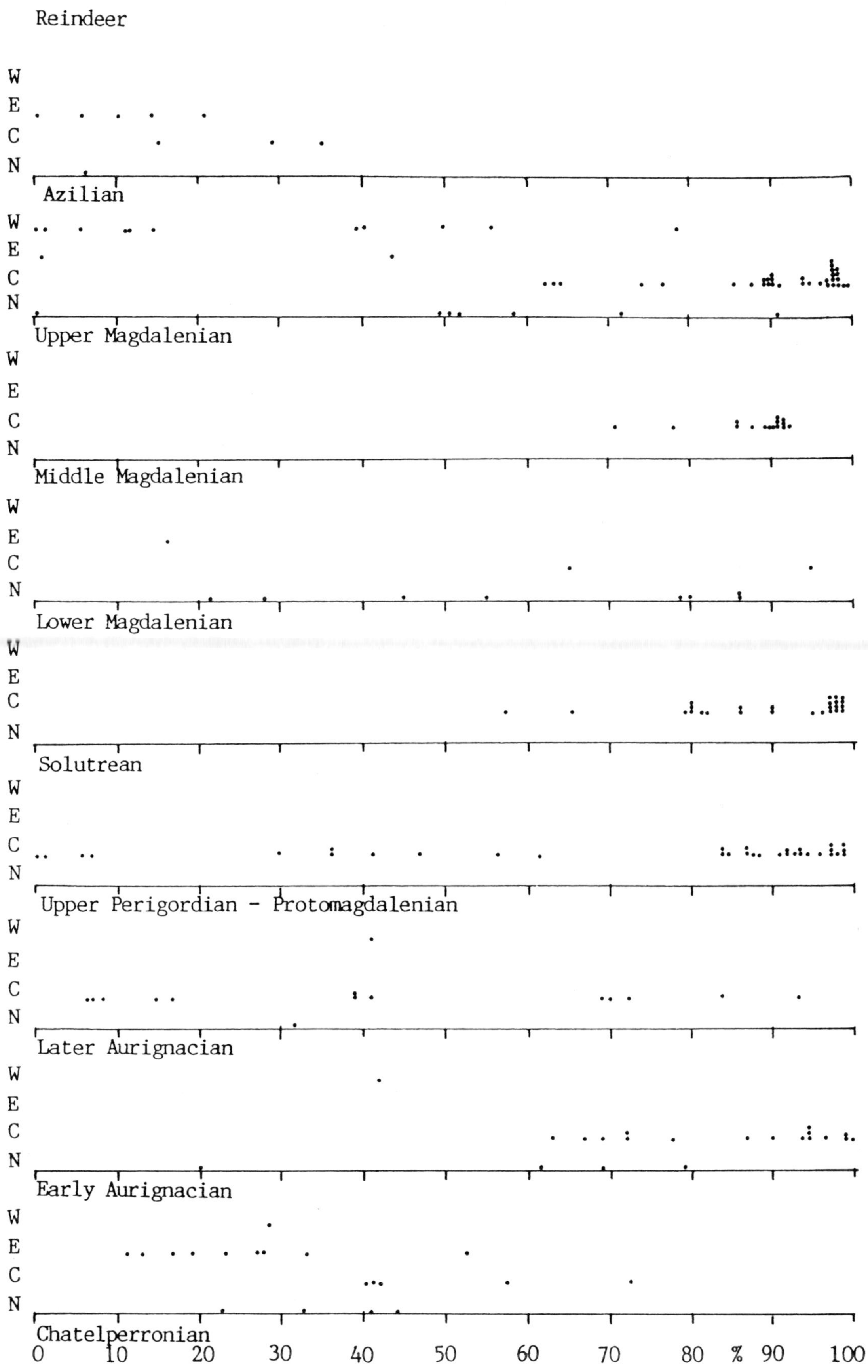
Reindeer
W
E
C
N
Azilian
Upper Magdalenian
Middle Magdalenian
Lower Magdalenian
Solutrean
Upper Perigordian - Protomagdalenian
Later Aurignacian
Early Aurignacian
Chatelperronian
0 10 20 30 40 50 60 70 80 % 90 100

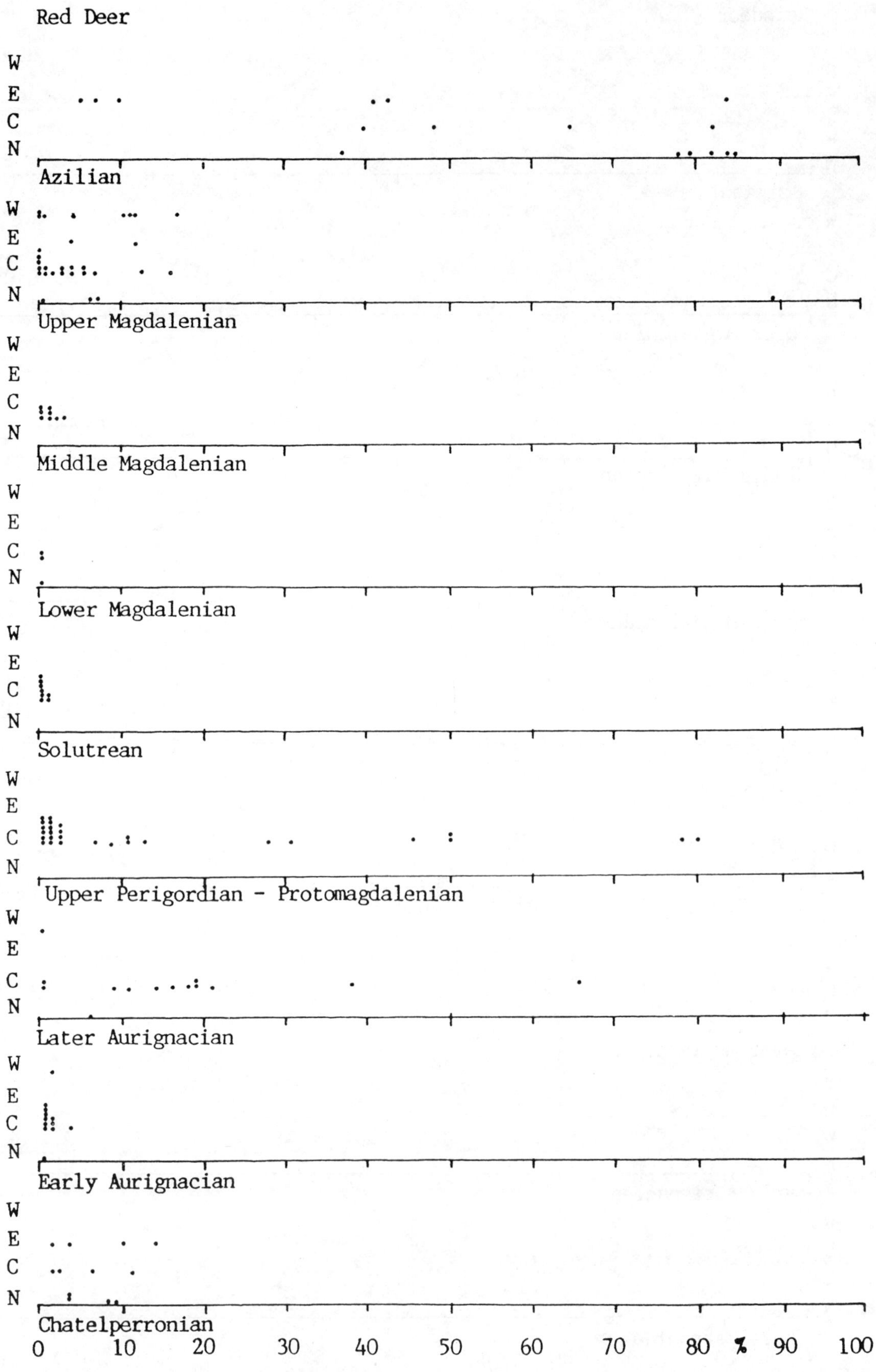
Red Deer
W
E
C
N
Azilian
Upper Magdalenian
Middle Magdalenian
Lower Magdalenian
Solutrean
Upper Perigordian - Protomagdalenian
Later Aurignacian
Early Aurignacian
Chatelperronian
0
10
20
30
40
50
60
70
80
%
90
100

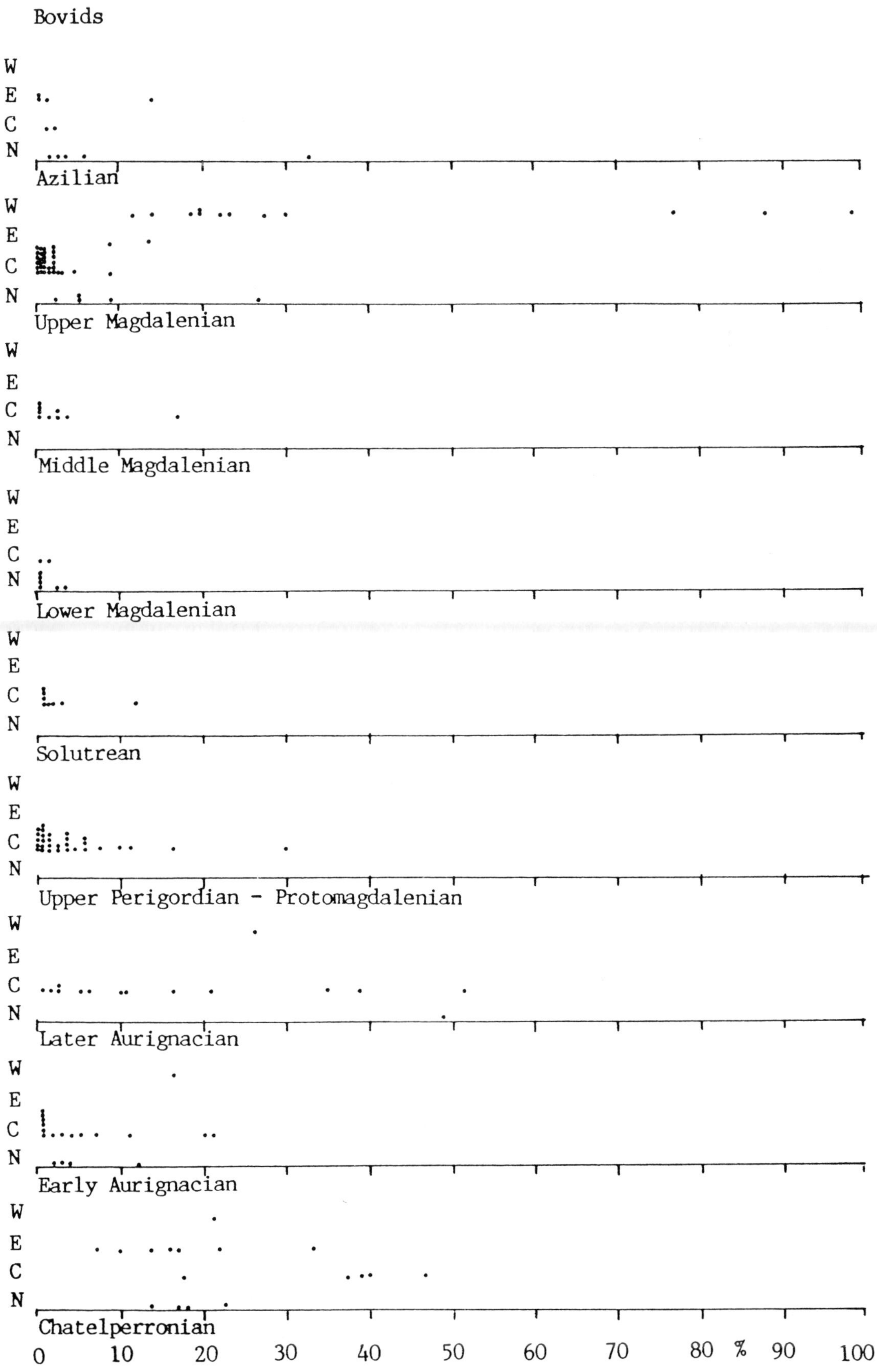
Bovids
W
E
C
N
Azilian
Upper Magdalenian
Middle Magdalenian
Lower Magdalenian
Solutrean
Upper Perigordian - Protomagdalenian
Later Aurignacian
Early Aurignacian
Chatelperronian
0
10
20
30
40
50
60
70
80
%
90
100

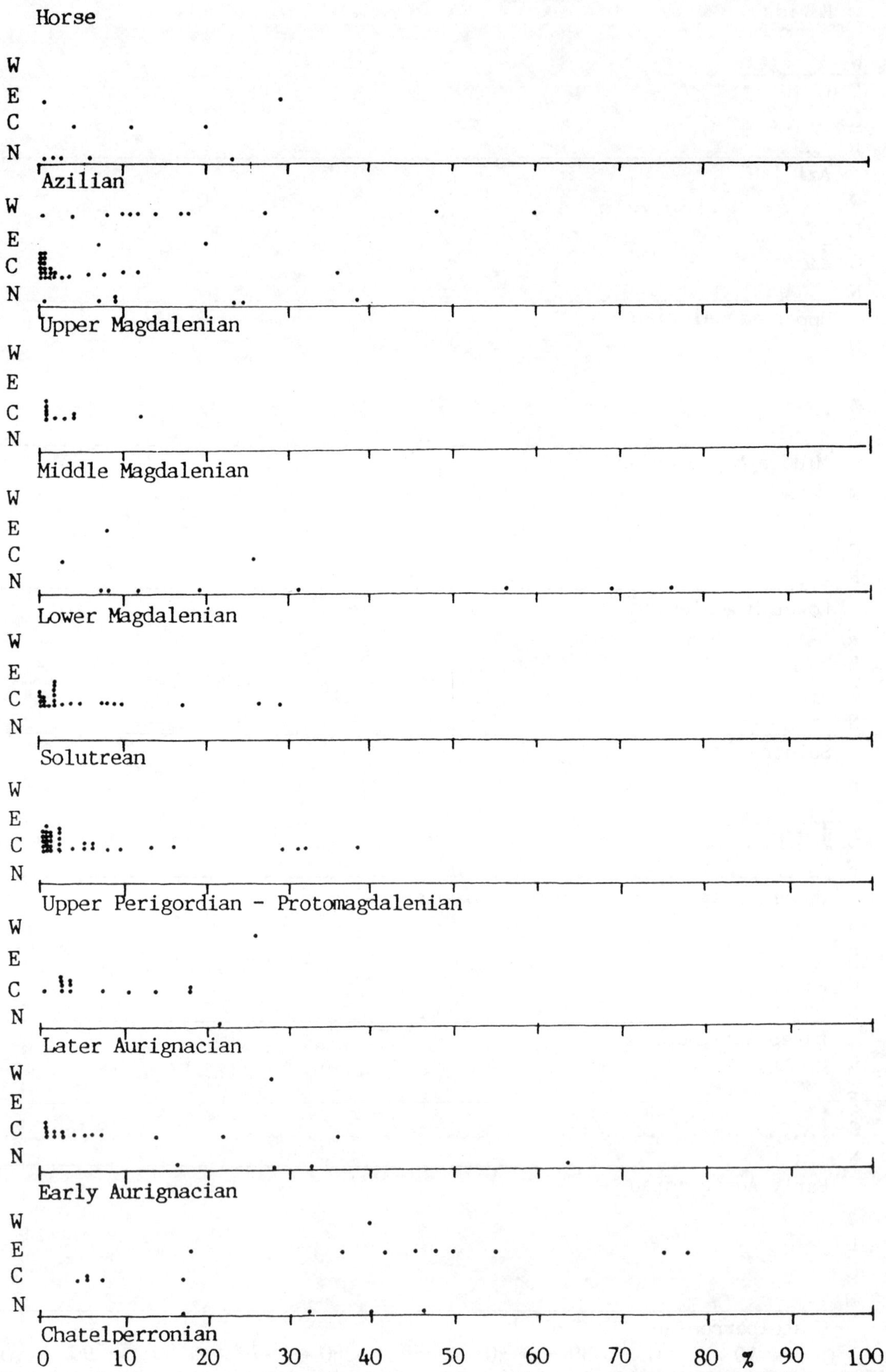
Horse
W
E
C
N
Azilian
Upper Magdalenian
Middle Magdalenian
Lower Magdalenian
Solutrean
Upper Perigordian - Protomagdalenian
Later Aurignacian
Early Aurignacian
Chatelperronian
0 10 20 30 40 50 60 70 80 % 90 100

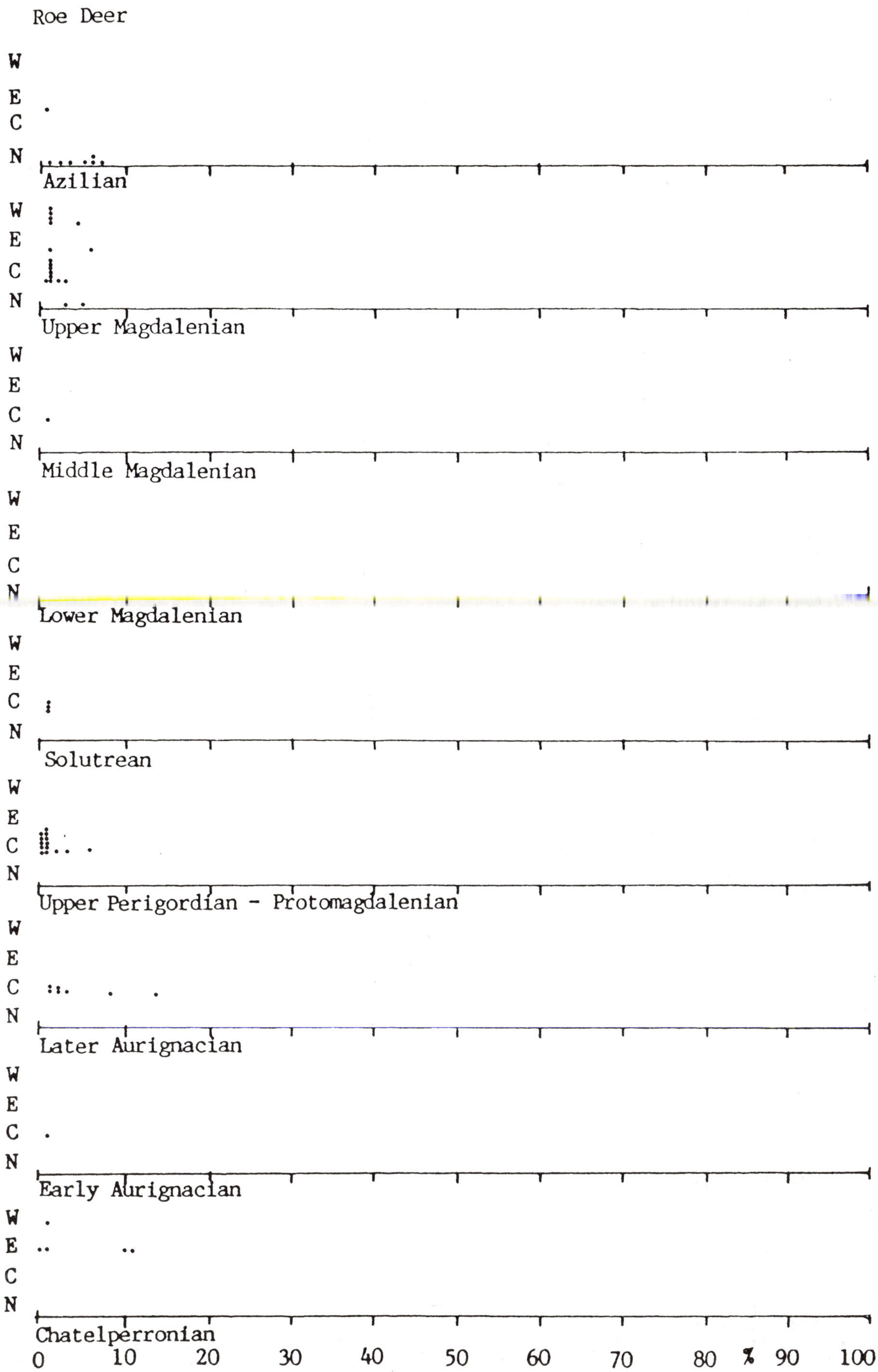
Roe Deer
W
E
C
N
Azilian
Upper Magdalenian
Middle Magdalenian
Lower Magdalenian
Solutrean
Upper Perigordian - Protomagdalenian
Later Aurignacian
Early Aurignacian
Chatelperronian
0
10
20
30
40
50
60
70
80
%
90
100

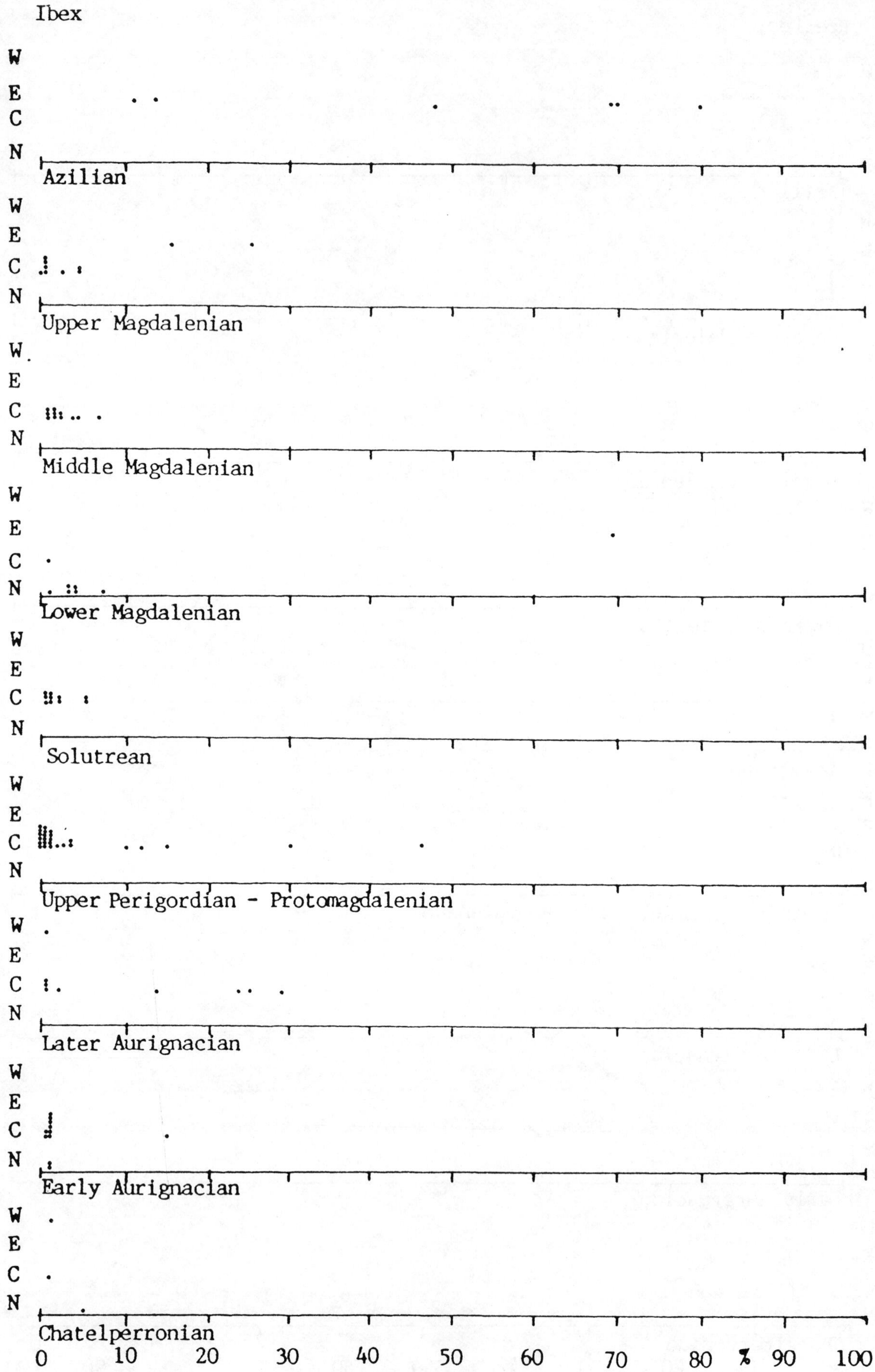
Ibex
W
E
C
N
Azilian
W
E
C
N
Upper Magdalenian
W
E
C
N
Middle Magdalenian
W
E
C
N
Lower Magdalenian
W
E
C
N
Solutrean
W
E
C
N
Upper Perigordian - Protomagdalenian
W
E
C
N
Later Aurignacian
W
E
C
N
Early Aurignacian
W
E
C
N
Chatelperronian
0
10
20
30
40
50
60
70
80
%
90
100

Chamois

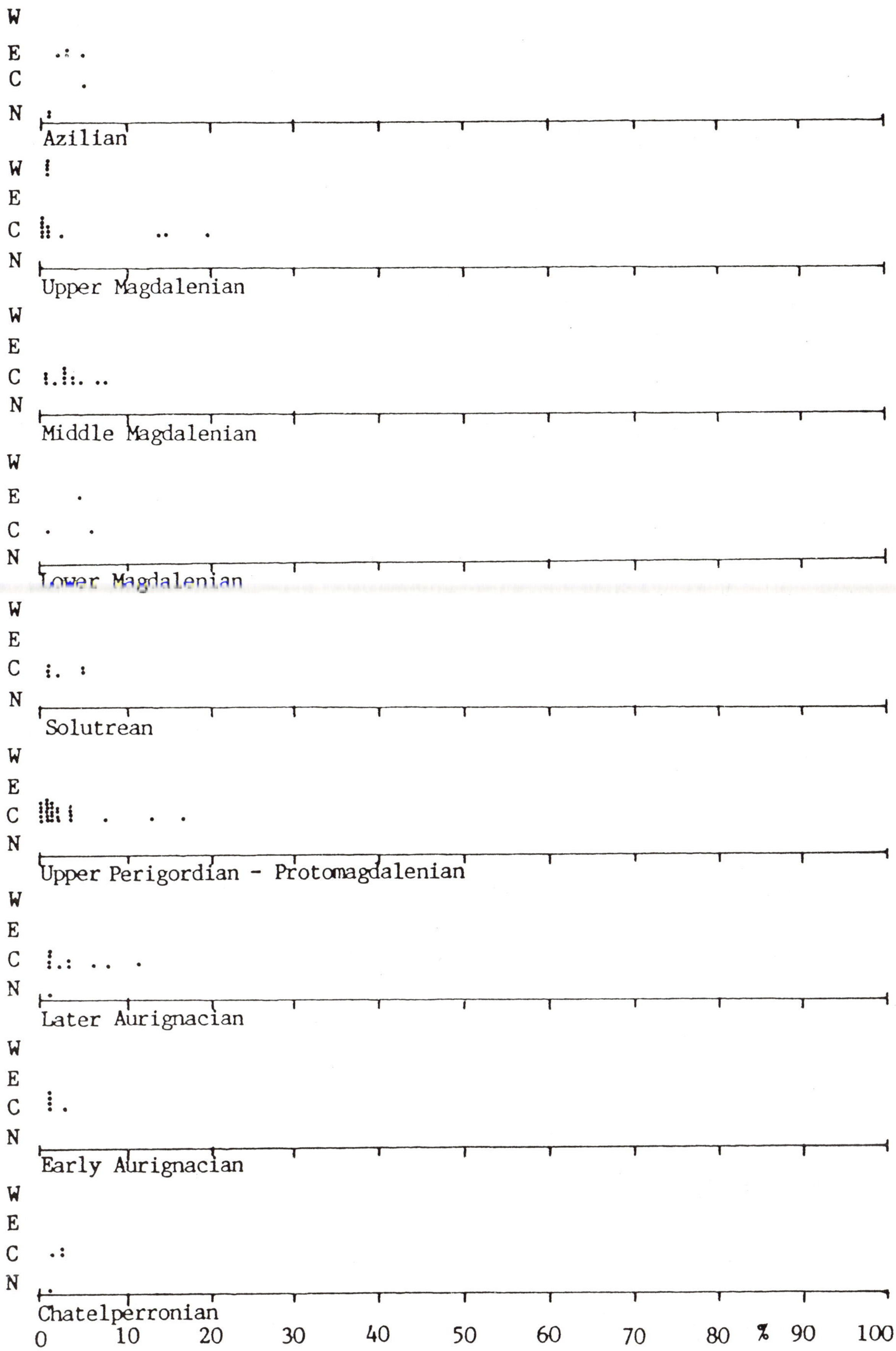
W
E
C
N
Azilian
Upper Magdalenian
Middle Magdalenian
Lower Magdalenian
Solutrean
Upper Perigordian - Protomagdalenian
Later Aurignacian
Early Aurignacian
Chatelperronian
0
10
20
30
40
50
60
70
80
%
90
100

Saiga

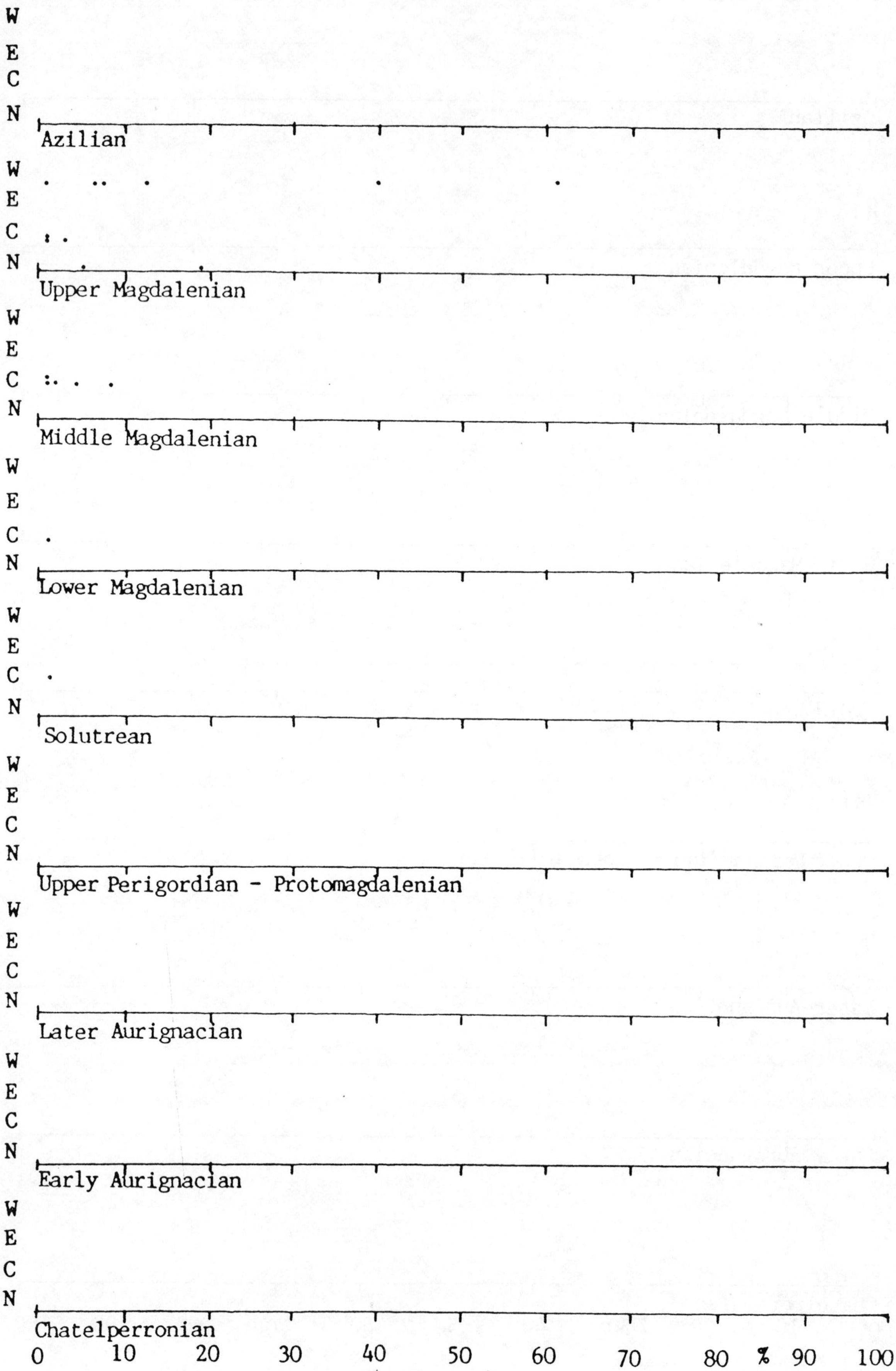
W
E
C
N
Azilian
W
E
C
N
Upper Magdalenian
W
E
C
N
Middle Magdalenian
W
E
C
N
Lower Magdalenian
W
E
C
N
Solutrean
W
E
C
N
Upper Perigordian - Protomagdalenian
W
E
C
N
Later Aurignacian
W
E
C
N
Early Aurignacian
W
E
C
N
Chatelperronian
0 10 20 30 40 50 60 70 80 % 90 100

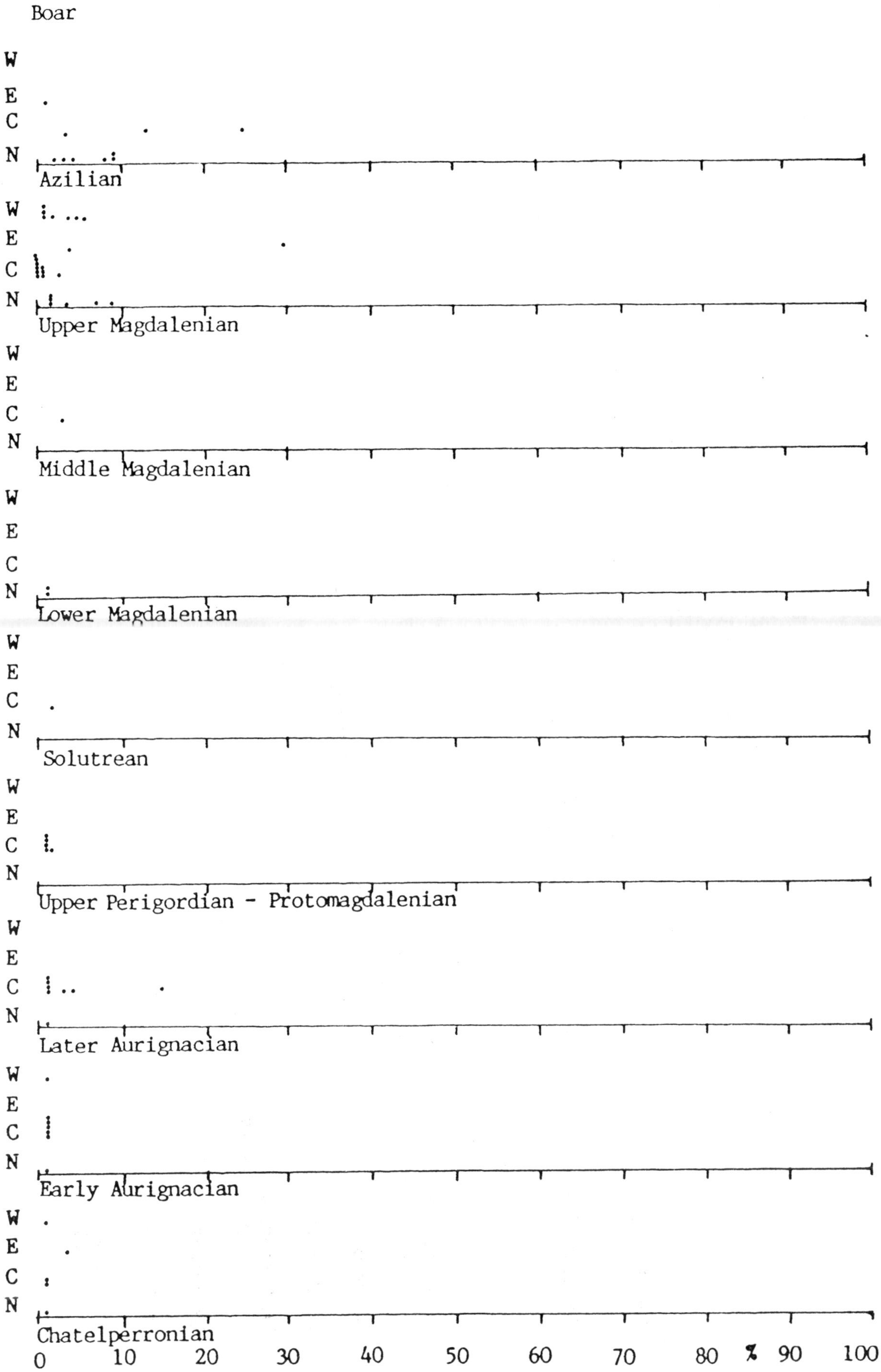
Boar
W
E
C
N
Azilian
W
E
C
N
Upper Magdalenian
W
E
C
N
Middle Magdalenian
W
E
C
N
Lower Magdalenian
W
E
C
N
Solutrean
W
E
C
N
Upper Perigordian - Protomagdalenian
W
E
C
N
Later Aurignacian
W
E
C
N
Early Aurignacian
W
E
C
N
Chatelperronian
0 10 20 30 40 50 60 70 80 % 90 100

Appendix IV

Supplementary Data (where sample size < 50 N.I.S.P.) Arranged in Approximate Chronological Order

	R.t.	C.e.	Bov.	E.c.	C.c.	C.i.	S.s.	R.r.	C.a.	M.p.
Roc de Combe 9	21.43		64.28	14.28						
7c	45.45	18.18	18.18	18.18						
Caminade Est G	42.86	14.29	42.86							
F	40.00	10.00	30.00	10.00				10.00		
E			28.57	57.14						
D2i	15.38	15.38	38.46	7.69			7.79			
D2s		11.11	44.44	11.11				11.11		
Caminade Ouest couche sup.	15.38	53.85		7.69	7.69	7.69	7.69			
Le Maldider 4b		91.00						8.33		
La Ferrassie I3/I2		29.17	33.33	16.67			16.67			
Le Flageolet I, XI	80.00			10.00		10.00				
TOT VIII/IX	27.59	34.48	6.89	17.24	3.35		3.35	3.35		
La Ferrassie B5-B3	100.00									
Battuts 12	100.00									
L.H.E. 33	73.33			13.33		13.33				
21	75.00			25.00						

	R.t.	C.e.	Bov.	E.c.	C.c.	C.i.	S.s.	R.r.	C.a.	M.p.
L.H.O.11	71.43			28.57						
Combe Cullier 4	90.00							10.00		
7	93.33			9.67						
10	81.81	9.09	9.09							
13b	68.42	5.26		5.26		2.63		18.42		
La Madeleine										
13/14	78.57		14.28	7.14						
10/12	100.00									
8/9	100.00									
3	96.15			3.85						
Le Flageolet II										
VIII	96.67			3.33						
II	100.00									
Pont d'Ambon 6	100.00									
3a & 3b		100.00								
2 & 3		100.00								

Equus hydruntinus = **Caminade Est** E (14.28%), D2i (7.69%), D2s (11.11%)
La Ferrassie I3/I2 (4.17%)
Le Flageolet I TOT VIII/IX (3.35%)